Wireless Networks

Series Editor

Xuemin Sherman Shen, University of Waterloo, Waterloo, Canada

The purpose of Springer's Wireless Networks book series is to establish the state of the art and set the course for future research and development in wireless communication networks. The scope of this series includes not only all aspects of wireless networks (including cellular networks, WiFi, sensor networks, and vehicular networks), but related areas such as cloud computing and big data. The series serves as a central source of references for wireless networks research and development. It aims to publish thorough and cohesive overviews on specific topics in wireless networks, as well as works that are larger in scope than survey articles and that contain more detailed background information. The series also provides coverage of advanced and timely topics worthy of monographs, contributed volumes, textbooks and handbooks.

Yulin Shao

AI-Empowered IoT Communications

Springer

Yulin Shao
Department of Electrical and Electronic Engineering
The University of Hong Kong
Hong Kong, China

ISSN 2366-1186 ISSN 2366-1445 (electronic)
Wireless Networks
ISBN 978-3-031-98815-8 ISBN 978-3-031-98816-5 (eBook)
https://doi.org/10.1007/978-3-031-98816-5

This Springer imprint is published by the registered company Springer Nature Switzerland AG
The registered company address is: Gewerbestrasse 11, 6330 Cham, Switzerland

To all those who have walked alongside me on this academic journey: mentors who illuminated the path, colleagues who challenged my thinking, friends who fueled my perseverance, and above all, my family, whose unwavering belief was the cornerstone that held me steady. Your support, in ways seen and unseen, made this possible. This is as much yours as it is mine.

Preface

Artificial Intelligence (AI) has rapidly evolved from a niche field of computer science into a pervasive force that powers countless innovations, from real-time speech recognition to self-driving vehicles. In parallel, the Internet of Things (IoT) has reshaped modern connectivity by embedding intelligence into everyday objects, enabling them to sense, communicate, and respond to the world around them. When AI and IoT converge, the potential becomes transformative: instead of merely collecting and transmitting data, IoT devices can learn from their environments, refine their operations, and autonomously coordinate with one another in real-time. It is this convergence – this marriage of sensing, networking, and intelligent processing – that forms the heart of *AI-Empowered IoT Communications*.

Despite the promise of this synergy, combining AI and IoT poses significant challenges. Traditional network designs must be overhauled to accommodate the demand for massive connectivity, ultra-reliable data transfer, and stringent latency requirements. Energy consumption, for instance, cannot simply be an afterthought; large-scale deployments of battery-powered IoT devices require highly optimized, feedback-enhanced communication mechanisms. Equally pressing are concerns such as data privacy, security, and ethical considerations, especially as AI-based models glean deeper insights from vast amounts of user-generated data. This book is a response to these emerging demands, offering a holistic view that spans foundational IoT principles, advanced AI algorithms, and their intersection in real-world applications.

The chapters that follow explore the core dimensions of AI-enabled IoT communications. We begin with a thorough grounding in the evolution of IoT architectures and the overarching principles of AI, establishing a framework for understanding the powerful synergy between the two fields. We then delve into specific mechanisms, such as AI-driven feedback communication, semantic communication, and deep reinforcement learning for multiple access, that drastically enhance the efficiency and adaptability of IoT systems. Emphasis is placed on both theoretical underpinnings and tangible proofs-of-concept, illustrating how techniques like feedback-based channel coding, semantic encoding, and federated learning can dramatically improve communication efficiency. Finally, the book discusses AI-enhanced network control mechanisms that leverage graph neural networks and large language models, pointing to a future where IoT operations become smarter, more decentralized, and even more collaborative.

Throughout these pages, our focus extends beyond mere technical detail. We aim to highlight the broader implications of AI-driven IoT, including scalability hurdles, regulatory frameworks, and societal responsibilities. By anticipating future trends, ranging from ethical oversight in data-driven systems to next-generation networking, we provide a roadmap for future research and development. This dual approach balances immediate, practical solutions with a vision of IoT ecosystems that continue to evolve in alignment with human values and global needs.

We hope this book serves as an in-depth guide for students, engineers, and researchers who share our enthusiasm for advancing AI and IoT technologies. Whether you are seeking to build next-generation communication protocols, refine machine learning architectures, or explore entirely new paradigms for distributed intelligence, *AI-Empowered IoT Communications* invites you on a journey through the latest innovations and the boundless possibilities they unlock. We look forward to seeing how our collective endeavors, across academia, industry, and beyond, shape an era in which IoT systems seamlessly learn, adapt, and thrive alongside us.

Hong Kong, China
May 2025

Yulin Shao

Acknowledgments I would like to express my deepest gratitude to my mentors – Prof. Soung Chang Liew, Prof. Deniz Gunduz, Prof. Yonina C. Eldar, and Prof. Vincent W. S. Chan – who have guided and supported me throughout my academic and research journey. Your wisdom, encouragement, and generous feedback have profoundly shaped my understanding of Artificial Intelligence and IoT communications.

I am especially indebted to my collaborators, whose insight, creativity, and persistence have been invaluable throughout my research career. I wish to thank: Prof. Kaibin Huang, Prof. Yuanwei Liu, Prof. Octavia A. Dobre, Prof. Rose Qingyang Hu, Prof. Krystian Mikolajczyk, Prof. Lu Lu, Prof. He Chen, Prof. Shaodan Ma, Prof. Liqun Fu, Prof. Qi Cao, Prof. Taotao Wang, Prof. Lizhao You, Prof. Qianqian Yang, Prof. Qing Yang, Prof. Yijin Zhang, Dr. Branislav Popovic, Dr. Alberto Perotti, Dr. Emre Ozfatura, Dr. Zhan Gao, Dr. Arman Rezaee, Dr. Haotian Wu, Dr. Chenghong Bian, Dr. Gongpu Chen, Dr. Haoyuan Pan, Dr. Jiaxin Liang, Dr. Jintao Wang, Dr. Pengfei Shen, Dr. Guangyu Zhu, Dr. Mengyao Ma, Dr. Ziyang Guo, Dr. Peng Liu, Dr. Jiajun Luo, Ms. Menghan Li, Mr. Xinyan Xie, Mr. Hongning Ruan, Ms. Pingping Zhang, Mr. Yuyang Du, Mr. Hongwei Cui, Mr. Qun Yang, Mr. Yucheng Cai, Ms. Yewen Cao, Mr. Yu Ding, Mr. Wenwei Lai, Mr. Fan Yang, Mr. Haoyang Di, Ms. Yimeng Li. Your collaboration and diverse perspectives have consistently challenged me to think beyond conventional boundaries and helped shape the direction of my work in meaningful ways.

I am also grateful to the editorial team at Springer Nature for their support and professionalism throughout the publication process. Your assistance helped bring this book to life with clarity and precision.

Finally, and most importantly, I extend my heartfelt thanks to my family, especially my parents and my wife, for your unwavering support, patience, and love. Your understanding throughout the years, through intense research periods, travel, and many missed personal moments, has meant more to me than words can express. This book is as much yours as it is mine.

Competing Interests The author has no competing interests to declare that are relevant to the content of this manuscript.

Contents

Acronyms

5G/6G	5th/6th Generation Mobile Networks
ACK	Acknowledgment
ACI	Artificial Complex Intelligence
AGI	Artificial General Intelligence
AI	Artificial Intelligence
AirComp	Over-the-Air Computation
AP	Access Point
AutoCA	Autonomous Channel Access
AWGN	Additive White Gaussian Noise
BER	Bit Error Rate
BLE	Bluetooth Low Energy
BLER	Block Error Rate
BSS	Basic Service Set
CDMA	Code Division Multiple Access
CSI	Channel State Information
CSMA/CA	Carrier Sense Multiple Access with Collision Avoidance
CSMA/CD	Carrier Sense Multiple Access with Collision Detection
D2D	Device-to-Device
DeepJSCC	Deep Joint Source-Channel Coding
DeepVLF	Deep Variable-Length Feedback
DL	Deep Learning
DMC	Discrete Memoryless Channel
DNN	Deep Neural Network
DRL	Deep Reinforcement Learning
DTAT	Discrete Time Amplitude Transmission
eDRX	Extended Discontinuous Reception
FDMA	Frequency Division Multiple Access
FEC	Forward Error Correction
FE-IoT	Feedback-Enhanced IoT
FL	Federated Learning
FSM	Finite-State Machine

GAN	Generative Adversarial Network
GBAF	Generalized Block Attention Feedback
GeLU	Gaussian Error Linear Unit
GNN	Graph Neural Network
GRU	Gated Recurrent Unit
HARQ	Hybrid Automatic Repeat Request
LDPC	Low-Density Parity-Check Code
LLM	Large Language Model
LoRaWAN	Long Range Wide Area Network
LPWAN	Low-Power Wide-Area Network
M2M	Machine-to-Machine
MA	Multiple Access
MAC	Medium Access Control
MADRL	Multi-Agent Deep Reinforcement Learning
ML	Machine Learning
MMSE	Minimum Mean Square Error
NB-IoT	Narrowband Internet of Things
NLP	Natural Language Processing
NR	New Radio (5G)
OFDM	Orthogonal Frequency Division Multiplexing
OFDMA	Orthogonal Frequency Division Multiple Access
PAM	Pulse Amplitude Modulation
PAPR	Peak-to-Average Power Ratio
PDF	Probability Density Function
PER	Packet Error Rate
PPP	Poisson Point Process
PSM	Power Saving Mode
QAM	Quadrature Amplitude Modulation
QoS	Quality of Service
QPSK	Quadrature Phase Shift Keying
ReLU	Rectified Linear Unit
RFID	Radio-Frequency Identification
RL	Reinforcement Learning
RNN	Recurrent Neural Network
RTS/CTS	Request to Send/Clear to Send
SNR	Signal-to-Noise Ratio
TDMA	Time Division Multiple Access
UE	User Equipment
URLLC	Ultra-Reliable and Low-Latency Communications
WOR	Wake on Radio
XAI	Explainable AI

Chapter 1
Introduction

To understand the world, we must connect it. To improve the world, we must make it intelligent.

1.1 Foundations of IoT Communication

The dramatic growth of the Internet of Things (IoT) has transformed how people and organizations connect physical objects, digital processes, and cloud-based services [1–6]. From the smallest wearable devices [7] to large-scale smart city infrastructures [8], IoT aims to bridge the physical and virtual worlds through seamless data exchange. Understanding the foundations of IoT communication is therefore critical in designing robust systems that can effectively harness data and deliver actionable insights. This section offers a comprehensive examination of how IoT is defined, identifies the fundamental components of an IoT system, explores the key principles that underpin IoT communication, and traces how this field has evolved over time.

1.1.1 Definition and Core Components of IoT

At its most basic level, IoT refers to a vast network of interlinked devices, often called "things" or "smart objects", that contain sensors, actuators, and software for transmitting and receiving data [1]. These devices may be as varied as home thermostats, wearable health monitors, industrial control sensors, and connected vehicles [3, 4]. They detect conditions in the physical environment (such as temperature, humidity, and motion) and share this information with other devices or centralized servers over a network. By analyzing and acting on this data, IoT systems create new opportunities for automation, efficiency gains, and decision-making based on real-time insights [9, 10].

Within an IoT ecosystem, several core components work together to facilitate these interactions. First, the physical devices themselves collect data about the

Y. Shao, *AI-Empowered IoT Communications*, Wireless Networks,
https://doi.org/10.1007/978-3-031-98816-5_1

surrounding environment and often respond to commands through actuators that enable physical changes [3, 11], such as adjusting a machine setting or turning on a light. Next, connectivity infrastructure links these devices to gateways or directly to the internet [12]. Depending on factors like the data rate required, power constraints, cost considerations, or coverage needs, various networking technologies can be employed, including short-range protocols like Bluetooth Low Energy [13] and Wi-Fi [14], or wide-area standards like LoRaWAN [15], NB-IoT [3], and cellular networks.

Edge devices or gateways frequently play a pivotal role in aggregating and preprocessing sensor data closer to the source. This approach, often referred to as "edge computing", can reduce latency and the bandwidth required to send raw data to the cloud [16, 17]. The cloud platform then serves as a centralized environment for storage, analytics, and application hosting, leveraging scalable computational resources for tasks such as big data analysis, machine learning, and overall device management. Finally, user-facing applications, including dashboards or mobile apps, allow individuals and organizations to visualize trends, receive alerts, and control IoT devices in an intuitive manner.

Security [18] and privacy [19] mechanisms overlay each layer of this ecosystem to safeguard data and ensure trustworthiness. Encryption, authentication, access control, secure firmware, and regular patching procedures are increasingly recognized as non-negotiable elements in modern IoT deployments, particularly as the number of connected devices continues to skyrocket.

1.1.2 Basic Principles of IoT Communication

Efficient exchange of data stands at the heart of IoT communication. Several fundamental principles guide the design and operation of IoT systems, beginning with the choice of network topology. In some settings, a simple star configuration [20], where devices connect directly to a central hub or gateway, is sufficient and cost-effective. In larger or more complex deployments, a mesh network, where each device can relay information on behalf of others [12], introduces greater resilience and range at the expense of more complex routing protocols. Hybrid structures also emerge in practice, combining the benefits of different topologies to match specific requirements related to range, power consumption, cost, and scalability.

Another important principle revolves around protocol selection. A wide variety of communication protocols support the diverse needs of IoT deployments, each offering a specific balance of range, power consumption, data rate, and complexity. For applications that require close-proximity connections and minimal power usage, short-range wireless protocols offer an ideal solution. Bluetooth Low Energy (BLE) [13], for example, is explicitly designed for low-power devices operating at short distances. Its efficient power profile and relatively straightforward implementation have made it ubiquitous in wearables, personal health monitors, and a wide range of consumer electronics, where battery longevity is a critical requirement and the range is typically only a few meters.

Zigbee [21], standardized under the IEEE 802.15.4 specification, also operates at low power and supports data rates sufficient for many IoT applications, especially in home automation and industrial sensor networks. By creating mesh networks where devices can relay messages to one another, Zigbee can extend coverage beyond point-to-point links. This capability makes it attractive for smart lighting, security systems, and other large-scale sensor deployments, even though its data throughput is modest compared to Wi-Fi.

Wi-Fi (IEEE 802.11) [14] stands out for delivering higher bandwidth and robust throughput, features that are advantageous when transferring larger amounts of data, such as media files or real-time sensor streams. Yet, because of its higher energy consumption, Wi-Fi is best suited for devices that have a stable power source or require frequent, high-speed data transfer. While it may not be the most power-efficient protocol, Wi-Fi's widespread global adoption and ease of integration into existing networks remain significant advantages.

Long-range wireless protocols target applications that must cover greater distances while still minimizing power usage. LoRa/LoRaWAN [15], for instance, is widely associated with applications in smart cities, agriculture, and remote environmental monitoring. Its key benefit lies in its ability to transmit data across several kilometers at ultra-low power, which enables end devices to operate on small batteries for extended periods. Similarly, Sigfox [22] is another low-power wide-area network (LPWAN) solution that emphasizes simplicity, low data rates, and cost-effective modules. Although it is proprietary, Sigfox offers an international network infrastructure that allows seamless device connectivity in many regions without requiring enterprises to set up and manage their own gateways.

Cellular-based IoT standards, such as NB-IoT [3] and LTE-M [23], emerge from the 3GPP ecosystem, leveraging existing cellular networks. These protocols support large-scale IoT rollouts, particularly in smart city applications, utility metering, and industrial IoT scenarios. By piggybacking on established mobile network infrastructure, NB-IoT and LTE-M can deliver broad coverage and dependable performance. They balance relatively low data rates and reduced power consumption with the benefit of wide availability, making them suitable for use cases that do not demand high-bandwidth communication but require reliable connectivity across dispersed locations.

In many industrial environments, wired connections are still the preferred choice, especially in scenarios demanding high data throughput or where wireless signals may be unreliable. Ethernet networks, for instance, are common in factories and facilities that rely on stable data rates for process automation, real-time monitoring, and high-bandwidth tasks such as machine vision. If a dedicated wiring infrastructure can be established, Ethernet provides dependable connectivity with minimal latency, as well as robust security features. Additionally, legacy fieldbus systems [24], such as CAN bus or Modbus, remain in use in some industrial installations. While these have traditionally dominated automation and process control, they are increasingly being replaced or supplemented by more modern, IP-based networks to unify data exchange and take advantage of emerging IoT platforms. Nonetheless, in many legacy plants, fieldbus systems persist because of their proven reliability, entrenched hardware investments, and specific industry certifications, underscoring the wide range of networking needs and solutions present in today's IoT landscape.

1.1.3 Evolution of IoT Communication

Although the phrase "Internet of Things" only gained popularity in the early 2000s, its roots stretch back several decades. Early research on distributed sensor networks, sponsored by government organizations and academia, laid the groundwork by demonstrating how multiple, low-power devices could collaborate to gather data about the environment. This concept found its first widespread commercial expression in the form of radio-frequency identification (RFID) technology [25], which allowed objects to be tagged and tracked via short-range wireless transmissions. Simultaneously, industrial sectors began experimenting with machine-to-machine (M2M) communication, where automated systems monitored and managed equipment remotely, reducing the need for constant human oversight.

The surge of consumer internet adoption in the late 1990s and early 2000s, alongside the development of home and office wireless standards like Wi-Fi and Bluetooth, ushered in a new era of connectivity. Suddenly, consumer products, from thermostats to security cameras, could be linked directly to home networks or the wider internet, transforming them into "smart" devices. Advances in microelectronics made sensors smaller, cheaper, and more energy-efficient. As the smartphone revolution took hold, a massive ecosystem of apps emerged, giving end users ever more intuitive ways to interact with connected devices.

By the 2010s, cloud computing had matured enough to offer cost-effective, on-demand resources for IoT systems. Companies took advantage of scalable cloud platforms to process enormous streams of sensor data, store logs, run analytics, and deploy sophisticated machine learning algorithms. This period also saw the rise of low-power wide-area networks (LPWANs), such as LoRaWAN and Sigfox, which targeted remote or dispersed applications that demanded battery-powered devices with long-range coverage. The introduction of 3GPP standards like NB-IoT and LTE-M built on existing cellular networks to deliver similarly low-power, wide-area capabilities, enabling the industry to integrate IoT solutions into a well-established telecommunications ecosystem.

Today, IoT communication is on the brink of another transformation as 5G rolls out across global markets. Promising higher bandwidth, ultra-low latency, and advanced features such as network slicing, 5G aims to broaden the scope of IoT to incorporate mission-critical use cases like autonomous vehicles, real-time robotics, and advanced healthcare applications. In parallel, the continuing convergence of IoT and artificial intelligence (AI) is adding powerful new capabilities in edge analytics, anomaly detection, and automated decision-making. These developments illustrate that IoT communication remains highly dynamic, evolving in response to technological innovations, market demands, and global connectivity trends.

With this groundwork established, readers can now appreciate why IoT communication forms a critical building block for interconnected solutions across industries. Subsequent sections of this chapter, as well as forthcoming chapters, will delve deeper into how AI both informs and benefits from IoT communication, and how the synergy between AI and IoT drives innovative, scalable, and secure systems that are poised to reshape modern life.

1.2 Overview of AI

AI has rapidly evolved from a theoretical pursuit among a handful of researchers into a powerful, transformative technology that underpins many of today's most innovative products and services [26]. While popular culture often depicts AI in the context of robots or futuristic machines that rival human intelligence, AI in practice is more nuanced, encompassing a broad range of computational techniques that enable computers to learn from data, interpret complex information, and make decisions with minimal human intervention. To understand how AI enriches the IoT, it is essential first to gain a clear overview of what AI is, how it emerged, what its core principles are, and how its most groundbreaking methodologies, particularly deep learning [26], came to be.

1.2.1 Emergence and Definition of AI

AI can be traced back to a simple question: Can machines think? This challenge, famously articulated by Alan Turing in 1950 [27], began a decades-long journey toward building computer systems capable of performing tasks that would usually require human-like intelligence. The formal field of AI research is often dated to the 1956 Dartmouth Workshop, where a group of scientists, including John McCarthy and Marvin Minsky, laid out an ambitious vision of creating machines that could replicate facets of human cognition. Although the term "intelligence" remains open to philosophical interpretation, AI generally denotes the ability of machines to interpret their environment, learn from experience, and use that learning to adapt to new inputs or accomplish specific goals.

AI is inherently interdisciplinary. It intersects with computer science, mathematics, psychology, linguistics, and neuroscience, among other fields. Early attempts at AI tended to rely on symbol manipulation and logic-based algorithms. Over time, and particularly in recent decades, AI has shifted its emphasis to more data-driven approaches, focusing on machine learning methods that uncover patterns in large datasets [28]. From natural language processing [29] to advanced robotics [30], AI's broad range of techniques underscores its central aim: to reproduce or augment the ways humans sense, reason, and respond to the world.

1.2.2 Principles of AI

At its core, AI relies on computational processes that mirror some aspects of human reasoning and problem-solving. Among the foundational ideas that guide AI research and development are:

1. Representation of Knowledge: AI systems require a meaningful way to encode information about the world. In early AI systems, knowledge was often represented in symbolic form: facts and rules that could be manipulated using logical inference. Although modern machine learning approaches often learn directly from raw data without extensive manual rule-coding, the concept of how to represent knowledge and context remains essential.
2. Learning from Data: A distinguishing feature of AI, especially in its modern incarnations, is the capacity to learn. Unlike traditional software that strictly follows programmed instructions, AI systems adapt their behavior based on patterns gleaned from examples or experience. This capacity encompasses supervised learning (learning with labeled data), unsupervised learning (finding patterns in unlabeled data), and reinforcement learning (improving performance based on trial-and-error feedback).
3. Reasoning and Inference: AI systems do not merely store information; they draw inferences and make decisions. Early AI models often used formal logic or handcrafted rules to deduce new facts from known premises. More recent methods, such as probabilistic models and neural networks, reason in a statistical or approximate manner, allowing them to cope better with uncertainty and real-world complexity.
4. Problem-Solving and Planning: Many AI applications revolve around solving complex tasks or planning sequences of actions. This capability includes everything from pathfinding for autonomous vehicles to scheduling and resource allocation within intricate systems. By breaking problems down into smaller steps and evaluating potential outcomes, AI systems aim to find efficient or even optimal solutions.
5. Interaction via Perception and Language: A remarkable aspect of human intelligence is its rich interaction with the environment through senses like vision, hearing, and touch, as well as language comprehension and production. AI researchers continuously strive to embed similar capabilities in machines: computer vision algorithms can identify objects in images, natural language processing techniques enable chatbots to understand user queries, and speech recognition allows digital assistants to interpret spoken instructions.

These principles form the backbone of AI systems, reflecting both historical insights and modern breakthroughs in computational methods. By blending these principles with domain-specific knowledge, AI can be customized to solve problems as varied as medical diagnosis, financial forecasting, and, crucially for this text, transforming and optimizing IoT communication.

1.2.3 Deep Learning

Among the most transformative developments in modern AI has been the rise of deep learning [26], a subset of machine learning that uses neural networks with many layers of computational nodes. While the basic concepts behind neural networks date back to the mid-twentieth century, deep learning only began to revolutionize AI once sufficient computational power and large datasets became available. In essence, a deep neural network tries to learn a hierarchy of features directly from data. For instance, in computer vision tasks, earlier layers of the network may learn to detect edges or simple shapes, while deeper layers capture more abstract features like textures or specific objects [31].

The power of deep learning lies in its ability to discover intricate patterns in data without relying on rigid, hand-crafted feature engineering. From image classification and object detection to speech recognition and language translation, deep learning has delivered major accuracy gains in tasks once deemed intractable for computers. This approach now underpins critical applications, such as recommendation engines on streaming services, advanced forecasting models in finance, and real-time language translation [29].

An important corollary to deep learning's success is the idea of transfer learning [32], wherein a model initially trained on a large general dataset is subsequently refined on a more specialized dataset, requiring far fewer data samples. This practice is particularly relevant in domains where gathering large amounts of labeled data can be costly or time-consuming, such as medical imaging or industrial defect detection.

Deep learning's success does not preclude other AI methods, nor is it without challenges. Training very large models can be computationally expensive, and deep neural networks typically behave as "black boxes," making it difficult to interpret their decision-making processes. Efforts in explainable AI (XAI) [33] and techniques for more efficient architectures (like pruning and quantization) are emerging to address these concerns. Nevertheless, the effectiveness of deep learning has expanded the boundaries of AI and fueled the industry's current wave of enthusiasm for machine intelligence.

1.2.4 Major Milestones of AI

The development of AI has not been a smooth, linear progression; rather, it has experienced periods of intense optimism punctuated by "AI winters" when funding and interest waned [34]. Key milestones help illustrate the field's evolving nature and its significant leaps forward:

- 1950: Alan Turing proposes the Turing Test in his paper "Computing Machinery and Intelligence" [27], setting a benchmark for machine intelligence.
- 1956: Dartmouth Conference organized by John McCarthy, Marvin Minsky, and others, marking the birth of AI as a formal field. The term "Artificial Intelligence" is coined.
- 1958: Frank Rosenblatt develops the Perceptron, an early neural network model.
- 1960s–1970s: Rule-based systems and symbolic AI dominate early projects; the first so-called "AI winter" ensues due to unmet promises.
- 1980s: Rise of Expert Systems like MYCIN (medical diagnosis) and XCON (computer configuration), driven by rule-based AI.
- 1986: Backpropagation algorithm is popularized, revitalizing neural network research (credited to Geoffrey Hinton, David Rumelhart, and others) [35].
- 1987–1993: Second AI Winter as expert systems prove costly and inflexible.
- 1997: IBM's Deep Blue defeats world chess champion Garry Kasparov, showcasing AI's capability in narrowly defined tasks.
- 2006: Geoffrey Hinton coins the term "Deep Learning" to describe neural networks with multiple layers.
- 2012: AlexNet (a deep convolutional neural network) dominates the ImageNet competition [36], triggering the deep learning boom.
- 2014: Ian Goodfellow introduces Generative Adversarial Networks (GANs) [37], enabling realistic synthetic data generation.
- 2015: TensorFlow (Google) and later PyTorch (Facebook) democratize deep learning tools.
- 2016: DeepMind's AlphaGo beats the Go champion Lee Sedol, underscoring the growing sophistication of AI in tackling even complex, intuition-based games [38].
- 2017: Google's Transformer Architecture [29] revolutionizes natural language processing (NLP).
- 2022: OpenAI's ChatGPT popularizes conversational AI, followed by DALL-E 2 (image generation) and Stable Diffusion.

These achievements reflect not only the fluctuating nature of AI research but also the pivotal role breakthroughs in computing hardware and data availability play in bringing theoretical concepts to life.

Overall, by understanding these key concepts of AI, including its origins, guiding principles, historical milestones, and deep learning breakthroughs, one can better appreciate how AI intersects with IoT. The synergy between IoT's ability to gather vast amounts of real-world data and AI's capacity to analyze and learn from this data holds significant promise for intelligent automation, predictive analytics, and adaptive systems that can transform how we live and work.

1.3 AI in IoT: The Synergy

By this point, it is clear that both IoT and AI have undergone significant developments. On one hand, IoT has succeeded in weaving intelligence into the fabric of everyday life by networking physical devices and collecting vast streams of data. On the other hand, AI has matured to the point where its advanced algorithms can sift through massive data sets to find meaningful patterns, predict outcomes, and automate complex decision processes. When these two fields converge, they give rise to intelligent environments that not only gather information from the world but also act on that information in real time, adjusting to changing conditions without the need for constant human oversight. The synergy between AI and IoT thus forms a powerful foundation for next-generation applications that promise unprecedented levels of efficiency, personalization, and automation.

1.3.1 Convergence of AI and IoT

The fusion of AI and IoT harnesses the strengths of both domains. IoT devices produce torrents of data – from simple temperature readings to video streams – capturing the state of everything from personal fitness to large-scale industrial systems. AI transforms this raw data into insights, detecting patterns that might elude traditional analytics or manual scrutiny. As a result, organizations can anticipate equipment failures, optimize resource usage, adapt services to individual preferences, and automate decisions that previously required human judgment.

Moreover, continuous improvements in hardware and software have reduced the barriers to deploying AI in resource-constrained environments. Embedded systems and edge computing platforms now have the capacity to run lightweight machine learning models, enabling real-time analytics and decision-making directly at the edge. This approach can lower latency, decrease bandwidth requirements, and preserve data privacy by limiting the transmission of sensitive information to the cloud. Whether in autonomous vehicles, smart healthcare devices, or industrial automation systems, the integration of AI at the edge represents an essential step in realizing the full potential of IoT.

Yet, bringing AI and IoT together also introduces new complexities. Data must be reliably collected, curated, and protected against security threats. AI models require careful design, from selecting appropriate algorithms to addressing potential biases in training data. In practical deployments, system architects must balance the computational costs of local versus cloud-based analytics, comply with evolving regulatory frameworks, and maintain user trust. Despite these hurdles, the growing synergy between AI and IoT is too compelling to ignore, offering both productivity gains in established industries and transformative possibilities in areas yet to be fully explored.

1.3.2 Overview of the Book

This book systematically addresses AI's transformative role in IoT communication, guiding the reader from fundamental principles to advanced, real-world applications. It opens with an introduction to the foundations of IoT communication and an overview of AI (including its historical milestones, core principles, and the emergence of deep learning). From there, each subsequent chapter explores a critical area where AI enhances the efficiency, reliability, and intelligence of IoT networks:

Chapter 2 delves into AI-enabled feedback communication for IoT, revealing how real-time adaptive transmission strategies can streamline data flow and make the most of limited resources at IoT devices. By learning from and reacting to the receiver's decoding status, AI-based feedback channel codes can intelligently generate the "parity" symbols to achieve ultra reliability.

Chapter 3 shifts the focus to AI-empowered semantic communication, demonstrating how meaningful, context-driven data exchange can dramatically improve both efficiency and interoperability among IoT devices. This chapter shows that by interpreting the semantics of a message, rather than merely transmitting raw bits, AI-powered systems can reduce data overhead while maintaining the integrity and usefulness of the information conveyed.

In Chap. 4, the spotlight moves to AI-driven multiple access, where techniques like multi-agent deep reinforcement learning (MADRL) and over-the-air computation (AirComp) are harnessed to optimize spectrum usage, mitigate interference, and enhance energy efficiency in crowded IoT environments. These approaches illustrate the next generation of wireless communication, where intelligent resource management ensures seamless, equitable access across numerous devices.

Chapter 5 tackles AI-enhanced network control, detailing how deep reinforcement learning (DRL), graph neural networks (GNNs), and large language models (LLMs) can overcome traditional scalability and adaptability barriers. From autonomously reconfiguring network topologies to contextualizing massive datasets of sensor information, these AI methods show significant promise in proactively shaping network behavior to meet ever-evolving demands.

Finally, Chap. 6 examines future trends and ethical considerations, offering insights into the emerging innovations that will continue to revolutionize IoT communication. Topics range from scaling issues in large-scale IoT deployments to governance and responsible AI principles, helping readers anticipate future challenges and develop solutions aligned with ethical best practices.

Throughout these chapters, readers will find both theoretical underpinnings and practical case studies, making this book a comprehensive guide for researchers, engineers, and practitioners working at the intersection of AI and IoT. By the end, one will be fully equipped with the concepts, strategies, and cutting-edge methods necessary to leverage AI for enhanced IoT communication.

References

1. S. Madakam, R. Ramaswamy, S. Tripathi, Internet of Things (IoT): A literature review. J. Comput. Commun. **3**(5), 164–173 (2015)
2. D.C. Nguyen, M. Ding, P.N. Pathirana, A. Seneviratne, J. Li, D. Niyato, O. Dobre, H.V. Poor, 6G Internet of Things: A comprehensive survey. IEEE Internet Things J. **9**(1), 359–383 (2021)
3. M. Kanj, V. Savaux, M. Le Guen, A tutorial on NB-IoT physical layer design. IEEE Commun. Surv. Tutorials **22**(4), 2408–2446 (2020)
4. R.S. Sinha, Y. Wei, S.-H. Hwang, A survey on LPWA technology: LoRa and NB-IoT. ICT Express **3**(1), 14–21 (2017)
5. H. Pan, S.C. Liew, J. Liang, Y. Shao, L. Lu, Network-coded multiple access on unmanned aerial vehicle. IEEE J. Sel. Areas Commun. **36**(9), 2071–2086 (2018)
6. Y. Shao, DEEP-IoT: Downlink-enhanced efficient-power Internet of Things. IEEE Trans. Wirel. Commun. **24**(2), 1722–1736 (2024)
7. I. Lee, K. Lee, The internet of things (IoT): applications, investments, and challenges for enterprises. Bus. Horiz. **58**(4), 431–440 (2015)
8. M. Marjani, F. Nasaruddin, A. Gani, A. Karim, I.A.T. Hashem, A. Siddiqa, I. Yaqoob, Big IoT data analytics: architecture, opportunities, and open research challenges. IEEE Access **5**, 5247–5261 (2017)
9. H. Cui, Y. Du, Q. Yang, Y. Shao, S.C. Liew, LLMind: Orchestrating AI and IoT with LLMs for complex task execution. IEEE Commun. Mag. **63**(4), 214–220 (2024)
10. Y. Shao, S.C. Liew, J. Liang, Sporadic ultra-time-critical crowd messaging in V2X. IEEE Trans. Commun. **69**(2), 817–830 (2020)
11. R. Zhang, Y. Shao, M. Li, L. Lu, Y.C. Eldar, Optical integrated sensing and communication with light-emitting diode. IEEE Internet Things J. **12**(9), 12896–12911 (2024)
12. Y. Shao, S.C. Liew, H. Chen, Y. Du, Flow sampling: Network monitoring in large-scale software-defined IoT networks. IEEE Trans. Commun. **69**(9), 6120–6133 (2021)
13. C. Bisdikian, An overview of the Bluetooth wireless technology. IEEE Commun. Mag. **39**(12), 86–94 (2001)
14. H.A. Omar, K. Abboud, N. Cheng, K.R. Malekshan, A.T. Gamage, W. Zhuang, A survey on high efficiency wireless local area networks: Next generation WiFi. IEEE Commun. Surv. Tutorials **18**(4), 2315–2344 (2016)
15. J.P. S. Sundaram, W. Du, Z. Zhao, A survey on LoRa networking: Research problems, current solutions, and open issues. IEEE Commun. Surv. Tutorials **22**(1), 371–388 (2019)
16. W. Shi, J. Cao, Q. Zhang, Y. Li, L. Xu, Edge computing: Vision and challenges. IEEE Internet Things J. **3**(5), 637–646 (2016)
17. Y. Mao, C. You, J. Zhang, K. Huang, K.B. Letaief, A survey on mobile edge computing: The communication perspective. IEEE Commun. Surv. Tutorials **19**(4), 2322–2358 (2017)
18. T. Xu, J.B. Wendt, M. Potkonjak, Security of IoT systems: Design challenges and opportunities, in *2014 IEEE/ACM International Conference on Computer-Aided Design (ICCAD)* (IEEE, Piscataway, 2014), pp. 417–423
19. L. Tawalbeh, F. Muheidat, M. Tawalbeh, M. Quwaider, IoT privacy and security: Challenges and solutions. Appl. Sci. **10**(12), 4102 (2020)
20. Y. Shao, D. Gündüz, S.C. Liew, Federated edge learning with misaligned over-the-air computation. IEEE Trans. Wirel. Commun. **21**(6), 3951–3964 (2021)
21. S. Safaric, K. Malaric, ZigBee wireless standard, in *Proc. ELMAR* (IEEE, Piscataway, 2006), pp. 259–262
22. A. Lavric, A.I. Petrariu, and V. Popa, Sigfox communication protocol: The new era of IoT?, in *International Conference on Sensing and Instrumentation in IoT Era (ISSI)* (IEEE, Piscataway, 2019), pp. 1–4
23. S.R. Borkar, Long-term evolution for machines (LTE-M), in *LPWAN Technologies for IoT and M2M Applications* (Elsevier, Cham, 2020), pp. 145–166

24. J.-P. Thomesse, Fieldbus technology in industrial automation. Proc. IEEE **93**(6), 1073–1101 (2005)
25. R. Weinstein, RFID: a technical overview and its application to the enterprise. IT Prof. **7**(3), 27–33 (2005)
26. Y. LeCun, Y. Bengio, G. Hinton, Deep learning. Nature **521**(7553), 436–444 (2015)
27. A.M. Turing, *Computing Machinery and Intelligence* (Springer, Dordrecht, 2009)
28. K. He, X. Zhang, S. Ren, J. Sun, Deep residual learning for image recognition, in *IEEE CVPR* (2016), pp. 770–778
29. A. Vaswani, N. Shazeer, N. Parmar, J. Uszkoreit, L. Jones, A.N. Gomez, L. Kaiser, I. Polosukhin, Attention is all you need. arXiv:1706.03762 (2017)
30. R.R. Murphy, *Introduction to AI Robotics* (MIT Press, Cambridge, 2019)
31. K. O'shea, R. Nash, An introduction to convolutional neural networks. Preprint. arXiv:1511.08458 (2015)
32. L. Torrey, J. Shavlik, Transfer learning, in *Handbook of Research on Machine Learning Applications and Trends: Algorithms, Methods, and Techniques* (IGI Global, Hershey, 2010), pp. 242–264
33. R. Dwivedi, D. Dave, H. Naik, S. Singhal, R. Omer, P. Patel, B. Qian, Z. Wen, T. Shah, G. Morgan et al., Explainable AI (XAI): Core ideas, techniques, and solutions. ACM Comput. Surv. **55**(9), 1–33 (2023)
34. A. Toosi, A.G. Bottino, B. Saboury, E. Siegel, A. Rahmim, A brief history of AI: how to prevent another winter (a critical review). PET Clin. **16**(4), 449–469 (2021)
35. T.P. Lillicrap, A. Santoro, L. Marris, C.J. Akerman, G. Hinton, Backpropagation and the brain. Nat. Rev. Neurosci. **21**(6), 335–346 (2020)
36. A. Krizhevsky, I. Sutskever, G.E. Hinton, Imagenet classification with deep convolutional neural networks, in *Advances in Neural Information Processing Systems*, vol. 25 (2012)
37. I. Goodfellow, J. Pouget-Abadie, M. Mirza, B. Xu, D. Warde-Farley, S. Ozair, A. Courville, Y. Bengio, Generative adversarial networks. Commun. ACM **63**(11), 139–144 (2020)
38. D. Silver, A. Huang, C.J. Maddison, A. Guez, L. Sifre, G. Van Den Driessche, J. Schrittwieser, I. Antonoglou, V. Panneershelvam, M. Lanctot et al., Mastering the game of Go with deep neural networks and tree search. Nature **529**(7587), 484–489 (2016)

Chapter 2
AI-Enabled Feedback Communication for IoT

He who learns to listen, masters the conversation.

2.1 IoT Communication with Feedback

The IoT has experienced unprecedented growth and adoption, driving the interconnection of billions of devices worldwide. With applications spanning from smart homes and cities to healthcare and industrial automation, the performance requirements placed on IoT communications have increased dramatically [1]. Among these requirements, energy efficiency and reliable communications stand out as critical. IoT devices, typically battery-powered, face significant constraints in power consumption [2], making it essential to innovate methods to enhance their operational lifespan without compromising their communication capabilities.

Feedback communication emerges as a potent solution in this context [3], providing a means to adaptively refine transmissions, thereby reducing unnecessary power usage and enhancing overall reliability. Unlike traditional IoT communication methods, which often rely on simple acknowledgment (ACK) or negative acknowledgment (NACK) feedback mechanisms, advanced feedback communication strategies [4–9] can dynamically adjust the coding and transmission power based on the receiver's decoding status and the instantaneous channel conditions.

2.1.1 Motivation

In conventional IoT communication models, the transmitter (IoT device) largely dominates the communication process, continuously sending data and receiving minimal feedback from the AP. This unidirectional approach, while straightforward, can be wasteful, especially in challenging environments where signals degrade due to interference, noise, or fading.

Feedback communication addresses this inefficiency by assigning the AP a more proactive role, allowing it to provide continuous or periodic information to IoT

Y. Shao, *AI-Empowered IoT Communications*, Wireless Networks,
https://doi.org/10.1007/978-3-031-98816-5_2

devices about channel quality, decoding success, and any adjustments required. With this knowledge, IoT devices can optimize their coding mechanism and transmission power in real time.

The primary advantage of incorporating feedback is the marked improvement in energy efficiency. In many IoT scenarios, devices operate under tight power constraints-batteries in remote sensors, wearables, or industrial monitors must last for months or even years. By listening for, and reacting to, receiver-side feedback, devices can:

- *Avoid Over-Transmission:* If the AP signals successful reception early, devices can halt further retransmissions, saving power [7].
- *Adapt Transmission Power:* Devices can lower power outputs when channel conditions are favorable, or revert to higher power (and potentially higher redundancy) only when needed [3].
- *Optimize Coding:* Using advanced feedback-aided coding schemes allows higher spectral efficiency under good channels and robust error protection under poor channels, without wasting energy over long code blocks [9].

From a reliability standpoint, feedback-based mechanisms enable near-instantaneous detection of decoding errors. This stands in contrast to classical systems that might rely on infrequent ACK/NACK mechanisms, which sometimes lead to delayed reactions and inefficient retransmissions. The near-real-time awareness of the receiver's status reduces the chance of unnecessary packet repetition and the corresponding energy overhead.

2.1.2 Comparison with Existing Schemes

Traditional IoT protocols such as NB-IoT or LoRa provide limited forms of feedback, primarily in the shape of simple ACK/NACK responses. While these minimal feedback systems can confirm correct reception, they do not facilitate substantial adaptation of transmission parameters. Current power-saving modes, such as enhanced Discontinuous Reception (eDRX) and Power Saving Mode (PSM) [2, 10], reduce energy consumption primarily by letting devices sleep for extended periods rather than by making active transmissions more power-efficient.

Hence, a device operating under legacy schemes may send the same data multiple times (up to a maximum repetition factor in NB-IoT [10]) just to combat deep fading in the channel. This repetitive approach can guarantee coverage but is inherently inefficient if channel conditions vary quickly. By contrast, more granular feedback enables devices to refine their coding mechanism and power levels on a per-packet or per-frame basis [8], drastically lowering the average transmission energy.

Furthermore, while hybrid automatic repeat request (HARQ) schemes [11] in some cellular standards use feedback to indicate whether a codeword is successfully decoded, they still rely on discrete repetition or incremental redundancy mechanisms, rather than truly adaptive, real-time adjustments of continuous amplitude signals and coding approaches. Leveraging deeper feedback beyond these legacy HARQ strategies opens new frontiers in optimizing each transmission dynamically, providing a continuous feedback loop that is especially relevant for IoT networks in harsh environments or with large coverage areas [12].

Roadmap

The remainder of this chapter is articulated in the following manner. Section 2.2 delves into the core information-theoretical results that underpin feedback's usefulness in communication systems. We present fundamental concepts, definitions, and theoretical bounds to highlight how and why feedback can drastically improve performance in IoT settings. Section 2.3 presents various AI-driven feedback codes. Section 2.4 envisions a feedback-enhanced IoT Communication framework with significantly enhanced energy efficiency and coverage. Section 2.5 concludes this chapter.

2.2 Information Theoretical Results

Information theory provides the bedrock on which modern communication systems are built [13]. In particular, channel coding, a central tenet of information theory, introduces carefully crafted redundancy to protect data against channel noise, ensuring reliable transfer of information even under imperfect transmission environments [14, 15]. Over the decades, coding theory has produced a variety of powerful schemes, many of them approaching or operating close to the Shannon limit, which indicates the maximum rate of error-free transmission a channel can achieve in the asymptotic regime [14]. This section discusses the fundamental definitions and concepts from information theory that guide feedback channel coding, outlining the known gains of feedback for enhancing reliability of data transmission.

2.2.1 Channel Coding

Communication system typically consists of four key components: a source (producing messages), an encoder, a channel, and a decoder, as illustrated in Fig. 2.1. The channel is described by a conditional probability distribution $P_{Y^n|X^n} : \mathcal{A}^n \to \mathcal{B}^n$, characterizing how the n-length channel input X^n in alphabet $\mathcal{A}^n$ is mapped (stochastically) to the output Y^n in alphabet $\mathcal{B}^n$.

- Source W: A random variable taking one of M possible symbols. Formally, $W \in \{1, 2, \ldots, M\}$.
- Encoder f: A function mapping each source message w to an n-length channel input sequence $x^n \in \mathcal{A}^n$. Thus, $f : \{1, 2, \ldots, M\} \to \mathcal{A}^n$.

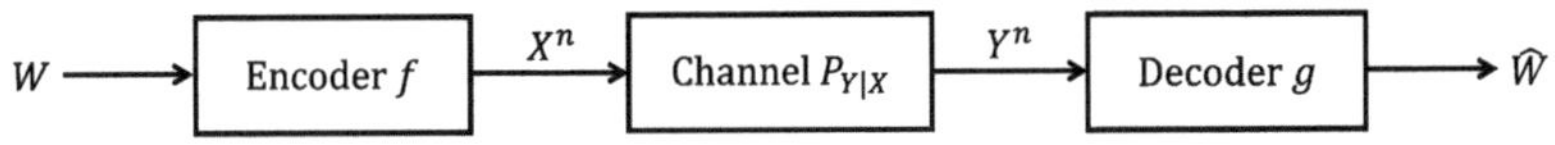

Fig. 2.1 A communication system operating in a discrete memoryless channel (DMC)

- Decoder g: A function mapping the channel output $y^n \in \mathcal{B}^n$ back to one of the M possible source messages, $g : \mathcal{B}^n \to \{1, 2, \ldots, M\}$.
- Block length n: The number of uses of the channel for each codeword.

Under a specific channel code, the overall information transmission can be modeled by the Markov chain

$$W \xrightarrow{f} X \xrightarrow{P_{Y|X}} Y \xrightarrow{g} \widehat{W}, \tag{2.1}$$

where W is the transmitted (true) message, X is the encoded channel input, Y is the channel output, and $\widehat{W}$ is the message estimate after decoding. Noise, interference, or other impairments may cause $\widehat{W} \neq W$ with some probability of error.

Definition 2.1 (Error Probability) We define the probability of error ϵ as

$$\epsilon \triangleq \Pr\big[W \neq \widehat{W}\big]. \tag{2.2}$$

To quantify how much information can be reliably conveyed in n channel uses under a certain constraint on ϵ, we introduce the notion of an (n, M, ϵ)-code.

Definition 2.2 ((n, M, ϵ)-code) An (n, M, ϵ)-code over a channel $P_{Y^n|X^n}$ is specified by a pair of functions (f, g) along with block length n, such that:

- $f : 1, 2, \ldots, M \to \mathcal{A}^n$ (the encoder),
- $g : \mathcal{B}^n \to 1, 2, \ldots, M$ (the decoder),

ensuring the overall probability of decoding error does not exceed ϵ.

We further define two fundamental limits related to channel codes [16]:

$$M^*(n, \epsilon) = \max\{M : \exists(n, M, \epsilon)\text{-code}\}, \tag{2.3}$$

$$\epsilon^*(n, M) = \inf\{\epsilon : \exists(n, M, \epsilon)\text{-code}\}. \tag{2.4}$$

The quantity $M^*(n, \epsilon)$ is the maximum number of distinct messages that can be transmitted reliably (with error probability not exceeding ϵ) using n channel uses.

The rate of a code is defined as

$$R \triangleq \frac{1}{n} \log_2 M, \tag{2.5}$$

capturing how many bits per channel use can be sent with acceptable error probability.

To further generalize performance to asymptotically large n, we define the ϵ-capacity.

Definition 2.3 The ϵ-capacity

$$C_\epsilon \triangleq \liminf_{n\to\infty} \frac{1}{n} \log M^*(n, \epsilon). \tag{2.6}$$

When $\epsilon \to 0^+$, one obtains the Shannon capacity C, i.e.,

$$C = \lim_{\epsilon\to 0^+} C_\epsilon, \tag{2.7}$$

which is the highest rate at which arbitrarily reliable communication is possible. Shannon's groundbreaking result showed that as n grows large, it becomes theoretically possible to transmit at rates below C with vanishing error probability [13]. Above C, the error probability remains bounded away from zero.

Yury Polyanskiy et al. [16] refined these ideas to finite block length settings, showing that for discrete memoryless channels (DMCs):

$$\log M^*(n, \epsilon) = nC - \sqrt{nV} Q^{-1}(\epsilon) + O(\log n), \tag{2.8}$$

where C is the channel capacity, V is the channel dispersion, and Q^{-1} is the inverse of the Q-function. This characterizes how code rate falls below capacity when n is finite.

2.2.2 *Feedback Channel Coding*

Now consider a channel augmented with feedback, as shown in Fig. 2.2, where the transmitter can access some function of the previous channel outputs Y_τ, $\tau < t$ before sending X_t. Thus, the input of the forward channel $P_{Y|X}$ at time t depends not only on the message W but also on all previously received outputs $Y_1, Y_2, \ldots, Y_{t-1}$.

An (n, M, ϵ)-code with feedback is then characterized by encoder mappings f_t that adapt as the transmission proceeds. We define [7]

$$M^*_{\text{fb}}(n, \epsilon) \triangleq \max\{M : \exists (n, M, \epsilon)\text{-code with feedback}\}, \tag{2.9}$$

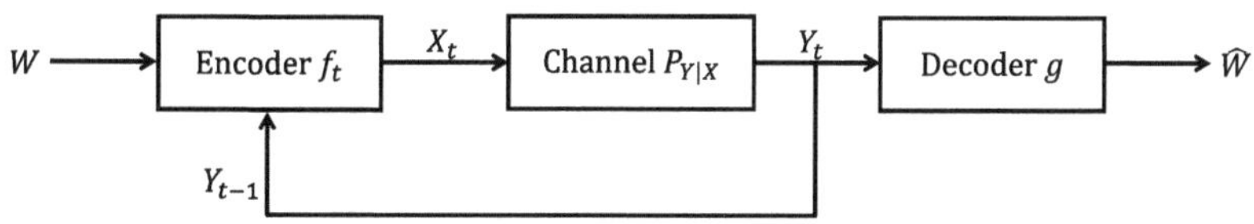

Fig. 2.2 A communication system operating in a discrete memoryless channel (DMC) with feedback

$$C_{\mathrm{fb},\epsilon} \triangleq \liminf_{n\to\infty} \frac{1}{n} \log M^*_{\mathrm{fb}}(n,\epsilon), \tag{2.10}$$

$$C_{\mathrm{fb}} \triangleq \lim_{\epsilon\to 0^+} C_{\mathrm{fb},\epsilon}. \tag{2.11}$$

Shannon proved as early as 1956 that feedback does not increase the capacity of discrete memoryless channels (DMC) [17], i.e., $C_{\mathrm{fb}} = C$, meaning feedback does not increase capacity in the strict asymptotic sense.

In the case of finite bock length, Yury Polyanskiy et al. demonstrated that for a class of weakly input symmetric DMC channels (e.g., additive noise, binary symmetric channel, binary erasure channel), the following formula still holds if we consider fixed block length coding [7]

$$\log M^*_{\mathrm{fb}}(n,\epsilon) = nC - \sqrt{nV}\,Q^{-1}(\epsilon) + O(\log n). \tag{2.12}$$

Thus, in fixed-block-length scenarios, feedback does not increase channel capacity or improve the convergence rate to capacity fundamentally; instead, it can only reduce higher-order (i.e., the third-order $O(\log n)$) terms. Nevertheless, feedback does offer dramatic advantages once these additional constraints, e.g., fixed-block-length and fixed-power, are relaxed.

2.2.2.1 Variable-Length Schemes

Traditional channel coding analyses typically assume fixed block lengths, thus limiting adaptability to varying channel conditions. In contrast, the introduction of feedback notably allows for dynamic adaptation of the coding rate in response to actual noise realizations encountered during transmission.

Specifically, for any channel with capacity C, employing feedback in conjunction with variable-length coding significantly improves achievable transmission rates. Such feedback-enabled variable-length coding can achieve an ϵ-capacity:

$$C_\epsilon = \frac{C}{1-\epsilon}. \tag{2.13}$$

Yury Polyanskiy et al. further quantified these benefits by establishing upper and lower bounds on the maximum number of reliably transmitted messages $M^*_{\mathrm{fb}}(n,\epsilon)$ for DMCs [7]:

$$\frac{nC}{1-\epsilon} - \log n + O(1) \leq \log M^*_{\mathrm{fb}}(n,\epsilon) \leq \frac{nC}{1-\epsilon} + O(1). \tag{2.14}$$

Moreover, if the transmitter is permitted to send an additional one-bit termination signal reliably to the receiver, signaling the conclusion of the message transmission, a further refinement to this result can be obtained [7]

$$\log M^*_{\text{fb}}(n, \epsilon) \leq \log M^*_{\text{fb,t}}(n, \epsilon) \leq \frac{nC + \log n}{1 - \epsilon} + O(1), \tag{2.15}$$

where $M^*_{\text{fb,t}}(n, \epsilon)$ denotes the fundamental limit of feedback codes augmented with such termination signals.

Comparing these results, we observe that incorporating feedback—especially with the capability of sending a reliable termination signal—significantly reduces the block length required to achieve a specific error probability. This advantage becomes especially pronounced when contrasting variable-length feedback codes with classical fixed-length schemes. Feedback thus enables practical IoT systems to achieve transmission rates much closer to the Shannon capacity at significantly reduced block lengths.

2.2.2.2 Variable-Power Schemes

For an additive white Gaussian noise (AWGN) channel, the effectiveness of feedback strongly depends on the nature of the power constraint imposed on the transmitter. Under strict instantaneous power constraints, feedback provides limited benefit. However, if the constraint is relaxed from instantaneous power control to an average power constraint, expressed mathematically as

$$\mathbb{E}\left[\|x^n\|^2\right] \leq nP, \tag{2.16}$$

the potential benefits of feedback become significantly more pronounced.

Utilizing this relaxation, Elias introduced an innovative coding scheme leveraging feedback. Elias's approach achieves substantial performance gains by iteratively refining the receiver's message estimate through feedback, progressively reducing the estimation error [18]. Subsequently, Schalkwijk and Kailath developed the Schalkwijk-Kailath (SK) scheme [4, 19], which extends Elias's ideas using pulse amplitude modulation (PAM). The SK scheme has become notable for its simplicity, achieving second-order exponential decay of the decoding error probability with linear encoding and decoding complexity. By using feedback to iteratively refine transmitted signals, the sender progressively corrects errors at the receiver, thus driving the receiver's estimates towards the true message and enhancing communication reliability.

Consider the communications between two nodes A and B, where node A aims to transmit a bitstream $\boldsymbol{b} \in \{0, 1\}^K$ to node B. The feedforward and feedback transmissions can be represented by

$$\boldsymbol{y} = \boldsymbol{x} + \boldsymbol{w}, \tag{2.17}$$

$$\boldsymbol{y}' = \boldsymbol{y} + \boldsymbol{w}' \triangleq \boldsymbol{x} + \overline{\boldsymbol{w}}, \tag{2.18}$$

Algorithm 1 The SK scheme

```
Input: b ∈ {0, 1}^K, N, P, P';
Initialization:
Node A maps b to a normalized 2^K-ary PAM constellation Θ;
Node A transmits x_1 = √P Θ to node B;
Node B receives y_1 = x_1 + w_1 and estimates the PAM constellation by Θ̂_1 = (1/√P) y_1;
Interaction:
while n = 1, 2, 3, ..., N − 1 do
    Node B feeds back Θ̂_n to node A via the noiseless feedback channel;
    Node A computes the estimation error of node B: u_n = Θ̂_n − Θ;
    Node A transmits x_{n+1} = √(P/ε_n^2) u_n to node B, where ε_n^2 = E(u_n^2);
    Node B receives y_{n+1} = x_{n+1} + w_{n+1};
    Node B updates its estimate by Θ̂_{n+1} = Θ̂_n − (√(P ε_n^2) / (σ̃_f^2 + P)) y_{n+1};
Decoding:
Node B decodes Θ from Θ̂_N using a minimum distance decoder.
```

where $\boldsymbol{x} \in \mathcal{R}^N$ consists of encoded symbols at node A (i.e., a codeword); $\boldsymbol{y}' \in \mathcal{R}^N$ consists of the received feedback from node B; $\boldsymbol{w} \in \mathcal{R}^N$ and $\boldsymbol{w}' \in \mathcal{R}^N$ are real AWGN vectors, and their elements follow $\mathcal{N}(0, \tilde{\sigma}_f^2)$ and $\mathcal{N}(0, \tilde{\sigma}_b^2)$, respectively; $\overline{\boldsymbol{w}} = \boldsymbol{w} + \boldsymbol{w}'$ consists of the aggregated noise realizations observed at node A. Nodes A and B are subjective to power constraints P and P', respectively.

The SK scheme is summarized in Algorithm 1, where the feedback channel is assumed to be noise-free. As can be seen, the SK scheme breaks the transmission into two phases.

In the initialization phase, node A maps the bitstream $\boldsymbol{b}$ to a constellation point Θ of a normalized M-ary PAM. In particular, $\Theta \in \{\pm 1\xi, \pm 3\xi, \pm(2^K - 1)\xi\}$, where $M = 2^K$ and $\xi = \sqrt{\frac{3}{2^{2K} - 1}}$ is the normalization factor to ensure that the average power of the PAM constellations is normalized to 1. Node A will first transmit Θ to node B via the noisy feedforward channel and node B will compute an estimate of Θ, i.e., $\hat{\Theta}_1 = \frac{1}{\sqrt{P}} y_1$.

In successive interactions, node B feeds back its current estimate $\hat{\Theta}_n$ to node A via the noiseless feedback channel, from which node A knows the current estimation error of node B, i.e., $u_n = \hat{\Theta}_n - \Theta$. To decrease node B's estimation error, node A transmits a scaled version of u_n to node B via the noisy feedforward channel, and node B updates its estimate by

$$\hat{\Theta}_{n+1} = \hat{\Theta}_n - \frac{\sqrt{P \epsilon_n^2}}{\tilde{\sigma}_f^2 + P} y_{n+1}, \tag{2.19}$$

where y_{n+1} is the observed noisy estimation error at node B.

Finally, after N interactions, node B decodes the transmitted PAM constellation by the minimum distance decoding. With the SK scheme, the authors proved that

Algorithm 2 The modulo-SK scheme

Input: $\boldsymbol{b} \in \{0, 1\}^K, N, P, P', \alpha, \{\gamma_n, \beta_n : n = 1, 2, \ldots, N\}$;
Initialization:
  Node A maps $\boldsymbol{b}$ to a normalized 2^K-ary PAM constellation Θ;
  Node A transmits $x_1 = \sqrt{P}\Theta$ to node B;
  Node B receives $y_1 = x_1 + w_1$ and estimates the PAM constellation by $\hat{\Theta}_1 = \frac{1}{\sqrt{P}} y_1$;
Interaction:
while $n = 1, 2, 3, \ldots, N-1$ **do**
  Given $\hat{\Theta}_n$, node B feeds back $x'_n = \mathbb{M}_d[\gamma_n \hat{\Theta}_n + V_n]$ to node A;
  Node A receives $y'_n = x'_n + w'_n$;
  Node A estimates the estimation error of node B by $u'_n = \mathbb{M}_d[y'_n - \gamma_n \Theta - V_n]$;
  Node A transmits $x_{n+1} = \alpha u'_n$ to node B;
  Node B receives $y_{n+1} = x_{n+1} + w_{n+1}$;
  Node B updates its estimate by $\hat{\Theta}_{n+1} = \hat{\Theta}_n - \beta_{n+1} y_{n+1}$;
Decoding:
  Node B decodes Θ from $\hat{\Theta}_N$ using a minimum distance decoder.

the decoding error of node B decays doubly exponentially fast in N for any rate below capacity.

The SK scheme was designed for the noiseless feedback setting, wherein node A perfectly observes the estimation error of node B. However, as a linear encoding scheme, the SK scheme fails when arbitrarily low noise is present in the feedback channel [19, 20]. In this context, the authors in [5] proposed a simple and efficient scheme, dubbed the modulo-SK scheme, for the AWGN-channel case to tackle noisy feedback channel.

The modulo-SK scheme is summarized in Algorithm 2. As can be seen, the initialization phase is the same as the SK scheme. Node A modulates the bitstream to a PAM constellation and transmits it to node B, from which node B generates the first estimate of Θ, i.e., $\hat{\Theta}_1$.

In the interaction phase, instead of $\hat{\Theta}_n$, node B feeds back a scaled version of $\hat{\Theta}_n$, modulo a fixed interval $[-\frac{d}{2}, \frac{d}{2})$, to node A. Specifically, the transmitted symbol from node B is given by

$$x'_n = \mathbb{M}_d[\gamma_n \hat{\Theta}_n + V_n], \tag{2.20}$$

where $\mathbb{M}_d[x] \in [-\frac{d}{2}, \frac{d}{2})$ denotes the modulo operation; $V_n \sim \text{Uniform}(-\frac{d}{2}, \frac{d}{2})$ is a common dither between nodes A and B (the dither can be ignored in the implementation); and γ_n is a scaling factor. In particular, d is set to $\sqrt{12P'}$ to ensure that node B's power constraint is satisfied.

Upon receiving y'_n, node A extracts a noisy scaled version of the estimation error $u'_n = \mathbb{M}_d[y'_n - \gamma_n \Theta - V_n]$, and transmits a scaled u'_n to node B. Specifically, the transmitted symbol from node A is $x_{n+1} = \alpha u'_n$, where α is chosen to meet node A's power constraint. Finally, node B receives y_{n+1} and updates its estimate of Θ by $\hat{\Theta}_{n+1} = \hat{\Theta}_n - \beta_{n+1} y_{n+1}$, where $\beta_{n+1} y_{n+1}$ is the minimum mean square error (MMSE) estimate of the estimation error.

After N interactions, node B decodes the transmitted PAM constellation by minimum distance decoding. The decoding error of the modulo-SK scheme is governed by (1) the probability that the remaining estimation noise in $\hat{\Theta}_n$ exceeds half the minimum distance of the PAM constellations, and (2) the event that the modulo aliasing [5] arises in one of the interactions. Compared with the SK scheme that completely fails when the feedback channel is noisy, the modulo-SK scheme works well when the feedback channel SNR is significantly larger than the feedforward channel SNR.

2.3 AI-Driven Feedback Codes

The information theory results showed that while capacity itself is not necessarily boosted by feedback in strict asymptotic limits, feedback can substantially lower blocklength requirements, reduce capacity gap, or accelerate error decay. Techniques like the SK scheme and its variants illustrate the powerful synergy of iteration, feedback, and intelligent power usage in AWGN channels.

When paired with AI-based learning methods, these feedback-driven ideas gain new potential. By incorporating real-time decoding results at the receiver, AI can guide the transmitter on how to encode or allocate power at each step. This enables practical IoT systems to use feedback not just for confirming receipt but to adaptively refine transmissions in a highly efficient manner. This section will expand upon these insights, and introduces a variety of flexible AI-enabled feedback code designs.

2.3.1 AI-Driven Symbol-Wise Feedback Codes

Early attempts in AI-driven feedback code focus on Shannon's immediate feedback channel model, where receiver feedback a symbol after received a symbol—we term these class of codes symbol-wise feedback codes.

2.3.1.1 DeepCode

DeepCode [6] is the first deep learning architecture to leverage AI techniques to design feedback code. Since the forward and backward channels are known to be AWGN channels with perfectly-known SNR, we can integrate the transmitter (node A), the channels, and the receiver (node B) into an overall system so that the interactions among the four parties become the internal logic of the integrated system. In this way, end-to-end supervised learning to jointly train the encoder and decoder is possible.

DeepCode considers passive feedback: node B will feedback the received noisy symbol verbatim to the transmitter. Both the encoder at node A and decoder at node B are parameterized to be deep neural networks (DNNs). For the passive feedback scheme, it has been proven in [20] that no linear code can achieve a positive communication rate when the feedback link is noisy. DNNs, on the other hand, are known for their abilities to approximate nonlinear functions, and hence, have the potential to learn good nonlinear feedback codes.

DeepCode consists of two transmission phases. The first phase is a BPSK transmission phase, in which the source bitstream $\boldsymbol{b}$ are BPSK modulated and transmitted to node B in K forward channel uses. The feedback information is not utilized in the first phase. The second phase, on the other hand, is a forward-coding phase with passive feedback: the forward transmission is coded and the feedback transmission is uncoded. Following (2.17) and (2.18), node A is equipped with a DNN encoder to generate the coded symbol in each interaction; node B is equipped with a DNN decoder to decode the source bitstream after collecting N noisy symbols.

Figures 2.3 and 2.4 illustrate the encoding and decoding processes in DeepCode, respectively, where the DNN encoder and decoder are chosen to be recurrent neural networks (RNNs) [21] with gated recurrent unit (GRU) [22]. Figure 2.3a shows the encoding process at node A in the second transmission phase of DeepCode. As can be seen, each encoding produces two coded symbols, denoted by $x_{1,k}$ and $x_{2,k}$. These two symbols will be transmitted to node B in one interaction and fed back to

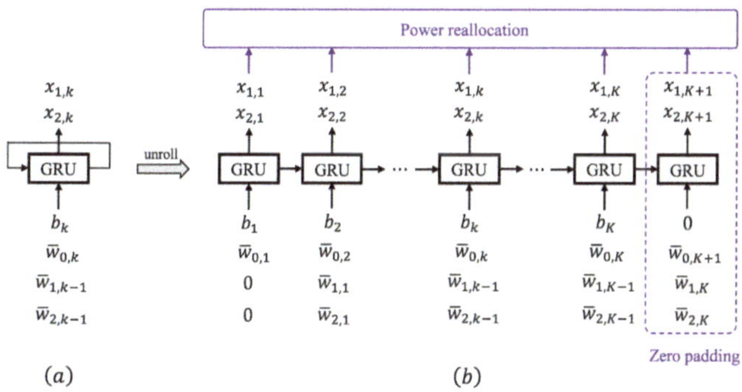

Fig. 2.3 DeepCode: the encoder design at node A

Fig. 2.4 DeepCode: the decoder design at node B

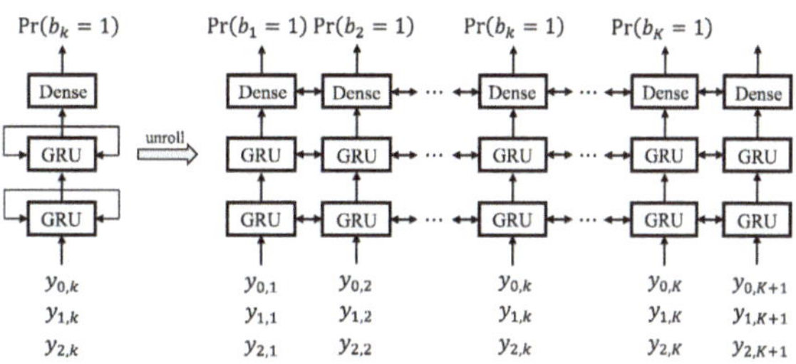

node A in a passive fashion. The feedback symbols received at node A are denoted by $y'_{1,k} \triangleq x_{1,k}+\overline{w}_{1,k}$ and $y'_{2,k} \triangleq x_{2,k}+\overline{w}_{2,k}$, where $\overline{w}_{1,k}$ and $\overline{w}_{2,k}$ are the aggregated noise realizations encountered by the first and second coded symbols, respectively. Since $x_{1,k}$ and $x_{2,k}$ are known at node A, $\overline{w}_{1,k}$ and $\overline{w}_{2,k}$ can be perfectly recovered and used for the encoding of $x_{1,k+1}$ and $x_{2,k+1}$.

In each encoding, the GRU takes four new symbols as input, in addition to the hidden states generated from the last encoding (i.e., the recurrent structure shown in Fig. 2.3a). The four new symbols are: the k-th bit b_k; the aggregated noise realization encountered by b_k in the first phase $\overline{w}_{0,k}$, where '0' in the subscript denotes the first phase; and the aggregated noise realizations encountered by $x_{1,k-1}$ and $x_{2,k-1}$ in the last transmission, i.e., $\overline{w}_{1,k-1}$ and $\overline{w}_{2,k-1}$. In particular, when $k = 1$, the authors set $\overline{w}_{1,0} = \overline{w}_{2,0} = 0$.

An unrolled version of the encoder is shown in Fig. 2.3b. To summarize, to transmit a stream of K bits, the first phase produces K real symbols and the second phase produces $2K$ real symbols. Therefore, the code rate of DeepCode is $K/3K = 1/3$.

In the final designs of DeepCode, the authors employed two techniques at node A, dubbed zero padding and power reallocation, to alleviate the unbalanced BER at different positions and further enhance the performance.

- **Zero padding**. As shown in Fig. 2.3b, an extra bit "0" is padded at the end of sequence $\boldsymbol{b}$. Zero padding consumes extra channel resources to transmit the original K bits, but brings a significant improvement to the BER performance.
- **Power reallocation**. The authors proposed to reallocate power for encoded symbols at different positions. The reallocation weights are learnable parameters and are determined over the course of training.

After all the interactions, node B receives $\{y_{0,k} : k = 1, 2, \ldots, K + 1\}$ in the first phase and $\{y_{1,k}, y_{2,k} : \forall k\}$ in the second phase, and starts to decode the bitstream. Unlike the encoder, the GRU in the decoder is chosen to be bi-directional since there is no causal constraint at node B. Given $\{y_{0,k}, y_{1,k}, y_{2,k} : \forall k\}$ as inputs, the bi-directional GRU aims to decode the transmitted bitstream.

2.3.1.2 AttentionCode

Following the footprint of DeepCode, [23] put forth the AttentionCode architecture. DeepCode relies on the recurrent encoder and decoder structures with GRU or LSTM as the recurrent units. AttentionCode, in contrast, explores a new network architecture, which is based purely on the attention mechanism [24], to design feedback codes. The architecture of AttentionCode is outlined in Fig. 2.5.

AttentionCode adopts the same two-phase transmission paradigm with passive feedback as DeepCode. As shown in Fig. 2.5, in the first phase, node A transmits quadrature phase shift keying (QPSK)-modulated symbols to node B in an uncoded fashion. Following (2.17) and (2.18), let us denote the received passive feedback at

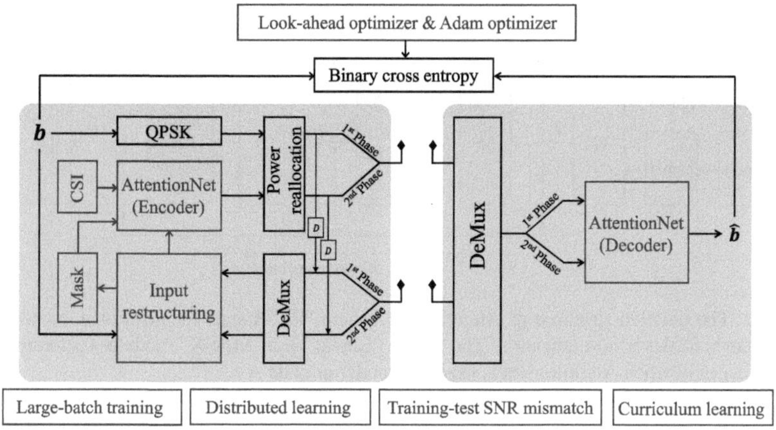

Fig. 2.5 The architecture of AttentionCode with two transmission phases

node A by $y'_{0,k} \triangleq x_{0,k} + \overline{w}_{0,k}$, $k = 1, 2, \ldots, K$, where '0' in the subscript denotes the first phase, $x_{0,k}$ is the QPSK symbol, and $\overline{w}_{0,k}$ is the aggregated noise realization encountered by $x_{0,k}$.

In the second phase, the encoder sequentially generates $2K$ real coded symbols and transmits them in $\widetilde{N} = K$ interactions. In the k-th interaction, two real coded symbols $x_{1,k}$ and $x_{2,k}$ are generated, and we denote the feedback symbols received at node A by $y'_{1,k} \triangleq x_{1,k} + \overline{w}_{1,k}$ and $y'_{2,k} \triangleq x_{2,k} + \overline{w}_{2,k}$, where $\overline{w}_{1,k}$ and $\overline{w}_{2,k}$ are the aggregated noise realizations encountered by the first and second coded symbols, respectively. Since $x_{1,k}$ and $x_{2,k}$ are known to node A, $\overline{w}_{1,k}$ and $\overline{w}_{2,k}$ can be obtained for the encoding of $x_{1,k+1}$ and $x_{2,k+1}$ in the next interaction.

After the two transmission phases, node B feeds all three received sequences $\{y_{0,k}, y_{1,k}, y_{2,k} : \forall k\}$ into the decoder. The encoder and decoder networks are jointly trained to minimize the binary cross-entropy loss between the transmitted bit stream $\boldsymbol{b}$ and the estimated probabilities:

$$\mathcal{L} = \frac{1}{K} \sum_{k=1}^{K} \Big[- b_k \log \Pr(b_k = 1) - (1 - b_k) \log \big(1 - \Pr(b_k = 1)\big) \Big]. \tag{2.21}$$

AttentionNet

Transformer is a sequence-to-sequence DNN model. It is composed of several encoding and decoding blocks [24]: the encoding blocks take the input sequence and map it into a higher-dimensional feature vector; the decoding blocks, on the other hand, take the feature vector as the input and decode it to the target sequence. AttentionCode utilizes a simplified version of the transformer, dubbed *AttentionNet*. In particular, AttentionNet is modified from the encoding blocks of the transformer

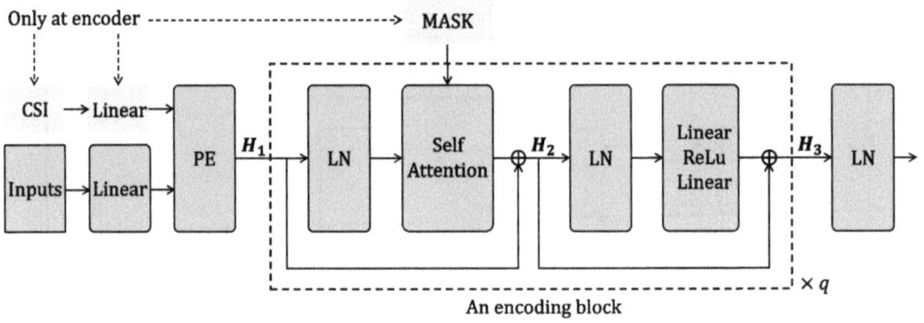

Fig. 2.6 The detailed structure of AttentionNet, where "PE" stands for positional encoding and "LN" stands for layer normalization. The "CSI", "Linear" and "MASK" modules (colored in light gray) exist only when AttentionNet is used as encoder at node A

and does away with the decoding blocks. In the literature, a similar model is known as bidirectional encoder representations from transformers (BERT) [25]. However, BERT often refers to the pre-trained language models using unsupervised learning. To avoid ambiguity, here we refer to the pure-attention-based DNNs with the encoding blocks of transformer as AttentionNet.

The detailed structure of AttentionNet is presented in Fig. 2.6. To illustrate the operations inside AttentionNet, we focus on AttentionNet as the encoder and assume that the input to AttentionNet is the same as DeepCode. The input matrix will be restructured later in Sect. 2.3.1.2.

Let us start from the "Inputs" module of Fig. 2.6. Denote by $\boldsymbol{S}(k)$ the input matrix to the encoder in the k-th interaction of the second phase. We have

$$\boldsymbol{S}(k) = \begin{bmatrix} b_1 & b_2 & b_3 & \dots & b_k \\ \overline{w}_{0,1} & \overline{w}_{0,2} & \overline{w}_{0,3} & \dots & \overline{w}_{0,k} \\ 0 & \overline{w}_{1,1} & \overline{w}_{1,2} & \dots & \overline{w}_{1,k-1} \\ 0 & \overline{w}_{2,1} & \overline{w}_{2,2} & \dots & \overline{w}_{2,k-1} \end{bmatrix}. \tag{2.22}$$

In the k-th interaction, $k = 1, 2, \dots, K+1$, the dimension of $\boldsymbol{S}(k)$ is $d_S \times k$, where $d_S = 4$ when AttentionNet is used as the encoder and $d_S = 3$ when AttentionNet is used as the decoder (because the input to the decoder consists of three sequences of received symbols: $\{y_{0,k}, y_{1,k}, y_{2,k} : \forall k\}$).

The input matrix $\boldsymbol{S}$ will first be passed through a linear layer for feature extraction and a positional encoding (PE) layer, yielding

$$\boldsymbol{H}_1 = \boldsymbol{L}_1 \boldsymbol{S} + \boldsymbol{W}_{\text{pos}}, \tag{2.23}$$

where $\boldsymbol{L}_1 \in \mathcal{R}^{d_m \times d_S}$ represents the weight matrix of the linear layer; $\boldsymbol{W}_{\text{pos}} \in \mathcal{R}^{d_m \times k}$ is a constant matrix, representing the positional information embedded in the input matrix (we use the same W_{pos} as [24]);[1] and d_m is a variate that determines the model size.

Then, H_1 will go through q encoding blocks. For each encoding block, we adopt a pre-normalization framework proposed in [26]. That is, input data are first layer-normalized [27] before they are fed into the self-attention module or the linear module, as shown in Fig. 2.6. Let us abstract a single encoding block as a function g, then the final output of the encoding layers is $H_3 = g^q(H_1)$. Since the encoding block is the most essential part of the AttentionNet, we will focus on how the encoding block works in the following.

To ease exposition, we assume $q = 1$ and explain how to get H_3 from H_1. The processing in one encoding block can be written as

$$H_2 = \text{SelfAttention}(\text{LN}(H_1)) + H_1, \tag{2.24}$$

$$H_3 = L_3\text{ReLU}(L_2(\text{LN}(H_2))) + H_2, \tag{2.25}$$

where "LN" stands for the layer normalization module; SelfAttention stands for the self-attention module, which will be detailed later; $L_2 \in \mathcal{R}^{4d_m \times d_m}$ and $L_3 \in \mathcal{R}^{d_m \times 4d_m}$ are the weight matrices of the linear layers, respectively. Note that LN and SelfAttention modules do not change the shape and size of their inputs.

The output $H_3 \in \mathcal{R}^{d_m \times k}$ will be layer normalized and used to generate the encoded symbols (at the encoder) or decode the bitstream (at the decoder). This can be realized simply by a linear transformation, i.e., a fully-connected layer with weights of size $\mathcal{R}^{2 \times d_m}$.

The essential component of the encoding block is the self-attention module. Let the input and output of the self-attention module be matrices J and J', respectively. Then, J' is the features extracted from J and they have the same size. In particular, the i-th column of J' is a linear combination of J's columns, and the combination weights is determined by the inter-correlations between the i-th column of J and the other columns of J.

More rigorously, let $J = \text{LN}(H_1) \in \mathcal{R}^{d_m \times k}$. We first generate a query matrix Q, a key matrix K, and a value matrix V [24] using three weight matrices $W^q \in \mathcal{R}^{d_m \times d_m}$, $W^k \in \mathcal{R}^{d_m \times d_m}$, $W^v \in \mathcal{R}^{d_m \times d_m}$, respectively, where $Q = W^q J$, $K = W^k J$ and $V = W^v J$. A correlation matrix capturing the inter-column correlations of J can be computed by $C = Q^\top K$.

For a given mask matrix $M \in \mathcal{R}^{k \times k}$, the masked correlation matrix is given by $C' = C \odot M$, where $\odot$ is the element-wise product. A commonly used mask in

[1] The self-attention model treats each column of $L_1 S$ as independent of the other and cannot recognize the sequential order of the columns. As a result, positional encoding explicitly adds the positional information to the input such that the knowledge about the order of columns is maintained in the input.

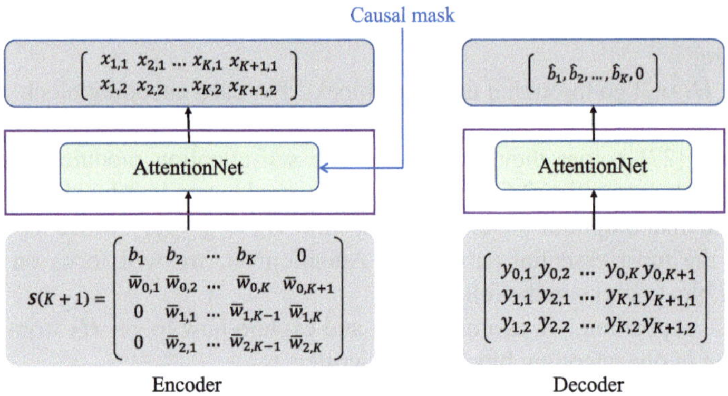

Fig. 2.7 Substituting AttentionNet for RNN in DeepCode

transformer is called the causal mask given by

$$\boldsymbol{M} = \begin{bmatrix} 1 & -\infty & -\infty & \dots & -\infty \\ 1 & 1 & -\infty & \dots & -\infty \\ 1 & 1 & 1 & \dots & -\infty \\ 1 & 1 & 1 & \dots & 1 \end{bmatrix}_{(K+1)\times(K+1)}. \tag{2.26}$$

The final output of the self-attention block is obtained as

$$\boldsymbol{J}' = \boldsymbol{V}\text{softmax}(\boldsymbol{C}'). \tag{2.27}$$

Note that after passing through the softmax operation in (2.27), $\boldsymbol{M}$ becomes a left triangular matrix.

The causal mask enables the parallel computation of coded symbols in the training phase. As shown in Fig. 2.7, we can feed all the information (i.e., the matrix $\boldsymbol{S}(K+1)$) simultaneously into the AttentionNet so that all the coded symbols can be computed in a parallel fashion. In this process, a causal mask is essential since it guarantees that a coded column $[x_{k,1}, x_{k,2}]^\top$ can only access the noise realizations that have already been observed, i.e., $\{\overline{w}_{0,k'}, \overline{w}_{1,k'-1}, \overline{w}_{2,k'-1} : k' = 1, 2, \dots, k\}$, but not those that will be observed in the future. In other words, albeit the real interactions among nodes A and B proceed sequentially, the training of Fig. 2.7 proceeds in parallel with the use of a "causal mask" to prevent future-information leakage, which makes training more efficient. This mechanism is similar to the unidirectional RNN in DeepCode.

About the Multi-Head Attention

Compared with different variations of transformer, AttentionNet does away with two widely-used techniques: one is the multi-head attention mechanism and the other is dropout, because it is empirically found that they hinder AttentionCode from achieving low BLER. The reason behind these empirical observations may be the fundamental difference between the channel coding problem and classical DL tasks that DL for channel coding does not always pursue the generalizability of DNNs.

Functions of AttentionNet

It is worth noting that AttentionNet serves two purposes: (1) as the encoder at node A to *create* temporal correlations among the encoded symbols; (2) as the decoder at node B to *exploit* the temporal correlations among coded symbols, and hence, combat channel impairments such as noise and fading.

Input Restructuring

As shown in Fig. 2.7, if we replace the RNN by AttentionNet, the information used to generate $x_{k,1}$ and $x_{k,2}$ is the first k columns of the input, i.e., the matrix $\boldsymbol{S}(k)$ given in (2.22). In particular, due to the delayed feedback, the last two rows of $\boldsymbol{S}(k)$ are delayed by one column compared with the first two rows.

In AttentionCode, we propose to reorganize $\boldsymbol{S}(k)$ so that the noise realizations encountered by a single bit in two transmission phases are aligned in one column of $\boldsymbol{S}(k)$. After restructuring, the new input matrix is given by

$$\widetilde{\boldsymbol{S}}(k)=\begin{bmatrix} b_1 & b_2 & b_3 & \cdots & b_{k-1} & b_k \\ \overline{w}_{0,1} & \overline{w}_{0,2} & \overline{w}_{0,3} & \cdots & \overline{w}_{0,k-1} & \overline{w}_{0,k} \\ \overline{w}_{1,1} & \overline{w}_{1,2} & \overline{w}_{1,3} & \cdots & \overline{w}_{1,k-1} & 0 \\ \overline{w}_{2,1} & \overline{w}_{2,2} & \overline{w}_{2,3} & \cdots & \overline{w}_{2,k-1} & 0 \end{bmatrix}. \tag{2.28}$$

This scheme, dubbed "input restructuring", turns out to be very effective. One possible explanation is that the restructured input matrix $\widetilde{\boldsymbol{S}}(k)$ matches the nature of self-attention that focuses on the column-wise correlations among the inputs.

Although input restructuring is a small modification to the input, it brings about a fundamental change in the system training from parallel to sequential. Specifically, without input restructuring, the difference between $\boldsymbol{S}(k)$ and $\boldsymbol{S}(k+1)$ is only the last column of $\boldsymbol{S}(k+1)$. Therefore, all $K+1$ columns can be fed into AttentionNet simultaneously, provided that a causal mask is used to prevent information leakage. In so doing, the overall training proceeds in parallel and all the coded symbols can be computed simultaneously.

With input restructuring, on the other hand, $\widetilde{\boldsymbol{S}}(k+1)$ differs from $\widetilde{\boldsymbol{S}}(k)$ in last two columns. More importantly, unlike (2.22), the generation of $\widetilde{\boldsymbol{S}}(k)$ in (2.28) is no longer recurrent: the generation of $\widetilde{\boldsymbol{S}}(k+1)$ does not implicitly generate $\widetilde{\boldsymbol{S}}(k')$, $\forall k' \leq k$. As a consequence, AttentionCode can only be trained sequentially: in each

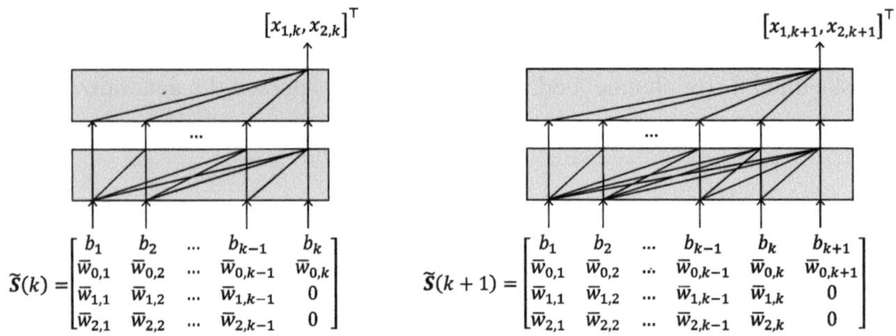

Fig. 2.8 AttentionCode: an illustration of input restructuring with masks. The grey blocks stand for the self-attention module, wherein the arrows stand for the unmasked correlations to be computed

interaction, we feed $\widetilde{\boldsymbol{S}}(k)$ into the AttentionNet and take the two output symbols of the last column as $[x_{1,k}, x_{2,k}]^\top$.[2] An illustration is given in Fig. 2.8.

Another ingredient we use in conjunction with input restructuring is the mask. As shown in Fig. 2.8, the self-attention layers are partially connected, subjected to a $k \times k$ dimensional mask

$$\boldsymbol{M}_k = \begin{bmatrix} 1 & -\infty & -\infty & \dots & -\infty \\ 1 & 1 & -\infty & \dots & -\infty \\ 1 & 1 & 1 & \dots & -\infty \\ 1 & 1 & 1 & \dots & 1 \end{bmatrix}_{k\times k} . \tag{2.29}$$

At first sight, $\boldsymbol{M}_k$ is a mask generated by the causal constraint since it is in the same form as (2.26). However, note that AttentionCode is trained sequentially and there is no causal constraint—$\widetilde{\boldsymbol{S}}(k)$ is fully known to node A when generating $[x_{1,k}, x_{2,k}]^\top$. In other words, it is legitimate to remove the mask and use fully-connected self-attention layers in Fig. 2.8. Nevertheless, we empirically found that masked self-attention layers lead to better feedback codes and we conjecture that this is thanks to the simplified correlations among the columns of $\widetilde{\boldsymbol{S}}(k)$ to generate $[x_{1,k}, x_{2,k}]^\top$. In the final design of AttentionCode, we use masked self-attention at the encoder and unmasked self-attention at the decoder.

The $q = 1$ Case

When the number of encoding blocks (which is also the number of self-attention modules) $q = 1$, there is no difference between masked and unmasked AttentionCode since the generated $[x_{1,i}, x_{2,i}]^\top$ are the same (see Fig. 2.8 for a visual explanation).

[2] Inside GPU, a concatenated computation graph is created for consecutive interactions.

Table 2.1 Hyper-parameter settings for AttentionCode

Parameters	Descriptions	Values
K	Bitstream length	50
R	Code rate	1/3
d_S	Input dim	4 (Enc), 3 (Dec)
q_t	#enc blocks in the encoder	2
q_r	#enc blocks in the decoder	3
d_m	Model size	32
B	Batch size	2000
V	Distributed learning factor	5
C	Look-ahead steps	4
Adam	Inner optimizer	$\beta_1=0.9$, $\beta_2=0.98$, $\epsilon=10^{-9}$
lr	Learning rate	10^{-3}

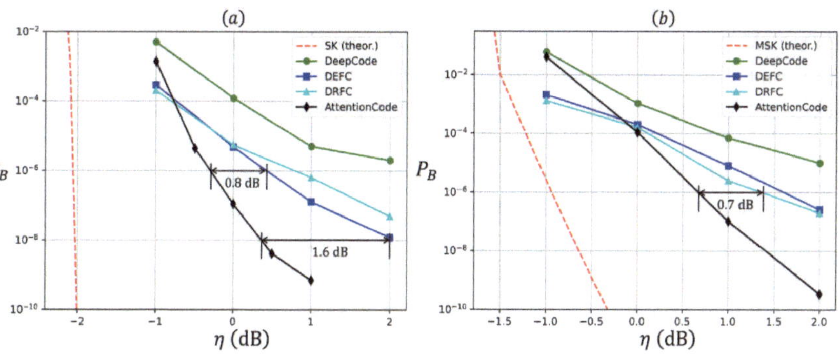

(a) (b)

Fig. 2.9 AWGN channels: BLER P_B versus feedforward channel SNR η (in dB) performance of various feedback codes. (**a**) noiseless feedback, where $\eta' = \infty$; (**b**) noisy feedback, where $\eta' = 20$ dB

Numerical Experiments

This section performs a numerical evaluation to compare the performance of DeepCode and AttentionCode. The hyper-parameter settings are summarized in Table 2.1, where the bitstream length is $K = 50$ and the code rate $R = 1/3$. The feedback channel can be noiseless and noisy ($\eta' = 20$ dB).

Figure 2.9 presents the BLER performance P_B versus the feedforward channel SNR η (in dB) for the SK scheme, the modulo-SK scheme, DeepCode, DEFC, DRFC, and AttentionCode. In particular, Fig. 2.9a consider a noiseless feedback channel and Fig. 2.9b considers a noisy feedback channel with a SNR $\eta' = 20$ dB. As can be seen, with the attention mechanism, AttentionCode significantly outperforms DeepCode and other RNN-based codes, including DEFC [28] and DRFC [29].

Computational Complexity of AttentionCode

The training and execution complexities of AttentionCode and RNN-based codes are compared as follows.

Let us first consider the number of computations in the neural networks. Consider a (K,N) feedback code. Existing DL-based feedback code designs, e.g., DeepCode, DEFS, and DRFS, leverage RNNs. The number of computations in the RNN is $O(Kd_{\text{RNN}}^2)$, where d_{RNN} is the size of latent variables, i.e., the number of latent features we use to represent one input. In prior works, the common settings are $K = 50$ and $d_{\text{RNN}} = 50$. AttentionCode, on the other hand, leverages the transformer with the self-attention mechanism. The number of computations in the self-attention mechanism is $O(K^2 d_{\text{Atten}})$. In the experiments, we set $K = 50$ and $d_{\text{Atten}} = 32$.

The training complexity is proportional to both the number of computations in the neural networks and the number of training epochs T. Thus, the training complexities of prior works and AttentionCode are $O(T_{\text{RNN}} K d_{\text{RNN}}^2)$ and $O(T_{\text{Atten}} K^2 d_{\text{Atten}})$, respectively. Since $T_{\text{Atten}} \gg T_{\text{RNN}}$, the training complexity of AttentionCode is much larger than prior works. Empirically, (1) if we set $T_{\text{Atten}} = T_{\text{RNN}}$, AttentionCode cannot outperform DeepCode; (2) if we set $T_{\text{RNN}} = T_{\text{Atten}}$, the performance of DeepCode does not improve further and is still much worse than AttentionCode.

However, it is worth noting that the training of feedback code is a one-shot cost. Once trained, the feedback code can be deployed and only the execution complexity matters. As stated above, the execution cost of existing feedback codes and AttentionCode are $O(Kd_{\text{RNN}}^2)$ and $O(K^2 d_{\text{Atten}})$, respectively. At a short block length (e.g., $K = 50$), the execution complexity of AttentionCode is on an equal footing with existing codes.

Forward Error Correction Code

As an additional benchmark to measure how much feedback improves the reliability of the feedforward channel, new radio (NR) LDPC, which does not exploit the feedback, achieves a BLER of 2×10^{-3} when the feedforward SNR is $\eta = 1$ dB; and a BLER of 3×10^{-5} when the feedforward SNR is $\eta = 2$ dB.

2.3.2 *AI-Driven Block-Wise Feedback Codes*

In this section, we introduce another AI-driven feedback code design, termed Generalized Block Attention Feedback (GBAF) codes [8, 30]. Unlike the AI-driven symbol-wise feedback codes discussed previously, which encode and decode each symbol individually based on instantaneous feedback, GBAF codes adopt a block-wise encoding strategy. Specifically, GBAF codes segment the information bits into blocks, encode each block collectively, and utilize feedback at the block level rather than symbol level. This block-wise approach leverages contextual dependencies across multiple symbols, allowing the feedback mechanism to incorporate richer context and exploit correlation among different blocks. Consequently, GBAF significantly enhances the performance and robustness of feedback coding, particularly beneficial for IoT communication scenarios where packet-based transmissions are common and computational efficiency and reliability are critical.

2.3.2.1 Node Interactions in GBAF Code

In each iteration of the GBAF code, node A transmits a frame of Q symbols to node B, and node B feeds a frame of Q symbols back to node A. Suppose there are G iterations between nodes A and B during one transmission. We denote by $\boldsymbol{x}_g \in \mathbb{R}^{Q\times 1}$ the vector of real channel symbols transmitted from node A to node B in the g-th frame, $g = 1, 2, \cdots, G$. Correspondingly, the received frame at node B is denoted by $\boldsymbol{y}_g$.

We further denote by $\widetilde{\boldsymbol{x}}_g$ the vector of real channel symbols fed back by node B in the g-th feedback frame to node A. Naturally, $\widetilde{\boldsymbol{x}}_g$ can be generated based on all the forward symbols previously received from node A, giving

$$\widetilde{\boldsymbol{x}}_g = f_B\left(\boldsymbol{y}_1, \boldsymbol{y}_2, \cdots, \boldsymbol{y}_g\right), \tag{2.30}$$

where f_B denotes the feedback encoder at node B. For passive feedback, we have $\widetilde{\boldsymbol{x}}_g = \boldsymbol{y}_g$.

After passing through the feedback channel, the received signal at node A is denoted by $\widetilde{\boldsymbol{y}}_g$. Given the feedback from node B, node A generates the $(g+1)$-th forward frame $\boldsymbol{x}_{g+1}$ based on all information bits, the previously transmitted coded symbols, and all symbols fed back from node B, yielding

$$\begin{cases} \boldsymbol{x}_1 = f_F\left(\boldsymbol{b}\right), \\ \boldsymbol{x}_{g+1} = f_F\left(\boldsymbol{b}; \boldsymbol{x}_1, \cdots, \boldsymbol{x}_g; \widetilde{\boldsymbol{y}}_1, \cdots, \widetilde{\boldsymbol{y}}_g\right), \end{cases} \tag{2.31}$$

where $\boldsymbol{b} \in \{0, 1\}^{K\times 1}$ is the source bit sequence, and f_F denotes the feedback encoder at the devices.[3]

At the end of the G-th forward transmission, node B receives G frames from node A and decodes $\boldsymbol{b}$ using a feedback decoder f_D:

$$\widehat{\boldsymbol{b}} = f_D\left(\boldsymbol{y}_1, \boldsymbol{y}_2, \cdots, \boldsymbol{y}_G\right). \tag{2.32}$$

[3] We point out that both feedback encoders f_B and f_F are no long channel encoders, but joint source-channel encoders.

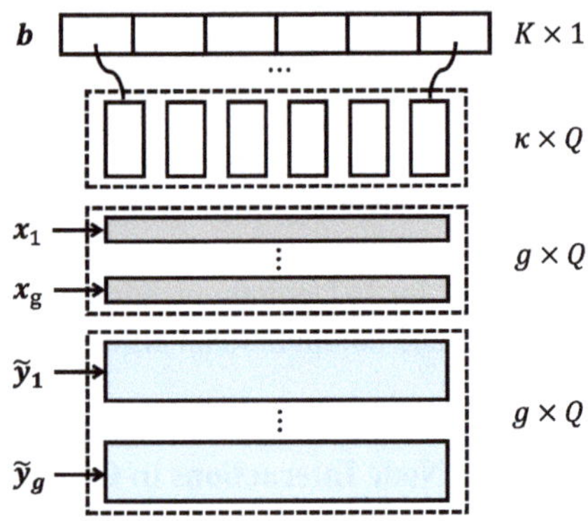

Fig. 2.10 The composition of feature matrix $\boldsymbol{M}_F$ for forward coding at node A

Overall, node A transmits $N \triangleq QG$ real channel symbols to node B. The channel coding rate, defined as the number of transmitted bits per coded channel symbol, can be written as

$$R \triangleq \frac{K}{N} = \frac{K}{QG}. \tag{2.33}$$

2.3.2.2 The GBAF Encoder

The primary function of the forward encoder f_F at node A is to generate the next forward frame, as in (2.31). Without loss of generality, we consider the generation of the $(g+1)$-th data frame $\boldsymbol{x}_{g+1}$. At that point, the available information at the ℓ-th device comprises the original information bits $\boldsymbol{b}$, the previously transmitted symbols $\{\boldsymbol{x}_1, \cdots, \boldsymbol{x}_g\}$, and all symbols fed back from node B $\{\widetilde{\boldsymbol{y}}_1, \cdots, \widetilde{\boldsymbol{y}}_g\}$.

To perform forward encoding, we re-organize the above information into a feature matrix $\boldsymbol{M}_F$ that serves as the input for the UL feedback encoder f_F. As depicted in Fig. 2.10, $\boldsymbol{M}_F$ is composed of three submatrices:

- The first part consists of the original bits $\boldsymbol{b}$. We divide K bits into Q groups, each containing $\kappa \triangleq K/Q$ bits. Subsequently, $\boldsymbol{b}$ is reshaped into a $\kappa \times Q$ matrix, forming the first submatrix.
- The second part consists of previously transmitted channel symbols $\{\boldsymbol{x}_1, \cdots, \boldsymbol{x}_g\}$. These are stacked together to form a $g \times Q$ submatrix.
- The third part consists of the received feedback $\{\widetilde{\boldsymbol{y}}_1, \cdots, \widetilde{\boldsymbol{y}}_g\}$. Given that each $\widetilde{\boldsymbol{y}}_g$ is of dimension $Q \times 1$, we reshape these feedback symbols into an $g \times Q$ submatrix.

The feature matrix is then fed into the forward encoder f_F, which comprises three main components: a noise suppression module, a self-attention mechanism module, and a power normalization module, as detailed in Fig. 2.11.

- The noise suppression module is constructed using three fully connected layers coupled with two ReLU activation functions. Its primary function is to clamp the

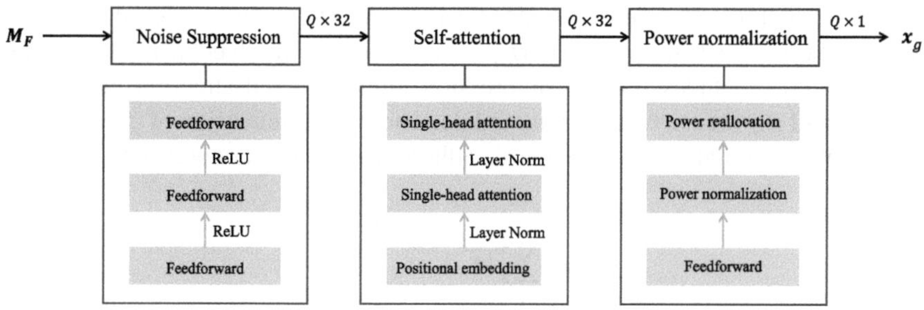

Fig. 2.11 The DNN architecture of the forward encoder f_F at node A

received signals, effectively reducing excessively large absolute values, which are introduced by noise [8]. This process ensures that the integrity of the signal is maintained by mitigating noise.

- The self-attention mechanism, a key component of feedback encoding, utilizes stacked encoders from the transformer architecture [24]. This module excels in selectively concentrating on different parts of the feature matrix, facilitating a dynamic and context-sensitive analysis of feedback. Essentially, it intelligently assesses node B's current decoding status using the feedback information and adaptively refines its encoding strategy to rectify any discrepancies at node B's end.
- The output of the self-attention mechanism then proceeds to the power normalization module [6]. This module is responsible for normalizing and redistributing power among the coded channel symbols, ensuring they adhere to the average power constraints.

The central element of the feedback encoder is the single-head attention layer, which correlate the encoding of Q input groups. To provide an analogy from coding theory, these Q input groups function as variable nodes, while the generated coded symbols in $\boldsymbol{x}_{\ell,g}^{(t)}$ act as check nodes. During each encoding cycle, the encoder prioritizes variable nodes that have not been successfully decoded, generating new variable nodes based on a learned degree distribution. This allows the decoder to correct any previously unsuccessful decoding attempts.

The output of f_F, $\boldsymbol{x}_g$, satisfies average power constraints:

$$\frac{1}{N}\mathbb{E}\left[\sum_{g=1}^{G}\left(\boldsymbol{x}_g\right)^\top \boldsymbol{x}_g\right] \leq 1. \tag{2.34}$$

The DL feedback encoder f_B and the decoder f_D are architecturally similar to f_F. Specifically, the feature matrix of node B can be written as

$$\boldsymbol{M}_B \triangleq \left[\boldsymbol{y}_1, \boldsymbol{y}_2, \cdots, \boldsymbol{y}_g\right]. \tag{2.35}$$

Compared with the feature matrix $\boldsymbol{M}_F$ at node A, $\boldsymbol{M}_B$ contains only the forward data symbols. The feature matrix $\boldsymbol{M}_B$ is then fed into the feedback encoder f_B, which, as f_F, includes a noise suppression module, a self-attention mechanism module, and a power normalization module. The output dimension of f_B is $1 \times Q$, with the post-normalization symbols $\widetilde{\boldsymbol{x}}_g$ adhering to the power constraint:

$$\frac{1}{N}\mathbb{E}\left[\sum_{g=1}^{G} \left(\widetilde{\boldsymbol{x}}_g\right)^{\top} \widetilde{\boldsymbol{x}}_g\right] \leq 1. \tag{2.36}$$

The coded symbols $\widetilde{\boldsymbol{x}}_g$ will be transmitted to node A in the g-th feedback frame. Upon receiving the feedback frame, node A extracts the data and integrates them into its feature matrix $\boldsymbol{M}_F$, as depicted in Fig. 2.10.

2.3.2.3 The GBAF Decoder

The transmission concludes upon completion of the G-th forward data frame. Node A, by this point, has transmitted $N = QG$ coded symbols to node B, all of which are stored in the feature matrix for decoding by node B. The DNN design of the decoder f_D is akin to that of f_B, except for the replacement of the power normalization module with a softmax function. Recall that the bit sequence $\boldsymbol{b}$ has been segmented into Q groups on the IoT device side. Mirroring the approach of QAM, node B engages in joint decoding of the bits within each group rather than individual bit decoding [8]. This approach effectively transforms the binary classification problem to a 2^{κ}-ary classification problem. Specifically, denote by $\boldsymbol{Y} \triangleq \{Y_{ij}\} \in \mathbb{R}^{2^{\kappa} \times Q}$ the output matrix of AttentionNet within the decoder f_D, the loss function can be written as

$$\mathcal{L} = \mathbb{E}\left[\sum_{j=0}^{Q-1}\left(-\sum_{i=0}^{2^{\kappa}-1} \delta(j = c_i) \log \frac{e^{Y_{ij}}}{\sum_{i=0}^{2^{\kappa}-1} e^{Y_{ij}}}\right)\right], \tag{2.37}$$

where the expectation is taken over a batch of bit sequences, and c_i is the ground truth of a group of bits, i.e., $c_i = \boldsymbol{b}^{\top}[2^{\kappa} - 1, 2^{\kappa} - 2, \cdots, 2^0]^{\top}$ for $\boldsymbol{b} \in \{0, 1\}^{\kappa \times 1}$.

2.3.2.4 Numerical Results

This section evaluates the performance of the GBAF code in comparison to symbol-wise feedback codes. We begin by considering the noiseless feedback scenario, assuming a fixed coding rate of $R = 1/3$. Figure 2.12a presents the BLER performance results across forward SNR values ranging from -1 dB to 2 dB. The results clearly indicate that the GBAF code achieves approximately an order-of-magnitude improvement over symbol-wise feedback codes.

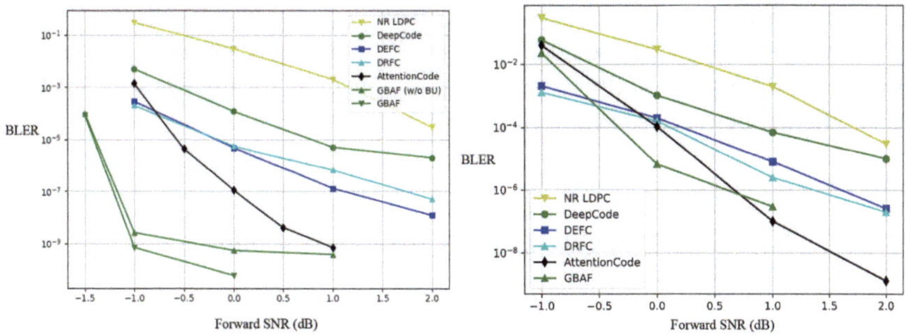

(a) (b)

Fig. 2.12 Performance comparison of GBAF with AttentionCode, DEFC, DRFC, DeepCode and NR-LDPC. (**a**) Noiseless feedback; (**b**) Noisy feedback

Table 2.2 BLER of GBAF codes at different code rates R

SNR/Rate	3/8	3/7	3/6	3/5
-1 dB	1.8×10^{-2}	–	–	–
0 dB	6.15×10^{-8}	2.8×10^{-3}	–	–
1 dB	2.7×10^{-8}	7.5×10^{-8}	1×10^{-2}	–
2 dB	–	1×10^{-9}	1.5×10^{-6}	6.5×10^{-2}
3 dB	–	–	2.7×10^{-8}	8.7×10^{-7}

Next, we analyze a scenario in which the feedback channel is subject to additive Gaussian noise, characterized by an SNR of 20 dB. Figure 2.12b demonstrate that, except at the lowest forward SNR of -1 dB, the GBAF code consistently outperforms DEFC, DRFC, DeepCode, and NR-LDPC. Although AttentionCode may surpass the GBAF code at higher SNR levels, it is important to emphasize that the GBAF code utilizes feedback approximately $6\times$ less frequently. This significantly reduces overhead, enabling GBAF to deliver comparable or superior performance with substantially lower communication requirements.

In the second set of simulations, we highlight the flexibility of the GBAF code in terms of achievable code rates. Unlike existing coding schemes, the GBAF framework readily accommodates different channel code rates simply by adjusting the number of transmitted parity symbols, without requiring any structural changes to the architecture. To illustrate this flexibility, we select values of $T = \{8, 7, 6, 5\}$ to achieve coding rates $R \in \{3/8, 3/7, 3/6, 3/5\}$, respectively, and evaluate the corresponding BLER performance for forward SNR values between -1 dB and 3 dB under noiseless feedback conditions. These results are summarized in Table 2.2. The simulation results demonstrate that, in high SNR regimes, it is possible to achieve acceptable BLER levels even at increased code rates. For example, a BLER target of 10^{-5}, which is sufficient for many practical applications [31], is attainable at rates $R \in \{3/8, 3/7, 3/6, 3/5\}$ depending on the SNR. This highlights the unique capability of GBAF for rate adaptation according to the channel conditions, unlike existing designs. We also observe from Table 2.2 that the BLER performance deteriorates rapidly as the code rate approaches the channel capacity at a given SNR. However, GBAF successfully reduces the error rate dramatically when the operating rate is slightly below capacity.

2.3.3 AI-Driven Variable-Length Feedback Codes

The AI-driven feedback codes presented thus far operate primarily at fixed lengths and rates. While these codes perform effectively at lower code rates, typically below $R = 1/3$, their performance degrades substantially when the rate is increased due to fewer available channel uses. This deterioration occurs because higher coding rates restrict the ability of these methods to adequately exploit the channel resources.

To overcome these limitations, this section introduces a AI-driven variable-length feedback code called the Deep Variable-Length Feedback (DeepVLF) code [9]. DeepVLF leverages multi-round feedback to achieve adaptive code rates with bit-group-level granularity, allowing the transmitter to adjust the amount of information transmitted based on the feedback. By dividing the message into bit groups and employing a threshold-based decoding strategy, each bit group can be decoded independently across different communication rounds. This approach reduces the total number of channel uses required to achieve the same performance as fixed-rate codes, particularly in high rate regimes.

2.3.3.1 DeepVLF Architecture

To start with, we outline the architecture of the DeepVLF code. We discuss the design and functionality of both the encoder and decode, highlighting the innovative aspects that differentiate DeepVLF from previous designs.

At the core of the DeepVLF architecture are several fundamental elements that guide its design.

Definition 2.4 (Bit Groups) As GBAF code, the encoding and decoding of DeepVLF are organized around groups of bits instead of individual bits. Given a vector of message bits $\boldsymbol{b}$, DeepVLF first divides $\boldsymbol{b}$ into Q equal-size groups, with each group containing $\kappa \triangleq K/Q$ bits:

$$\boldsymbol{b} = [\boldsymbol{b}_1^\top, \boldsymbol{b}_2^\top, \boldsymbol{b}_3^\top, \cdots, \boldsymbol{b}_Q^\top]^\top, \tag{2.38}$$

where each group $\boldsymbol{b}_q \in \mathcal{A} \triangleq \{0, 1\}^{\kappa \times 1}$.

The total number of possible patterns per group is $|\mathcal{A}| = 2^\kappa$. By reducing the effective length of the input sequence for the DNN, this method efficiently reduces the computational and memory demands on the network, facilitating more effective learning and processing.

Definition 2.5 (Belief Matrix) In the DeepVLF framework, node B formulates a belief matrix $\boldsymbol{P}$ to represent each transmitted bit group:

$$\boldsymbol{P} = [\boldsymbol{p}_1, \boldsymbol{p}_2, \boldsymbol{p}_3, \cdots, \boldsymbol{p}_Q], \tag{2.39}$$

where column $\boldsymbol{p}_q$ corresponds to bit group $\boldsymbol{b}_q$. Specifically, $\boldsymbol{p}_q$ is a 2^K-dimensional probability distribution, with the j-th element indicating the likelihood that node B interprets the transmitted bit group $\boldsymbol{b}_q$ as the j-th element of set $\mathcal{A}$.

Definition 2.6 (Threshold Decoding) In DeepVLF, node B utilizes a threshold decoding strategy to decode each bit group individually. Specifically, a bit group $\boldsymbol{b}_q$ is considered successfully decoded by node B if any element of the belief vector $\boldsymbol{p}_q$ surpasses a predefined threshold γ, i.e.,

$$\|\boldsymbol{p}_i\|_\infty \geq \gamma, \tag{2.40}$$

where $\|\cdot\|_\infty$ denotes the infinity norm, which evaluates the maximum absolute value among the elements of a vector.

Contrary to other AI-driven feedback code designs, where decoding is performed only after the complete sequence of transmissions are received by the decoder, DeepVLF employs threshold decoding at the end of each communication round, and iteratively updates its belief with the help of the feedback link.

Definition 2.7 (Feedback Information) In DeepVLF, during each communication round, node B provides user A with two types of feedback information:

- The packet of received symbols $\boldsymbol{y}_g$.
- The indices of the bit groups that have been successfully decoded within the last communication round.

Encoder

The encoding process of DeepVLF is iterative, unfolding across several rounds. Each round leverages feedback to refine the generation of parity symbols, thereby enhancing the robustness and efficiency of the transmission. We next detail the mechanics of how feedback is incorporated within each round and the subsequent generation of parity symbols.

During each communication round g, $g = 1, 2, \ldots, G$, the generation of parity symbols $\boldsymbol{x}_g$ can be represented by a bipartite graph with variable nodes $\mathcal{V}_{g,i}$, $i = 1, 2, \ldots, Q$, check nodes $C_{g,i}$, $i = 1, 2, \ldots, Q$, and a set of edges $\mathcal{E}_g$.

- Each i-th variable node and check node is uniquely associated with the i-th bit group, designed specifically to cater to the encoding and decoding needs of that group. This ensures targeted processing of each bit group throughout the communication rounds.

- Each check node contains a parity symbol intended for transmission to node B. As stated in Definition 2.6, DeepVLF actively decodes the bit groups in all interactions. If bit group i is decoded successfully during communication round g, the parity symbol in $C_{g',j}$, $g' > g$, will no longer be transmitted to node B.
- The variable nodes contain the information to generate the check nodes. Initially, in the first round, we have $\mathcal{V}_{1,i} = \{\boldsymbol{b}_i\}$. With each subsequent communication round, $\mathcal{V}_{g,i}$ is updated to incorporate additional feedback from node B, enriching the information used for future encoding decisions and enhancing the system's adaptability to channel noise realizations.

The check nodes and variable nodes are interconnected through a set of edges. Let e_{ij} represent a possible edge linking the i-th variable node with the j-th check node. During communication round g, the set of active edges is denoted by $\mathcal{E}_g$. The configuration of $\mathcal{E}_g$ is dynamically adjusted based on the decoding status of bit groups as follows:

- Initially, when no bit group has been decoded, all possible connections are active, thus $\mathcal{E}_g = \{e_{ij}, \forall i, j\}$.
- If a bit group j' was successfully decoded in the previous round $g-1$, the edges connecting to the corresponding check node j' are deactivated in the current round. Therefore, $\mathcal{E}_g = \mathcal{E}_{g-1} \backslash \{e_{ij'}, \forall i\}$.
- Each edge $e_{ij} \in \mathcal{E}_g$ is associated with a weight ρ_{ij}. For each C_j, we have $\sum_{i=1}^{Q} \rho_{ij} = 1$.

Given the bipartite graph, the relationship between the variable and check nodes can be written as

$$C_j = f_c\bigg(\sum_{i=1}^{Q} \rho_{ij} f_v(\mathcal{V}_i)\bigg), \tag{2.41}$$

where the encoding functions f_c and f_v are realized by DNNs. The coefficients ρ_{ij} are learnable parameters, which are optimized to effectively determine the influence of each variable node on the corresponding check node. This architecture of DNNs and their role in mapping variable nodes to check nodes is detailed further in Sect. 2.3.3.2.

The updating process for variable nodes after each communication round is crucial for the overall system's adaptability and efficiency. This update is governed by an auto-regressive mechanism as described below:

$$\mathcal{V}_{g,i} = \begin{cases} \{\boldsymbol{b}_i\}, & \text{if } g = 1; \\ \mathcal{V}_{g-1,i} \cup C_{g-1,i} \cup \{\boldsymbol{y}_{g-1,i}\}, & \text{if } \|\boldsymbol{p}_{g-1,i}\|_\infty < \gamma; \\ \mathcal{V}_{g-1,i}, & \text{otherwise.} \end{cases} \tag{2.42}$$

In the initial round ($g = 1$), each variable node is simply initialized to the corresponding bit group. In subsequent rounds, the update depends on the decoding success: if the infinity norm of the probability vector $\boldsymbol{p}_{g-1,i}$ is below the threshold γ, indicating unsuccessful decoding, $\mathcal{V}_{g,i}$ is expanded to include the parity information from the last round and the received symbols. If the threshold is met or exceeded, further updates are unnecessary, and $\mathcal{V}_{g,i}$ retains its prior state from $\mathcal{V}_{g-1,i}$. Nevertheless, this "frozen" node will continue to contribute to the generation of other parity symbols for the remaining rounds.

Decoder

At node B, the decoder engages in decoding operations during each communication round upon receiving $\boldsymbol{y}_g$. The design principles of the decoder mirror those of the encoder. In the following, we succinctly outline the decoder's functionality.

The core functionality of the decoder within DeepVLF is estimating the belief matrix $\boldsymbol{P}$. This process can also be modeled by a bipartite graph representation, consisting of variable nodes $\widetilde{\mathcal{V}}_{g,i}$, and check nodes $\widetilde{\mathcal{C}}_{g,j}$, akin to the encoder's structure in (2.41). We have

$$\widetilde{\mathcal{C}}_j = \widetilde{f}_c\left(\sum_{i=1}^{Q} \widetilde{\rho}_{ij} \widetilde{f}_v(\widetilde{\mathcal{V}}_i)\right). \tag{2.43}$$

In (2.43), the variable node $\widetilde{\mathcal{V}}_i$ stores the received parity symbols from the previous communication rounds and the current belief about the i-th bit group $\boldsymbol{p}_{g-1,i}$. The check node $\widetilde{\mathcal{C}}_j$ estimates the updated belief $\boldsymbol{p}_{g,j}$ for the j-th bit group. The functions $\widetilde{f}_c$ and $\widetilde{f}_v$ are realized by DNNs, and $\widetilde{\rho}_{ij}$ are learnable parameters, mirroring the learnable weights ρ_{ij} used in the encoder.

If $\|\boldsymbol{p}_{g,i}\|_\infty \geq \gamma$, the i-th bit group is deemed successfully decoded. Consequently, the corresponding $\widetilde{\mathcal{V}}_{g,i}$ and $\widetilde{\mathcal{C}}_{g,j}$ are frozen and will not undergo further updates in future communication rounds, thereby preserving the state achieved at the point of successful decoding.

Definition 2.8 (Code Rate) DeepVLF decodes bit groups individually across different communication rounds, thus introducing variability in the decoding timeline for each bit group. We define g_q^* as the communication round during which the q-th bit group is decoded:

$$g_q^* \triangleq \left\{g : \|\boldsymbol{p}_{g-1,q}\|_\infty < \gamma, \|\boldsymbol{p}_{g,q}\|_\infty \geq \gamma\right\}. \tag{2.44}$$

The overall code rate R of DeepVLF can be written as

$$R = \frac{K}{\sum_{q=1}^{Q} g_q^*}. \tag{2.45}$$

Given the DeepVLF framework described above, our primary objective is to maximize the code rate R while ensuring that the block error rate (BLER) $\xi \triangleq \Pr(\boldsymbol{b} \neq \widehat{\boldsymbol{b}})$, where $\widehat{\boldsymbol{b}}$ denotes the decoded bit sequence, remains below a specified target ξ^*. This optimization is performed under an average power constraint. Formally, we have

$$\begin{aligned}
\max_{f_c, f_v, \widetilde{f}_c, \widetilde{f}_v, \{\rho_{ij}, \widetilde{\rho}_{ij}\}} \quad & R \\
s.t. \quad & \xi \leq \xi^*, \\
& \mathbb{E}\left[\frac{1}{\sum_{q=1}^{Q} g_q^*} \sum_{q=1}^{Q} \sum_{g=1}^{g_q^*} \left(x_{g,q}\right)^2\right] \leq 1.
\end{aligned}$$

2.3.3.2 Realization of DeepVLF

This section presents our realization of the DeepVLF methodology using a transformer-based DNN architecture [23, 24].

Parity Symbol Generation

The encoding process at user A is guided by the principles outlined in (2.41), wherein the variable nodes $\mathcal{V}_{g,i}$ are updated in each communication round g based on the feedback received from node B following (2.42). The knowledge vectors contained in the variable nodes are first projected onto a high-dimensional latent space through the feature extractor function f_v, implemented via a variable-depth feature extractor (VDFE). Subsequently, these latent vectors are aggregated using coefficients ρ_{ij} generated by self-attention layers. The aggregated latent representations are then mapped to the check nodes $C_{g,j}$ via a header function f_c, resulting in the generation of parity symbols for transmission.

VDFE (f_v) In DeepVLF, the feature extractor f_v plays a crucial role in projecting the knowledge vectors of the variable nodes into a latent space suitable for processing by self-attention. We implement f_v using fully-connected layers with ReLU activation functions. Notably, the depth of the fully-connected layers varies depending on the communication round g through the introduction of a VDFE, where the number of fully-connected layers in f_v is adjusted based on whether the current communication round g is before or after a predefined threshold τ_{VD}, while a shallower feature extractor suffices in the earlier rounds of communication ($g \leq \tau_{\text{VD}}$). In contrast, during the later rounds ($g > \tau_{\text{VD}}$), a deeper feature extractor is employed to model the increasing complexity and nonlinearity of the encoding function.

Self-attention Coefficients (ρ_{ij}) The aggregation coefficients $\{\rho_{ij}\}$ determine the impact of each variable node on the check nodes during the encoding process. In DeepVLF, we employ a self-attention mechanism to dynamically generate these coefficients, leveraging the ability of attention mechanisms to capture dependencies and interactions between elements in a sequence. The effectiveness of using self-attention in feedback coding has been demonstrated in [8, 23, 30].

In this approach, each coefficient ρ_{ij} is computed based on the inner product of the latent representations of the variable nodes:

$$\rho_{ij} = \text{softmax}\left(f_v(\mathcal{V}_{g,i})^\top f_v(\mathcal{V}_{g,j})\right), \tag{2.46}$$

where softmax ensures that the coefficients are non-negative and sum to one for each check node.

To realize the threshold decoding mechanism and account for the decoding status of bit groups, we utilize masking within the self-attention layers. Specifically, we define a mask $\mathcal{M}_g \in \{0, 1\}^Q$ for the Q bit groups, representing the decoding status at communication round g:

$$\mathcal{M}_{g,j} = \begin{cases} 1, & \text{if } \|\boldsymbol{p}_{g,j}\|_\infty < \gamma \\ 0, & \text{otherwise} \end{cases} \tag{2.47}$$

Thanks to the feedback channel, this mask is known to both the encoder and decoder. In the self-attention computation, we mask the check nodes whose associated bit group has been successfully decoded.

Feedforward Network (f_c) After the latent representations are aggregated using the self-attention coefficients, they are mapped to the check nodes through the header function f_c. We design f_c as a two-layer fully connected network with Gaussian error linear unit (GeLU) activation functions [32].

At the encoder side, the header function f_c maps each aggregated latent vector to a single parity symbol $x_{g,i}$, which is then transmitted over the forward channel to node B. At the decoder side, the corresponding header function $\widetilde{f_c}$ generates a belief vector $\boldsymbol{p}_{g,j}$ for the j-th bit group by adding an additional linear layer followed by a softmax activation function. This setup allows the decoder to produce a probability distribution over the possible bit group patterns, facilitating effective threshold decoding as per Definition 2.6.

Decoding and Loss Function Design

At node B, the decoder utilizes a similar transformer-based architecture to process the received signals and update the belief matrix $\boldsymbol{P}$. The decoder's variable nodes $\widetilde{\mathcal{V}}_{g,i}$ store the received parity symbols from previous communication rounds and the current belief about each bit group. The check nodes $\widetilde{C}_{g,j}$ are responsible for generating the updated belief vectors $\boldsymbol{p}_{g,j}$, as described in (2.43).

The decoder employs the same VDFE and self-attention mechanism, ensuring consistency in processing and facilitating effective learning of the encoding-decoding relationship. The mask vector $\mathcal{M}_g$ is also utilized at the decoder side to mask decoded bit groups.

Upon receiving the parity symbols $\boldsymbol{y}_g$, the decoder updates the belief vectors and applies the threshold decoding criterion. If $|\boldsymbol{p}_{g,j}|_\infty \geq \gamma$, the j-th bit group is declared successfully decoded, and its corresponding variable and check nodes are frozen in subsequent rounds. This iterative process continues until all bit groups are decoded.

Given that DeepVLF attempts to decode the bit groups after each communication round, the training process must accommodate the dynamic nature of the decoding accuracy, which improves over successive rounds. In the following, we elaborate on the loss function designed for DeepVLF and discuss the strategies employed to enhance its training efficacy.

DeepVLF is trained in a supervised manner, aiming to minimize the discrepancy between the transmitted bit groups $\boldsymbol{b}_i$ and the estimated probabilities $\boldsymbol{p}_{g,i}$ at each round g. A straightforward loss function design is to sum the cross-entropy losses for every bit group across all communication rounds:

$$\mathcal{L} = -\sum_{q=1}^{Q}\sum_{g=1}^{g_q^*}\sum_{j=1}^{2^K} \mathbb{I}_{(\boldsymbol{b}_q=\mathcal{A}_j)} \cdot \log(p_{g,q,j}), \tag{2.48}$$

where $\mathbb{I}$ is the indicator function, and $\mathcal{A}_j$ denotes the j-th element in the set $\mathcal{A}$. This loss function aims to minimize the cross-entropy between the true bit groups and the estimated probabilities at each round, encouraging the model to improve its decoding accuracy progressively.

However, due to the interactive nature of DeepVLF, the BLER naturally decreases as the communication rounds progress. In the initial rounds, the decoder's estimates are less accurate, resulting in higher cross-entropy losses. Consequently, when averaging the cross-entropy equally across all rounds, the loss function in (2.48) becomes dominated by the high losses from the early rounds. This dominance can hinder the training process, as the model may focus disproportionately on minimizing the loss in the early rounds, potentially at the expense of performance in later rounds where significant improvements occur.

To mitigate this problem and enhance the performance of DeepVLF, we introduce two advancements in our training.

First, we set a threshold g^+, and initiate decoding attempts only after communication round g^+. Communication rounds before g^+ are treated as undecodable, acknowledging that the decoder's estimates during these early rounds are not sufficiently reliable. By excluding the high-loss contributions from the initial rounds, we prevent them from dominating the training process. The value of g^+

is determined based on the Shannon limit [13], providing a rough guideline for the minimum number of transmissions required for reliable communication given the channel conditions:

$$g^{+} \triangleq \max\left\{\mu, \left\lfloor \frac{2m}{\log_2(1+\eta))} \right\rfloor\right\}, \tag{2.49}$$

where $\eta = \frac{1}{\sigma^2}$ is the SNR of the forward channel and μ is served as a baseline threshold for g^{+}, adjusted based on γ.

Second, we incorporate an exponential weighting factor into the loss function to balance the contributions of losses from different rounds. By assigning higher weights to losses from later rounds, where the decoder's estimates are more accurate, we encourage the model to focus on improving performance in these critical stages. The final loss function for training DeepVLF is thus formulated as:

$$\mathcal{L} = -\sum_{q=1}^{Q}\sum_{g=g^{+}}^{g_q^*} \vartheta^{g-\epsilon} \sum_{j=1}^{2^{\kappa}} \mathbb{I}_{(b_q=\mathcal{A}_j)} \cdot \log(p_{g,q,j}), \tag{2.50}$$

where ϑ is the base of the exponential weighting factor while ϵ denotes a fixed offset parameter. By carefully selecting ϑ and ϵ, we can effectively balance the influence of each communication round on the total loss, ensuring that later rounds – more indicative of the model's ultimate performance – have a greater impact on the training process. An ablation study comparing different loss functions is given in Sect. 2.3.3.3.

2.3.3.3 Numerical Results

This section evaluates the performance of DeepVLF benchmarked against other feedback coding schemes.

Training DeepVLF

In our training procedure, we employ the AdamW optimizer, a refined version of the Adam optimizer that decouples weight decay from the gradient-based updates [33], and a learning rate scheduling strategy where the learning rate progressively decreases as the training steps advance.

The average code rate of DeepVLF varies with the decoding threshold γ, which determines when bit groups are considered successfully decoded based on the belief matrix. To optimize DeepVLF's performance across different code rates and BLER requirements, we structure the training process into two phases:

Table 2.3 Hyperparameter settings for DeepVLF

Parameters	Descriptions	Value
K	Bitstream length	51
Q	Number of bit groups	17
κ	Size of a bit group	3
B	Batchsize	8192
lr	Initial learning rate	10^{-3}
λ	Weight decaying factor	10^{-3}
ϑ	Base of the exponential factor	10
ϵ	Offset parameter of the exponential factor	9

- **Pretraining:** The DeepVLF model undergoes pretraining with varying values of the decoding threshold γ. By exposing the model to a range of decoding conditions, it learns to capture diverse feature representations and adapts to different requirements of belief for decoding. This phase enhances the model's generalization capabilities and prepares it for various operational scenarios.
- **Fine-tuning:** Following pretraining, we fine-tune the model with a target value of γ that aligns with the desired balance between the code rate and BLER. This phase focuses on optimizing the model's parameters to control DeepVLF's average code rate and to improve its BLER performance.

The hyper-parameters settings for the training of DeepVLF are summarized in Table 2.3. To dynamically adjust its average code rate, we vary the decoding threshold γ from $1 - 10^{-3}$ to $1 - 10^{-7}$. The parameter μ in (2.49) is set to

$$\mu = \begin{cases} 5, & \text{if } \gamma <= 1 - 10^{-5}; \\ 6, & \text{if } 1 - 10^{-5} < \gamma <= 1 - 10^{-6}; \\ 7, & \text{if } \gamma > 10^{-6}. \end{cases} \tag{2.51}$$

Performance

Considering a short block length $K = 51$ and noiseless feedback, we present the BLER versus code rate performances of DeepVLF in Fig. 2.13a and b, where the forward SNRs are set to 1 and 0 dB, respectively. As can be seen, DeepVLF outperforms existing DL-aided feedback coding schemes, establishing a new state-of-the-art performance in high code rate scenarios. Specifically, under a code rate of $1/2$ and a forward SNR of 1 dB, DeepVLF achieves a BLER below 10^{-5}, whereas the best existing DL-based code, GBAF, reaches only up to a BLER of 10^{-2}. This demonstrates DeepVLF's superior reliability and efficiency.

To further evaluate DeepVLF's performance, we extend our analysis to scenarios with noisy feedback. DeepVLF relies on two types of feedback: the received symbols and the indices of successfully decoded bit groups. As the indices comprise

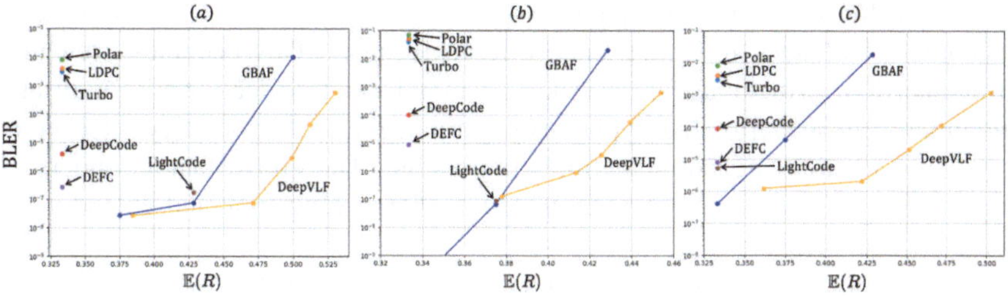

Fig. 2.13 The BLER versus average code rate performances of DeepVLF: (**a**) forward SNR is 1 dB, noiseless feedback; (**b**) forward SNR is 0 dB, noiseless feedback; (**c**) forward SNR is 1 dB, feedback SNR is 20 dB

a minimal amount of data, we assume they are transmitted reliably over a dedicated control channel. Only the received symbol packets are sent over the noisy feedback channel. The results are presented in Fig. 2.13c, where the forward SNR is set to 1 dB, and the feedback SNR is set to 20 dB, consistent with prior studies. The experimental findings reveal that even under noisy feedback conditions, DeepVLF consistently outperforms other DL-based feedback coding schemes, especially in the high code rate regime. This demonstrates the robustness of DeepVLF to noise in the feedback channel, affirming its effectiveness in practical scenarios with feedback imperfections.

2.4 Feedback-Enhanced IoT Communication

AI-driven feedback codes have the potential to dramatically enhance practical IoT communication systems by enabling a new communication paradigm, termed feedback-enhanced IoT (FE-IoT). This paradigm leverages proactive and intelligent feedback from the receiver (often the access point, AP) to significantly improve the uplink performance of IoT devices, thereby addressing critical challenges such as limited energy reserves and constrained uplink coverage. In what follows, we discuss two major practical advantages brought by feedback coding: the notable energy saving at IoT devices and the significant enhancement of uplink coverage.

2.4.1 Energy-Efficient IoT Through Feedback

IoT devices are typically battery-powered with limited capacity, yet they are deployed over vast areas and are required to continuously update their sensing data to a remote AP. The demand for extensive and deep coverage necessitates increased energy consumption for wireless transmission. For instance, in Narrowband IoT (NB-IoT) [10], IoT devices have to repetitively transmit a data packet up to 128 times to achieve a maximum path loss of 164 dB. Moreover, IoT devices often encounter challenges in timely energy replenishment. This becomes even more pronounced in critical use cases such as monitoring in deep wells or disaster surveillance in scenarios like typhoons or earthquakes, where battery replacement is not just challenging but sometimes impossible. The above factors underscore the importance of enhancing communication energy efficiency and extending the lifespan of IoT devices.

The predominant approach for reducing the communication power expenditure of IoT devices is to extend their sleep periods [2, 10, 34]. Take NB-IoT, for example, which implements two main power-saving strategies: PSM and eDRX [10]. In PSM, IoT devices transition into a partial sleep state after a specified duration of inactivity, significantly reducing the power expenditure of their radio chips. While in this mode, the devices remain registered on the network but cease listening for paging from the AP. On the other hand, eDRX allows devices to remain idle before entering PSM yet receptive to AP paging for some time, ensuring the reachability of downlink data. A similar concept exists in the LoRa IoT protocol with the Wake on Radio (WOR) mechanism [2]. WOR optimizes the communication energy expenditure while maintaining network connectivity by keeping IoT devices in a sleep state and periodically activating them to listen for any wake-up signals. Overall, prolonging the sleep cycles in NB-IoT and LoRa protocols has notably extended the lifetimes of IoT devices. Nevertheless, this method does not fundamentally modify the essence of communication, as it does not inherently increase the energy efficiency of the data transmission process itself: the energy expenditure for transmitting each unit of data to a remote AP remains the same, and the total amount of data an IoT device can transmit is still constrained by its power limitations.

FE-IoT, in contrast, introduces a pioneering energy-saving transmission strategy [3] by innovating the channel coding techniques used in IoT wireless communications. The genesis of FE-IoT is rooted in analyzing the energy consumption patterns of IoT devices during transmission. Taking NB-IoT as an example, a typical NB-IoT chip's transmit power, receive power, and sleep power are approximately 135mA, 4mA, and 4μA, respectively. Traditionally, the transmitter, possessing information unknown to the receiver, plays a dominant role in communications—it transmits data continuously while the receiver passively responds with a 1-bit acknowledgement (ACK) or negative ACK (NAK). In this setup, the major energy consumption occurs at the transmitting end.

However, recognizing that (1) the transmit power consumption of IoT devices far exceeds their receive power consumption, (2) the AP has a comparatively more abundant energy supply than IoT devices, we advocate for a fundamental paradigm shift: empowering the receiver with a more proactive role by enabling it to provide real-time feedback on its decoding status to the transmitter. With access to this feedback, the transmitter can dynamically adjust its coding strategy to match the receiver's current conditions. This iterative process of feedback and adaptation allows for a more efficient use of the transmitter's energy, embodying a "listen more, transmit less" approach: by listening more feedback than traditional systems

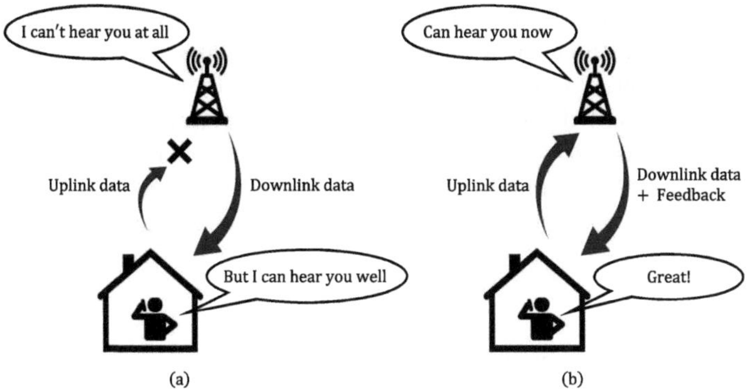

Fig. 2.14 (**a**) Typical asymmetric communication scenario, where downlink is successful while uplink fails. (**b**) Enhanced uplink with real-time feedback, where IoT devices leverage feedback to improve uplink coverage

from the AP, FE-IoT can transmit the same amount of information bits as traditional systems but at a significantly lower power. Adopting this strategy, FE-IoT essentially facilitates an energy exchange between IoT devices and the AP, leading to a marked decrease in energy consumed for transmissions and a notable extension of the devices' operational lifespan.

2.4.2 Connecting the Unconnectable Through Feedback

In the rapidly expanding realm of the IoT, reliable connectivity for devices at coverage edges remains a persistent challenge. Figure 2.14a depicts a frequently encounter situation where an IoT device at the cell edge is still able to "hear" downlink transmissions from its AP, yet the AP struggles to detect the device's uplink signals. This disparity arises from the limited transmit power and single-antenna design typical of many IoT devices, resulting in a pronounced imbalance between downlink and uplink performance. Overcoming this "unconnectable" gap is essential for unlocking the full potential of IoT applications, particularly those spanning smart cities, healthcare, and industrial automation, that demand wide and deep coverage.

To mitigate such uplink coverage limitations, conventional strategies have involved increasing uplink transmission power, employing repetitive transmissions (e.g., NB-IoT [35]), or exploiting multi-antenna diversity (e.g., the Transmission Mode (TM) 2 of LTE and TM 9 of 5G NR [36]). However, these methods are not favorable or impractical for energy-constrained, single-antenna IoT devices.

In this context, FE-IoT provides an alternative that boosts uplink connectivity without raising the transmission power of IoT devices [12]. FE-IoT is built on the innate asymmetry between downlink and uplink communications of IoT cells.

Specifically, APs typically have multiple antennas and virtually unlimited power resources, allowing them to achieve extensive downlink coverage, as shown in Fig. 2.14a. By exploiting this capability, the AP can transmit real-time feedback, indicating its current decoding status, to the IoT device, thereby boosting the IoT device's uplink coding efficiency through feedback channel codes [7, 17]. As illustrated in Fig. 2.14b, this feedback mechanism effectively extends the uplink coverage radius at the same device power consumption, thereby connecting the unconnectable.

2.4.2.1 Uplink Coverage Analysis

We consider an IoT environment comprising distributed wireless APs. Focusing on a single IoT device, our primary interest lies in its uplink transmission coverage, specifically, the number of APs that fall within the communication range of the device with the real-time feedback from the APs.

We model the spatial distribution of the APs using a Poisson Point Process (PPP) with intensity λ, representing the average number of APs per unit area. The uplink path loss from the IoT device to an AP at a distance R can be written as

$$L_U(R) = C_U R^{-\alpha_U}, \tag{2.52}$$

where C_U is the uplink path loss intercepts and α_U is the uplink path loss exponents.

The transmit power of the IoT device and the AP are denoted by P_U and P_D, respectively. Due to energy constraints typical in IoT devices, we have $P_D \gg P_U$. The uplink SNR at the AP can be expressed as

$$\eta_U(R) = P_U \frac{G_t G_r L_U(R)}{\sigma_U^2} |h_U|^2, \tag{2.53}$$

where G_t and G_r denote the transmit and receive antenna gains, respectively; σ_U^2 is the AWGN power; h_U is the uplink small-scale fading coefficient. We model h_U by Rayleigh distribution, i.e., $|h_U|^2 \sim Exp(\mu_U)$. We will also write SNR in decibels, in which case a 'dB' will be added in the subscript, e.g., $\eta_{U,\mathrm{dB}} \triangleq 10 \lg \eta_U$.

Downlink path loss is similarly modeled but with different parameters $L_D(R) = C_D R^{-\alpha_D}$, where α_D, C_D are the respective downlink path loss exponents and intercepts. The downlink SNR received at the IoT device can be written as

$$\eta_D(R) = P_D \frac{G_t G_r L_D(R)}{\sigma_D^2} |h_D|^2, \tag{2.54}$$

where σ_D^2 is AWGN power; h_D is the downlink Rayleigh fading coefficient, and $|h_D|^2 \sim Exp(\mu_D)$.

To transmit a packet of K bits, the IoT device conventionally employs a forward error correction code $C(K, N, \epsilon^*)$ to protect the information bits, where N is the number of uplink channel uses, and ϵ^* is the target packet error rate (PER), reflecting the target throughput. By the finite length coding theorem [16], the PER ϵ is determined by the uplink SNR η_U:

$$\sqrt{NV}Q^{-1}(\epsilon) = \frac{N}{2}\log(1+\eta_U) - K + O(\log N), \tag{2.55}$$

where $V = \frac{\eta_U(\eta_U+2)}{2(\eta_U+1)^2}\log^2 e$ is the channel dispersion, and Q represents the Q-function.

We refer to this conventional approach, where only forward channel coding is employed, as the forward mode. In forward mode, the critical uplink SNR required to achieve the target PER ϵ^* is denoted by Ω_c. Therefore, an AP is said to be connectable for the IoT device, or the IoT device is under the coverage of the AP, if the uplink SNR $\eta_U \geq \Omega_c$.

When the AP provides real-time feedback to the IoT device, the channel coding efficiency can be significantly improved. This enhancement effectively lowers the critical uplink SNR required to achieve the target PER, thereby increasing the probability that an AP at any distance R is connectable. We refer to this enhanced communication approach as the feedback mode.

In feedback mode, the IoT device employs a feedback error correction code $C_f(K, N, N', \epsilon^*)$, where K, N, ϵ^* are as defined in $C(K, N, \epsilon^*)$, while N' is the number of downlink channel uses dedicated to feedback. In this analysis, we consider the feedback coding architecture in [3, 8], in which case $N' = aN$, where $a \in \mathbb{Z}^+$ is a positive integer. The critical uplink SNR in the feedback mode required to meet the target ϵ^*, denoted as Ω_f, follows a logistic function:

$$\Omega_{f,\text{dB}} = \frac{1}{\exp\left(u_0\eta_{D,\text{dB}} + u_1 a + u_2\eta_{D,\text{dB}}a + u_3\right) + u_4} + u_5, \tag{2.56}$$

where $\{u_0, u_1, u_2, u_3, u_4, u_5\}$ are constants, determined by the target PER ϵ^*. An AP is deemed connectable for the IoT device if the uplink SNR $\eta_U \geq \Omega_f$.

As can be seen from (2.56), unlike the critical SNR Ω_c in the forward mode, which depends solely on the target PER ϵ^*, the critical SNR Ω_f in the feedback mode also depends on the feedback channel quality η_D and allocated feedback channel resources a. Consequently, Ω_f becomes a random variable due to the inherent randomness of the downlink channel quality η_D. This dual dependency – on both uplink and downlink channel conditions – introduces a unique coupling effect between the two channels. Specifically, the uplink channel determines the IoT device's ability to meet the SNR threshold for a given distance to the AP, while the downlink channel impacts the efficiency of feedback communication, which in turn affects the required critical SNR in the uplink.

In the following, we analyze the number of APs that are connectable in the feedback mode and quantify the performance gains achieved by leveraging real-time feedback. Specifically, we focus on a circular coverage area centered around the IoT device with radius D. The number of connectable APs in this feedback scenario is denoted as $M_f(D)$.

In feedback mode, the interplay between the uplink and downlink channels introduces a coupled dual-channel fading behavior. This coupling complicates the analysis of feedback communication systems, as the statistical properties of both the uplink and downlink channels must be jointly considered. Accurately characterizing this interdependence is critical for evaluating the system's coverage performance.

We define the probability that an AP at a distance R is connectable under feedback mode as $\varphi_f(R) \triangleq \Pr\left(\eta_U(R) \geq \Omega_f\right)$. Note that both $\eta_U(R)$ and Ω_f are random variables: the randomness of $\eta_U(R)$ arises due to uplink small-scale fading, while the randomness of Ω_f is introduced through the downlink SNR $\eta_D(R)$, which influences the feedback quality.

Substituting the expression for $\eta_U(R)$ into $\varphi_f(R)$, the coverage probability can be reformulated as

$$\varphi_f(R) = \Pr\left(|h_U|^2 > \frac{\Omega_f \sigma_U^2}{P_U G_t G_r C_U R^{-\alpha_U}}\right). \tag{2.57}$$

To simplify the notation, we introduce a function $g(R, |h_D|^2)$:

$$g(R, |h_D|^2) \triangleq \frac{\Omega_f \sigma_U^2}{P_U G_t G_r C_U R^{-\alpha_U}},$$

where Ω_f, as given in (2.56), is governed by the downlink received SNR η_D.

Since both $|h_U|^2$ and $|h_D|^2$ follow exponential distributions, the coverage probability can be rewritten as

$$\begin{aligned}
\varphi_f(R) &= \Pr\left(|h_U|^2 > g(R, |h_D|^2)\right) \\
&= \int_0^{+\infty} [1 - (1 - e^{-\mu_U g(R, |h_D|^2)})] f(|h_D|^2) d|h_D|^2 \\
&= \int_0^{+\infty} e^{-\mu_U g(R, x)} \mu_D e^{-\mu_D x} dx \\
&\overset{(a)}{\approx} \sum_{k=1}^{K} w_k e^{-\mu_U g\left(R, \frac{x_k}{\mu_D}\right)},
\end{aligned} \tag{2.58}$$

where $f(x)$ is the probability density function (PDF) of the exponential distribution; (a) follows from the Gauss-Laguerre quadrature [37], in which x_k represents the roots of the Laguerre polynomials and w_k are the weights associated with these roots.

To derive the number of connectable APs, an accurate approximation of $g(R, |h_D|^2)$ is necessary. We start by approximating the downlink SNR in decibels, denoted as $\eta_{D,dB}\left(\frac{x_k}{\mu_D}\right)$, as follows:

$$\begin{aligned}
\eta_{D,dB}\left(\frac{x_k}{\mu_D}\right) &= 10\lg\left(\frac{P_D G_t G_r C_D x_k R^{-\alpha_D}}{\mu_D \sigma_D^2}\right) \\
&= 10\lg\left(Z_{1,k} R^{-\alpha_D}\right) = 10\lg Z_{1,k} + 10\lg R^{-\alpha_D} \\
&= 10\lg Z_{1,k} + \frac{10}{\ln 10}\ln R^{-\alpha_D} \\
&\approx 10\lg Z_{1,k} + \frac{10}{\ln 10}(R^{-\alpha_D} - 1),
\end{aligned} \tag{2.59}$$

where $Z_{1,k} \triangleq \frac{P_D G_t G_r C_D x_k}{\mu_D \sigma_D^2}$ and $Z_2 \triangleq \frac{\sigma_U^2}{P_U G_t G_r C_U}$ encapsulates system parameters for the downlink and uplink, respectively.

Substituting (2.59) into (2.56), the critical SNR in decibels $\Omega_{f,dB}$ can be approximated as

$$\Omega_{f,dB} \approx \frac{1}{\exp\left(B_{1,k} + B_2 R^{-\alpha_D} - 1 - u_4\right) + u_4} + u_5, \tag{2.60}$$

where $B_{1,k} = 1+u_1a+u_3+u_4+10(u_0+u_2a)(\lg Z_{1,k} - \frac{1}{\ln 10})$ and $B_2 = \frac{10}{\ln 10}(u_0+u_2a)$.

Using the first-order Taylor series expansion $e^x = 1+x+R_1^{(1)}$, the approximation of $\Omega_{f,dB}$ becomes

$$\Omega_{f,dB} \approx \frac{1}{B_{1,k} + B_2 R^{-\alpha_D} + R_1^{(1)}} + u_5. \tag{2.61}$$

Applying another Taylor series expansion $\frac{1}{1+x} = 1 - x + R_1^{(2)}$, we further simplify $\Omega_{f,dB}$ as

$$\Omega_{f,dB} \approx \frac{1}{B_{1,k}}\left(1 - \frac{B_2}{B_{1,k}} R^{-\alpha_D} - \frac{R_1^{(1)}}{B_{1,k}} + R_1^{(2)} + u_5 B_{1,k}\right). \tag{2.62}$$

Here, the bounds of the Lagrange remainder terms satisfy $R_1^{(1)} \geq 0$ and $R_1^{(2)} \geq 0$.

We now approximate $g\left(R, \frac{x_k}{\mu_D}\right)$, defined as the effective threshold function:

$$g\left(R, \frac{x_k}{\mu_D}\right) = Z_2 R^{\alpha_U} 10^{\frac{1}{10}\Omega_{f,dB}} \tag{2.63}$$

$$= Z_2 R^{\alpha U} e^{\frac{\ln 10}{10}\Omega_{f,dB}} = Z_2 R^{\alpha U}(1 + \frac{\ln 10}{10}\Omega_{f,dB} + R_1^{(3)}),$$

where $R_1^{(3)}$ represents the remainder of the Taylor expansion.

Substituting (2.62) into (2.63), we obtain

$$g\left(R, \frac{x_k}{\mu_D}\right) \approx Z_2 R^{\alpha U}[1 + \frac{\ln 10}{10 B_{1,k}}(1 - \frac{B_2}{B_{1,k}} R^{-\alpha D} - \frac{R_1^{(1)}}{B_{1,k}} + R_1^{(2)} + u_5 B_{1,k}) + R_1^{(3)}]. \tag{2.64}$$

In particular, the Lagrange form of the remainder $R_1^{(3)}$ can be approximated by

$$\begin{aligned} R_1^{(3)} &= \frac{1}{2!} e^{\frac{\ln 10}{10 B_{1,k}}(1 - \frac{B_2}{B_{1,k}} c^{-\alpha D} - \frac{R_1^{(1)}}{B_{1,k}} + R_1^{(2)} + u_5 B_{1,k})} \\ &\times \left[\frac{\ln 10}{10 B_{1,k}}(1 - \frac{B_2}{B_{1,k}} R^{-\alpha D} - \frac{R_1^{(1)}}{B_{1,k}} + R_1^{(2)} + u_5 B_{1,k})\right]^2 \\ &\approx \frac{1}{2}\left[\frac{\ln 10}{10 B_{1,k}}(1 - \frac{B_2}{B_{1,k}} R^{-\alpha D} - \frac{R_1^{(1)}}{B_{1,k}} + R_1^{(2)} + u_5 B_{1,k})\right]^2. \end{aligned} \tag{2.65}$$

Since $R_1^{(1)} \geq 0$, $R_1^{(2)} \geq 0$, and $(1 + x)^{\alpha} \approx 1 + \alpha x$, $R_1^{(3)}$ can be further refined as

$$\begin{aligned} R_1^{(3)} &\approx \frac{\ln^2 10}{200 B_{1,k}^2}(1 + u_5 B_{1,k})^2 \left[1 - \frac{B_2 R^{-\alpha D}}{B_{1,k}(1 + u_5 B_{1,k})}\right]^2 \\ &\approx \frac{\ln^2 10}{200 B_{1,k}^2}(1 + u_5 B_{1,k})^2 \left[1 - \frac{2 B_2 R^{-\alpha D}}{B_{1,k}(1 + u_5 B_{1,k})}\right]. \end{aligned} \tag{2.66}$$

Substituting (2.66) into (2.64) yields

$$g\left(R, \frac{x_k}{\mu_D}\right) \approx J_{1,k} R^{\alpha U} - J_{2,k}, \tag{2.67}$$

where $J_{1,k} \triangleq Z_2 + \frac{Z_2(1+u_5 B_{1,k}) \ln 10}{10 B_{1,k}} + \frac{Z_2(1+u_5 B_{1,k})^2 \ln^2 10}{200 B_{1,k}^2}$ and $J_{2,k} \triangleq \frac{Z_2 B_2 \ln 10}{10 B_{1,k}^2} + \frac{Z_2 B_2 (1+u_5 B_{1,k}) \ln^2 10}{100 B_{1,k}^3}$.

Given the coverage probability $\varphi_f(R)$, the number of connectable APs $M_f(D)$ within a circular area of radius D can be expressed as

$$\begin{aligned} M_f(D) &= \int_0^D \varphi_f(R)\lambda 2\pi R dR \\ &\approx \int_0^D \sum_{k=1}^{K} w_k e^{-\mu_1 g\left(R, \frac{x_k}{\mu_D}\right)} \lambda 2\pi R dR \\ &= 2\pi\lambda \sum_{k=1}^{K} w_k \int_0^D e^{-\mu_1 g\left(R, \frac{x_k}{\mu_D}\right)} R dR, \end{aligned} \tag{2.68}$$

Substituting the approximation of $g\left(R, \frac{x_k}{\mu_D}\right)$ in (2.67), we finally have

$$\begin{aligned} M_f(D) &\approx 2\pi\lambda \sum_{k=1}^{K} w_k e^{\mu_U J_{2,k}} \int_0^D e^{-\mu_U J_{1,k} R^{\alpha_U}} R dR \\ &= 2\pi\lambda \sum_{k=1}^{K} w_k e^{\mu_U J_{2,k}} \frac{\gamma(\frac{2}{\alpha_U}, \mu_U J_{1,k} D^{\alpha_U})}{(\mu_U J_{1,k})^{\frac{2}{\alpha_U}} \alpha_U}. \end{aligned} \tag{2.69}$$

where $\gamma(\cdot,\cdot)$ is lower incomplete gamma function.

To provide a benchmark for comparison with the feedback mode, we derive the number of connectable APs in the forward mode, where no feedback is employed, and the coverage performance depends solely on the uplink channel conditions. Let $M_c(D)$ denote the number of connectable APs within a circular area of radius D. This can be similarly expressed as $M_c(D) = \int_0^D \varphi_c(R)\lambda 2\pi R dR$, where $\varphi_c(R)$ represents the probability that a AP at a distance R is connectable in forward mode. Specifically, we have

$$\varphi_c(R) = \Pr\left(\eta_U(R) \geq \Omega_c\right) = e^{-AR^{\alpha_U}}, \tag{2.70}$$

where $A \triangleq \frac{\mu_U \Omega_c \sigma_U^2}{P_U G_t G_r C_U}$. Substituting (2.70) into $M_c(D)$, the number of connectable APs in forward mode is given by

$$M_c(D) = 2\pi\lambda \int_0^D e^{-AR^{\alpha_U}} R dR = \frac{2\pi\lambda \cdot \gamma(\frac{2}{\alpha_U}, AD^{\alpha_U})}{\alpha_U A^{\frac{2}{\alpha_U}}}. \tag{2.71}$$

Coverage Probability Improvement with Feedback

The comparison between the coverage probabilities $\varphi_c(R)$ in forward mode and $\varphi_f(R)$ in feedback mode highlights the substantial improvement brought by feedback.

In forward mode, the coverage probability decays exponentially with distance R, as captured by the fixed term $e^{-AR^{\alpha U}}$, which depends solely on the uplink channel quality. In contrast, feedback mode leverages both uplink and downlink channels, where the feedback gains are encapsulated in the terms $J_{1,k}$ and $J_{2,k}$. These terms effectively reduce the required critical SNR, thereby slowing down the decay in coverage probability as R increases.

This improvement is particularly pronounced at larger distances, such as at the cell edge, where the uplink SNR is typically weak, and connectivity is more prone to failure. By increasing the coverage probability through feedback, the connectivity in these edge regions can be significantly enhanced.

2.4.2.2 Numerical Results

This section evaluates the performance of FE-IoT through numerical simulations. To establish an evaluation framework, we configure the system parameters as follows. The density of APs is set to $\lambda = 6 \times 10^{-3}$, with path loss exponents and intercepts set as $C_U = C_D = 10^{-4.7}$ and $\alpha_U = \alpha_D = 4$ [38]. The uplink and downlink Rayleigh fading channels are modeled with parameters $\mu_U = \mu_D = 2$. We consider a short block length $K = 48$, a code rate of $1/3$, resulting in $N = 144$ channel uses, and a target uplink PER $\epsilon^* = 10^{-4}$. In the forward mode, Polar and Turbo codes with the same block length and coding rate are utilized as benchmarks, enabling a direct comparison of their performance against the proposed feedback-aided approach. The downlink transmit power is fixed at $P_D = 50$ mW, while the uplink transmit power is varied across $P_U \in [0.5, 1, 2]$ mW to examine the impact of different power levels on system performance.

We begin by verifying the accuracy of the analytical models established for feedback-aided connectivity. To this end, we compare the derived expressions for the number of connectable APs with simulation results, as illustrated in Fig. 2.15.

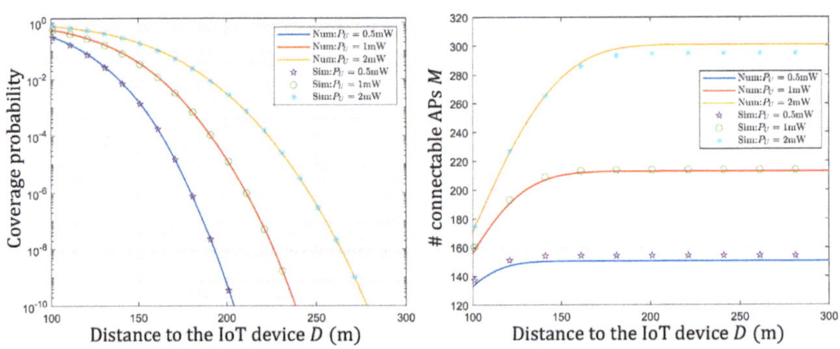

Fig. 2.15 Comparison of analytical and simulation results for validating the accuracy of the analytical expressions, where $P_U \in [0.5, 1, 2]$ mW, $P_D = 50$ mW, $a = 1$, and $\lambda = 6 \times 10^{-3}$

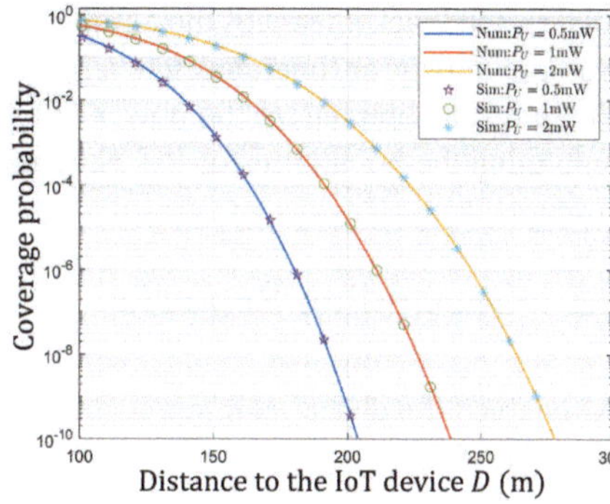

Fig. 2.16 The coverage probability gains of the feedback mode versus the forward mode with Polar and Turbo codes

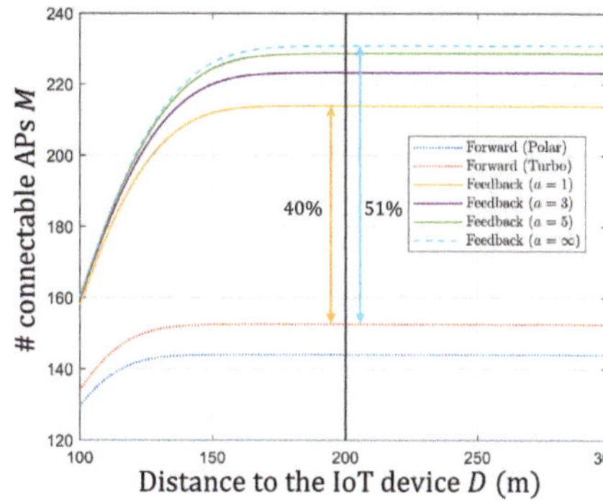

Fig. 2.17 The gains in the number of connectable APs achieved by the feedback mode versus the forward mode using Polar and Turbo codes

This comparison focuses on two key steps: the approximations of coverage probability and the computation of connectable APs. The results in Fig. 2.15 reveal a strong alignment between the analytical predictions and simulation outcomes across varying uplink transmit power levels and distances to the IoT device. This close match confirms the validity and robustness of our analytical method, demonstrating its effectiveness as a reliable tool for evaluating feedback-aided IoT coverage performance.

Next, we explore the transformative impact of feedback on connecting previously unreachable APs. Figure 2.16 illustrates the substantial enhancement in coverage probability achieved through feedback at varying distances. Notably, its benefits are most pronounced at larger distances, particularly for IoT devices situated at the cell edge. At 200 meters, for instance, feedback boosts the coverage probability by three orders of magnitude, effectively bridging the gap where traditional forward-only communication fails. The extent of this improvement is closely tied to the amount of feedback, a. More feedback allows the user to precisely assess the decoding status of the AP, significantly extending the coverage range. Without feedback, Turbo codes ensure a coverage probability of 10^{-4} up to 166 meters. In stark contrast, feedback extends this reliability to 205 meters, a remarkable 24% increase in range, enabling connections far beyond the traditional limits.

The benefits of feedback extend beyond probability gains. Figure 2.17 showcases the number of connectable APs, M, at varying distances. At a distance of 200 meters, feedback enhances the number of reachable APs by 51%, transforming what was previously an unconnected region into a significantly more connected zone. These results emphasize the power of feedback in "connecting the unconnectable", redefining the boundaries of IoT network performance.

2.5 Concluding Remarks

AI-enabled feedback communication represents a paradigm shift in the design and optimization of IoT networks, offering significant improvements in energy efficiency, reliability, and adaptability. This chapter has explored the theoretical underpinnings of feedback in communication systems, highlighting its potential to enhance transmission performance beyond conventional acknowledgment-based approaches. By integrating artificial intelligence into feedback mechanisms, IoT devices can dynamically adjust their coding strategies in response to real-time decoding status of the AP, reducing unnecessary retransmissions and optimizing power consumption.

The FE-IoT framework embodies the principle of "listen more, transmit less", enabling IoT devices to make intelligent coding based on feedback. We examined information-theoretical results that quantify the gains of feedback in finite block-length regimes, as well as AI-driven approaches such as DeepCode, AttentionCode, and GBAF codes, which leverage deep learning to improve feedback efficiency. These methods demonstrate that AI can not only refine feedback-based adaptation but also surpass traditional feedback coding schemes, particularly in noisy and resource-constrained IoT environments.

As IoT continues to scale, the need for efficient, scalable, and intelligent communication mechanisms becomes more pressing. Future research should focus on adaptive feedback architectures that can seamlessly integrate with emerging IoT standards, the development of low-complexity AI models for real-time feedback optimization, and the exploration of hybrid digital-analog feedback systems. Additionally, extending AI-driven feedback schemes to multi-device collaborative scenarios and heterogeneous IoT networks could further unlock the potential of feedback in large-scale deployments.

By harnessing AI-enhanced feedback mechanisms, the next generation of IoT networks will be more energy-efficient, resilient, and capable of supporting intelligent interactions among IoT devices, paving the way for robust and scalable IoT ecosystems.

References

1. D.C. Nguyen, M. Ding, P.N. Pathirana, A. Seneviratne, J. Li, D. Niyato, O. Dobre, H.V. Poor, 6G Internet of Things: A comprehensive survey. IEEE Internet Things J. **9**(1), 359–383 (2021)
2. J.P.S. Sundaram, W. Du, Z. Zhao, A survey on LoRa networking: Research problems, current solutions, and open issues. IEEE Commun. Surv. Tutorials **22**(1), 371–388 (2019)

3. Y. Shao, DEEP-IoT: Downlink-enhanced efficient-power Internet of Things. IEEE Trans. Wirel. Commun. **24**(2), 1722–1736 (2024)
4. J. Schalkwijk, T. Kailath, A coding scheme for additive noise channels with feedback I: No bandwidth constraint. IEEE Trans. Inf. Theory **12**(2), 172–182 (1966)
5. A. Ben-Yishai, O. Shayevitz, Interactive schemes for the AWGN channel with noisy feedback. IEEE Trans. Inf. Theory **63**(4), 2409–2427 (2017)
6. H. Kim, Y. Jiang, S. Kannan, S. Oh, P. Viswanath, Deepcode: Feedback codes via deep learning, in *Advances in Neural Information Processing Systems*, vol. 31 (2018)
7. Y. Polyanskiy, H.V. Poor, S. Verdú, Feedback in the non-asymptotic regime. IEEE Trans. Inf. Theory **57**(8), 4903–4925 (2011)
8. E. Ozfatura, Y. Shao, A.G. Perotti, B.M. Popović, D. Gündüz, All you need is feedback: Communication with block attention feedback codes. IEEE J. Sel. Areas Inf. Theory **3**(3), 587–602 (2022)
9. W. Lai, Y. Shao, Y. Ding, D. Gunduz, Variable-length feedback codes via deep learning, in *IEEE ICC* (2024)
10. M. Kanj, V. Savaux, M. Le Guen, A tutorial on NB-IoT physical layer design. IEEE Commun. Surv. Tutorials **22**(4), 2408–2446 (2020)
11. P. Frenger, S. Parkvall, E. Dahlman, Performance comparison of HARQ with chase combining and incremental redundancy for HSDPA, in *IEEE Vehicular Technology Conference*, vol. 3 (IEEE, Piscataway, 2001), pp. 1829–1833
12. Y. Li, Y. Shao, Connecting the unconnectable through feedback. Preprint. arXiv:2501.02335 (2025)
13. C.E. Shannon, A mathematical theory of communication. Bell Syst. Tech. J. **27**(3), 379–423 (1948)
14. W. Ryan, S. Lin, *Channel Codes: Classical and Modern* (Cambridge University Press, Cambridge, 2009)
15. J.H. Bae, A. Abotabl, H.-P. Lin, K.-B. Song, J. Lee, An overview of channel coding for 5G NR cellular communications. APSIPA Trans. Signal Inf. Process. **8**, e17 (2019)
16. Y. Polyanskiy, H.V. Poor, S. Verdú, Channel coding rate in the finite blocklength regime. IEEE Trans. Inf. Theory **56**(5), 2307–2359 (2010)
17. C. Shannon, The zero error capacity of a noisy channel. IRE Trans. Inf. Theory **2**(3), 8–19 (1956)
18. R.G. Gallager, B. Nakiboğlu, Variations on a theme by Schalkwijk and Kailath. IEEE Trans. Inf. Theory **56**(1), 6–17 (2009)
19. J. Schalkwijk, A coding scheme for additive noise channels with feedback II: Band-limited signals. IEEE Trans. Inf Theory **12**(2), 183–189 (1966)
20. Y.-H. Kim, A. Lapidoth, T. Weissman, The Gaussian channel with noisy feedback, in *IEEE ISIT* (2007), pp. 1416–1420
21. M. Schuster, K.K. Paliwal, Bidirectional recurrent neural networks. IEEE Trans. Signal Process. **45**(11), 2673–2681 (1997)
22. K. Cho, B. Van Merriënboer, C. Gulcehre, D. Bahdanau, F. Bougares, H. Schwenk, Y. Bengio, Learning phrase representations using RNN encoder-decoder for statistical machine translation. arXiv:1406.1078 (2014)
23. Y. Shao, E. Ozfatura, A. Perotti, B. Popovic, D. Gunduz, Attentioncode: Ultra-reliable feedback codes for short-packet communications. IEEE Trans. Commun. **71**, 4437 (2023)
24. A. Vaswani, N. Shazeer, N. Parmar, J. Uszkoreit, L. Jones, A.N. Gomez, L. Kaiser, I. Polosukhin, Attention is all you need. arXiv:1706.03762 (2017)
25. J. Devlin, M.-W. Chang, K. Lee, K. Toutanova, BERT: pre-training of deep bidirectional transformers for language understanding. arXiv:1810.04805 (2018)
26. R. Xiong, Y. Yang, D. He, K. Zheng, S. Zheng, C. Xing, H. Zhang, Y. Lan, L. Wang, T.-Y. Liu, On layer normalization in the transformer architecture. arXiv:2002.04745 (2020)
27. J.L. Ba, J.R. Kiros, G.E. Hinton, Layer normalization. arXiv:1607.06450 (2016)
28. A.R. Safavi, A.G. Perotti, B.M. Popovic, M.B. Mashhadi, D. Gunduz, Deep extended feedback codes. ITU J. Future Evolving Tech. **2**(6), 33–41 (2021)

29. M.B. Mashhadi, D. Gunduz, A.G. Perotti, B.M. Popovic, DRF codes: Deep SNR-robust feedback codes. arXiv (2021)
30. E. Ozfatura, Y. Shao, A. Ghazanfari, A. Perotti, B. Popovic, D. Gündüz, Feedback is good, active feedback is better: Block attention active feedback codes, in *IEEE International Conference on Communications* (2023), pp. 6652–6657
31. M. Bennis, M. Debbah, H.V. Poor, Ultra reliable and low-latency wireless communication: tail, risk, and scale. Proc. IEEE **106**(10), 1834–1853 (2018)
32. D. Hendrycks, K. Gimpel, Gaussian error linear units. Preprint. arXiv:1606.08415 (2016)
33. I. Loshchilov, Decoupled weight decay regularization. arXiv:1711.05101 (2017)
34. R.S. Sinha, Y. Wei, S.-H. Hwang, A survey on LPWA technology: LoRa and NB-IoT. ICT Express **3**(1), 14–21 (2017)
35. M. Kanj, V. Savaux, M. Le Guen, A tutorial on NB-IoT physical layer design. IEEE Commun. Surv. Tutorials **22**(4), 2408–2446 (2020)
36. S. Parkvall, E. Dahlman, A. Furuskar, M. Frenne, NR: The new 5G radio access technology. IEEE Commun. Stand. Mag. **1**(4), 24–30 (2017)
37. N.I. Ioakimidis, The Gauss-Laguerre quadrature rule for finite-part integrals. Commun. Numer. Methods Eng. **9**(5), 439–450 (1993)
38. T. Bai, R.W. Heath, Coverage and rate analysis for millimeter-wave cellular networks. IEEE Trans. Wirel. Commun. **14**(2), 1100–1114 (2015)

Chapter 3
AI-Empowered Semantic Communication for IoT

Words are but symbols; meaning is the true essence.

3.1 Understanding Semantic Communication

In the burgeoning landscape of the IoT, the efficiency and effectiveness of communication have become paramount. Traditional communication systems have been primarily concerned with the accurate transmission of data. However, as IoT devices become increasingly pervasive and integral to our daily lives, the need for these devices to understand and process the significance or meaning behind the data, known as semantic communication, has emerged as a critical research area [1–3].

Semantic communication transcends the mere exchange of data to ensure that the information conveyed is understood and actionable by the recipient, be it human or machine. This chapter delves into the core of semantic communication within the IoT paradigm, exploring how it could revolutionize interactions by ensuring that devices not only communicate but also comprehend each other's messages in context [4].

The concept of semantic communication is not new; it was first hinted at by Weaver in 1949 [5], categorizing communication problems into three distinct levels: technical, semantic, and effective problems. In the realm of IoT, these distinctions become increasingly significant as devices must act autonomously and in concert based on the exchange of meaningful and contextually relevant information. This chapter builds upon the foundational theories of semantic communication. By integrating advanced AI techniques with traditional communication theories, we aim to outline a framework whereby IoT devices can achieve a mutual understanding and perform tasks more efficiently.

3.1.1 Historical Context of Semantic Communication

One year after the birth of Shannon's information theory [6], Weaver classified a broad communication problem into three levels [5]:

Y. Shao, *AI-Empowered IoT Communications*, Wireless Networks,
https://doi.org/10.1007/978-3-031-98816-5_3

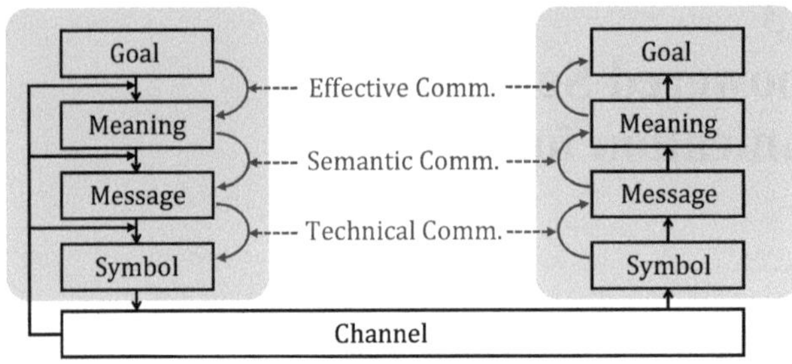

Fig. 3.1 A broad communication process with three levels of problems

- **Level A** (The technical problem): How accurately can the symbols of communication be transmitted?
- **Level B** (The semantic problem): How precisely do the transmitted symbols convey the desired meaning?
- **Level C** (The effectiveness problem): How effectively does the received meaning affect conduct in the desired way?

Following Weaver's formulation, we depict a communication process in Fig. 3.1 and expound on some key concepts to be used throughout this chapter. We will provide general descriptions of these concepts here, and relegate their more precise mathematical definitions in our context to later sections.

- **Goal**. Communication serves for a cooperative task between the transmitter and receiver. The task is the ultimate goal of communication.
- **Meaning**. At different phases of the task, the transmitter can generate different meanings to be conveyed to the receiver for achieving the ultimate goal. The intended meaning can be an observation of the environment, an instruction, a state update, etc.
- **Message**. Messages are the carriers of meaning and are constructed following an agreed-upon language between the transmitter and receiver.
- **Symbol**. Symbols refer to the coded channel symbols to be transmitted to the receiver via the physical channel.

As shown in Fig. 3.1, effective/pragmatic communication deals with the problem of generating the right meaning for achieving the ultimate goal, considering the current states of the transmitter, receiver, and the progress of the task; semantic communication deals with the problem of how to construct the right message to accurately convey the meaning based on the agreed language; technical communication studies how to design channel symbols for different messages such that the messages can be reconstructed at the receiver as accurately as possible.

Shannon provided an elegant solution to the technical problem [6], which further spawned the field of information theory. In his solution, the amount of information in a message is measured by the notion of "uncertainty", which is inversely proportional to the statistical probability of the message. Thus, it is the probabilities, but not the meanings of messages matter in Shannon's theory. In contrast, semantic communication is more concerned with the meaning that can be conveyed by a message. We give two examples below to show the difference between semantic and technical communications.

Example 3.1 (Successful Technical Communication but Failed Semantic Communication) Consider the communication between a farmer and his grandchild. The farmer says: "The apple looks good". The intended meaning of the farmer is that the fruit "apple" looks good. His grandchild, however, may interpret "apple" as the mobile phone. In this example, technical communication succeeds since the message has been perfectly transmitted to the receiver. Semantic communication, however, fails due to misinterpretation – the same message can convey different meanings in a language.

Example 3.2 (Failed Technical Communication but Successful Semantic Communication) Consider the message transmission from Bob to Ted, and then to Alice. Bob says: "Carol does not like carrots". Ted relays the message to Alice: "Carol dislikes carrots". In this example, technical communication fails since the received message of Alice is different from the original message transmitted by Bob. Nevertheless, semantic communication succeeds because Alice knows "does not like" is the same as "dislikes" in English – the intended meaning of Bob is successfully conveyed.

3.1.2 Fundamental Problems of Semantic Communication

The cornerstone of communication is an agreed language among the parties involved. In Shannon's theory, for example, two key procedures for transmitting a message are source coding and channel coding. In source coding, each message is assigned a unique bit sequence (more generally, a sequence of alphabet symbols) according to the statistical probability of the message. In channel coding, structured redundancies are added to the source-coded bit sequences to combat physical channel impairments, yielding channel symbols to be transmitted to the receiver. At the receiver, first the channel input is resolved then the message is decoded.

The successful operation in Shannon's approach relies on the transmitter and receiver's consensuses on (1) the statistical distribution of messages; (2) the source code, e.g., the one-to-one mappings between messages and bit sequences in lossless source coding; and (3) the channel code, e.g., the algebraic structures among coded symbols, before transmission. These serve as the common ground for technical communication, and we refer to them as *technical language* in the sequel.

Likewise, semantic communication relies on a common *semantic language* between the transmitter and receiver for conveying the meaning. A typical example of semantic language is the human language, e.g., English, as in Examples 3.1 and 3.2. Unlike hand-crafted technical language, semantic language is often formed naturally over the course of interactions and is much richer in expression and

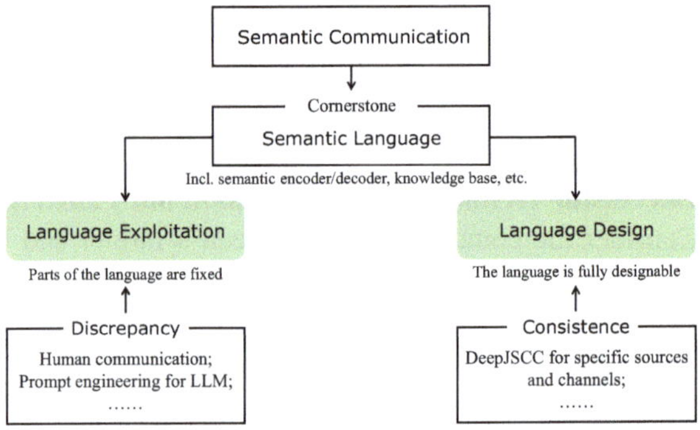

Fig. 3.2 The framework of semantic communications. The language design problem focuses on crafting a new set of languages for efficient communication between the transmitter and receiver for specific sources, aligning with classical communication principles. In contrast, the language exploitation problem addresses the nuanced challenges arising from discrepancies between the transmitter and receiver, extending beyond the traditional communication framework

interpretation: a meaning can be expressed by multiple messages and a message can be interpreted as multiple meanings. The reason for such richness can be attributed to the fact that semantic language has to be robust to the mismatched or unagreed prior information between the transmitter and receiver [7]. That is, the prior distribution of the intended meaning may not be known to the receiver before transmission.

Based on the above understanding of languages, we formulate two fundamental problems of semantic communication, as outlined in Fig. 3.2.

Definition 3.1 (The Language Exploitation Problem) The transmitter and receiver have agreed on semantic and technical languages. When conveying an intended meaning, how to minimize the misinterpretations of the receiver from the transmitter's perspective (semantic encoding) or the receiver's perspective (semantic decoding)?

- **Semantic encoding**: How can the transmitter generate the message such that the intended meaning can be recovered at the receiver as accurately as possible while the communication cost is minimized?
- **Semantic decoding**: Given a received message, how can the receiver decode the intended meaning of the transmitter without accurate prior information about the meaning?

As noted by Weaver:

> The semantic problems are concerned with the interpretation of meaning by the receiver, as compared with the intended meaning of the sender.

The first fundamental problem emphasizes how to leverage the agreed language to reduce misinterpretations of the receiver. An implication of the language exploitation problem is that there is no negotiation before the transmission. The only things that the transmitter and receiver agree on are the semantic and technical languages.

Note that semantic encoding and decoding can also be used at the same time, which we refer to as combined semantic encoding and decoding (CSED). The language exploitation problem resembles human communications.

Definition 3.2 (The Language Design Problem) Assuming that the transmitter and receiver are allowed to negotiate before the transmission, how can the semantic and technical languages be designed to efficiently convey the meaning of a semantic source?

The second fundamental problem concerns how to design common languages or codebooks between the transmitter and receiver to efficiently convey the meaning. Since the transmitter and receiver are given full freedom to design the languages, we can directly design the mappings between meanings and channel symbols, to which joint source-channel coding (JSCC) theory naturally applies. In other words, language design in semantic communication falls into the scope of the classical communication problem that researchers have been striving to solve for decades—designing the communication system to minimize a prescribed distortion measure between the transmitted and reconstructed data. In technical communication, we focus on sequences of equal likely bits and consider the block error rate (BLER) as the measure of reconstruction quality. Semantic communication, on the other hand, considers more general sources (e.g., text [1], image [8], video [9], point cloud [10]) and end-to-end distortion measures (e.g., classification loss [1, 11], perceptual loss [12], goal-oriented distortions [13, 14]).

The language design problem can be further divided into two classes of subproblems: joint semantic and technical language design, and technical language design under a given semantic language. A popular trend nowadays is leveraging data-driven techniques, in particular deep learning (DL), to learn a common language [1, 8, 15, 16]. In this approach, there is a training phase and an evaluation phase. The training phase is essentially a process of language negotiation/design between the transmitter and receiver for a specific kind of source. When the training is done, the agreed language, manifested as the neural encoder and decoder, is used in the evaluation phase for semantic transmission. In general, data-driven solutions to the language design problem are particularly useful in the case of analytically intractable distortion measures, a semantic source with memory, and resource-constrained channels, to name a few.

Related Work

In the literature, there are relatively few research efforts [17–20] devoted to formulating the theory of semantic communication, perhaps owing to the lack of clear definitions of semantics and languages. Among these few studies, the most notable works are Carnap and Bar-Hillel's characterization of semantic information [17] and the follow-up attempt at formulating semantic communication [18].

Carnap and Bar-Hillel's work focused on how to define semantic information and how to measure the amount of semantic information contained in a message. To this end,

1. They defined a language system with messages being limited to inductive logic. That is, messages can only be declarative statements or propositions.
2. They measured the amount of semantic information carried in a message by the probability that the message is logically true.

In this context, the semantic entropy of a message s is defined as

$$H_{\text{sem}}(s) = -\log_2 \sum_{w \models s} p(w), \tag{3.1}$$

where w denotes a world model (e.g., a meaning under Weaver's formulation), $\models$ denotes the propositional satisfaction relation, and $\sum_{w \models s} p(w)$ is the probability of all world models in which s is true.

As can be seen, in Shannon's theory, the information of a message is governed by its statistical probability but not its meaning (e.g., whether the message itself is true or false). By contrast, in Carnap and Bar-Hillel's formulation, the statistical probability $p(s)$ is irrelevant, and the semantic information of s is determined by its logical probability $\sum_{w \models s} p(w)$ under their language system.

Following Carnap and Bar-Hillel's characterization of semantic information, the authors of [18] defined a complete procedure of semantic communication and established semantic source coding and semantic channel coding by mimicking Shannon's theory. They assumed the same logical language system as [17]. In source coding, for example, the authors proposed to merge semantically equivalent messages in the language into a single message, thereby constructing a minimum subset of the original set of messages with no semantic loss. In particular, the minimum subset exhibits the minimum entropy, and hence, the minimum expected coding length.

Although established under the specific logical language system, the formulations and insights obtained in [17, 18] shed light on forming a generic theory of semantic communication, and also inspire the framework in this chapter.

On the other hand, for the language design problem, the transmitter and receiver are allowed to negotiate both semantic and technical languages. For now, the problem that receives wide attention is the joint design of semantic and technical languages using DL techniques.

Joint Semantic and Technical Language Design via DL

In the basic form, the objective of DL-based semantic communication is transmitting a specific kind of source beyond a simple bitstream (e.g., text [1], image [8], video [9]), for which DL techniques are utilized to design a common language. DL-based approaches consist of a training phase for language design and an evaluation phase for data transmission. The main research challenge is how to learn an efficient language in the training phase tailored for the considered source. To this end, various schemes, such as DeepJSCC [1, 8], discrete-time analog transmission [16], the information bottleneck approach [13], the goal-oriented principle [21], the nonlinear transform [22], etc., have been proposed. We refer readers to survey papers [23–25] for a more detailed treatment of this line of work.

Technical Language Design Under a Given Semantic Language
In addition to the joint design of semantic and technical languages, another interesting problem is the design of technical language under a given semantic language. This line of work is relatively few.

In [26], the authors proposed to optimize the communication system performance by taking the semantic content of messages into account. There are three parties in their framework: a transmitter, a receiver, and an influencer. The transmitter aims to transmit a message (more precisely, the meaning of the message) to the receiver, which, in turn, decodes the message with the side information provided by the influencer. To quantify the semantic distortion in the system, the authors defined a semantic measure that quantifies the similarity of messages. The transmitter and receiver have a fixed set of encoding and decoding functions, respectively. In this semantic communication system, the semantic language and the semantic distortion among messages are given. To deliver the meaning of a message, the transmitter and receiver design the technical language, i.e., choose between the set of encoding and decoding functions, thereby minimizing the semantic distortion in the presence of the influencer.

In [27], the authors considered a semantic source with an intrinsic state and an extrinsic observation, where the transitions between the state and observation are determined. The receiver is interested in both state and observation, thus there are two distortions at the receiver with respect to the reconstructed state and observation, respectively. The authors characterized the trade-off between the two distortions under a given code rate. Translating into our formulation, the intrinsic state and extrinsic observation can be viewed as meaning and message, respectively, and the semantic language is the fixed (and known) mapping between state and observation. The authors designed the technical language (codebook) to strike a balance between the two distortions using techniques from indirect rate-distortion theory and rate-distortion under multiple distortion measures.

3.1.3 Semantic Communication in IoT Contexts

This chapter offers insights into semantic communication's role in enhancing IoT communications by addressing the unique challenges and requirements of IoT systems.

- **Complexity and heterogeneity of IoT devices.** IoT environments typically consist of a diverse array of devices, each with varying capabilities and roles. The semantic communication framework provides a method for these devices to understand and interpret messages in a context that makes sense to their specific functions and capabilities. For example, in a smart factory, different machines and sensors might interpret data differently based on their operational parameters and roles. Semantic communication ensures that messages such as "increase production rate" or "adjust temperature settings" are interpreted correctly by each device according to its functionality and current state.

- **Need for precision in commands and data.** IoT systems often operate in environments where precision and accuracy in command execution and data interpretation are crucial. The semantic communication helps ensure that messages are not just received but are understood correctly, minimizing the chances of errors that could lead to inefficient operations or safety hazards. For instance, in automated healthcare systems, a command like "administer 10 ML of medication" must be interpreted with absolute precision to avoid medical errors.
- **Energy efficiency.** In IoT environments, particularly those involving remote or challenging locations such as wildlife habitats or difficult-to-access monitoring stations, minimizing energy usage is crucial. These devices often operate under stringent power constraints. Semantic communication offers a significant advantage in these settings by enhancing the efficiency of data transmission. By employing advanced AI-based data compression techniques that prioritize the transmission of meaningful information, semantic communication reduces the frequency of transmissions needed.
- **Scalability and system evolution.** As IoT networks grow, the complexity of managing communications increases. The principles of semantic communication, such as those involving language exploitation and design discussed in the chapter, provide a scalable way to manage communications in large, diverse IoT networks. This ensures that as systems evolve and new types of devices are added, the fundamental communication processes remain robust and effective.
- **Enhancing collaborative IoT interactions.** Semantic communication supports better collaboration between devices by enabling a shared understanding of messages. In scenarios where multiple devices must work together, such as coordinated responses in smart cities or collaborative manufacturing processes, ensuring that all devices interpret the messages within the same semantic context is crucial for smooth operation.

By addressing these specific needs through a structured approach to semantic understanding and data transmission, the semantic communication model discussed in the chapter is ideally suited to meet the complex demands of IoT communications, enhancing overall system performance, reliability, and scalability.

Roadmap

The remainder of this chapter is articulated in the following manner. Section 3.2 explores the language utilization problem via semantic coding, including semantic encoding in Sect. 3.2.2, semantic decoding in Sect. 3.2.3, and CSED in Sect. 3.2.4. Section 3.2.5 presents an example that illustrates these concepts in action. Section 3.3 explores the language design problem, utilizing the framework of deep joint source-channel coding. Section 3.4 addresses a practical concern, the peak-to-average power ratio (PAPR) problem, in semantic communication systems.

Section 3.5 concludes the chapter, summarizing the key insights and implications for the future of IoT communications.

3.2 Language Utilization via Semantic Coding

This section formulates the language utilization problem rigorously. We shall go through some basic semantic-related definitions, such as language, channel, distortion, cost, etc., and define the problems of semantic encoding and decoding, respectively.

3.2.1 *Definitions*

Let us set out to define semantic language, the prerequisite of semantic communication. A semantic language is a structured system that defines the transformation between meanings and messages. We characterize a semantic language by four main ingredients: word, syntax, expression, and interpretation. To ease exposition, henceforth "language" refers specifically to semantic language, unless specified otherwise.

Definition 3.3 (Words and Syntax) Words are the smallest elements of a message in a language. Words can be used on their own or together to form a message. Syntax is a set of rules that determine the arrangement of words in a message.

Definition 3.4 (The Set of Messages) The word and syntax of a language determine the set of all possible messages, denoted by $\mathcal{S}$. Suppose $\mathcal{S}$ is finite or countably infinite, we define $\mathcal{S} \triangleq \{s_m : m \in [M]\}$, where s_m denotes a message, M the number of all possible messages, and we define $[M] \triangleq \{1, 2, \ldots, M\}$.

Definition 3.5 (The Set of Meanings) Suppose the messages in $\mathcal{S}$ can convey a finite or countably infinite number of meanings. We define the set of all possible meanings as $\mathcal{W} \triangleq \{w_n : n \in [N]\}$, where w_n denotes one meaning and N denotes the number of meanings. Furthermore, we denote by $p(w_n)$ the probability that the intended meaning of the transmitter is w_n.

Definition 3.6 (Expression) The expression of a language defines a mapping from the set of meanings to the set of messages, denoted by

$$\left\{ p(s|w) \in [0, 1] : w \in \mathcal{W}, s \in \mathcal{S}, \sum_s p(s|w) = 1 \right\}. \tag{3.2}$$

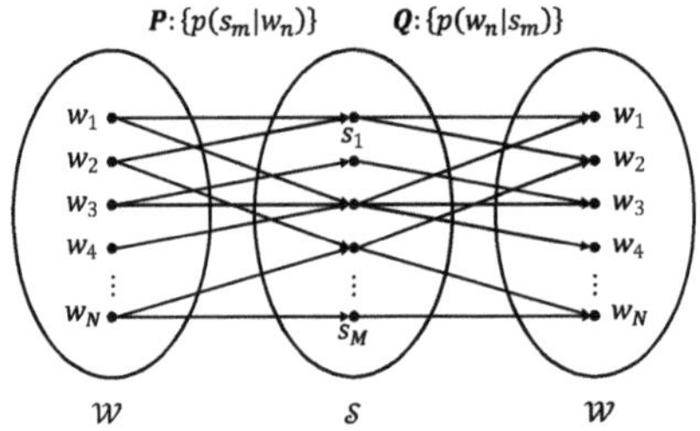

Fig. 3.3 The main ingredients of a semantic language $(\mathcal{W}, \mathcal{S}, \boldsymbol{P}, \boldsymbol{Q})$

To simplify the notation, we write the mappings into a matrix form as $\boldsymbol{P} \in \mathbb{R}^{N\times M}$: the element on the n-th row and m-th column of $\boldsymbol{P}$ is $p(s_m|w_n)$ and each row of $\boldsymbol{P}$ sums to 1.

Definition 3.7 (Interpretation) The interpretation of a language defines a mapping from the set of messages to the set of meanings, denoted by

$$\left\{q(w|s) \in [0, 1] : w \in \mathcal{W}, s \in \mathcal{S}, \sum_w q(w|s) = 1\right\} \tag{3.3}$$

or the matrix form $\boldsymbol{Q} \in \mathbb{R}^{M\times N}$. Each column of $\boldsymbol{Q}$ sums to 1.

Based on the above definitions, we denote a semantic language by a 4-tuple $(\mathcal{W}, \mathcal{S}, \boldsymbol{P}, \boldsymbol{Q})$, as illustrated in Fig. 3.3. As can be seen, the words and syntax of the language formulate the message set $\mathcal{S}$. A meaning $w \in \mathcal{W}$ can be expressed by a subset of messages $\{s : s \in \mathcal{S}, p(s|w) > 0\}$, the cardinality of which reflects the expressive redundancy of the language. On the other hand, a given message s can be interpreted as a subset of meanings $\{w : w \in \mathcal{W}, q(w|s) > 0\}$, the cardinality of which reflects the interpretation ambiguity of the language.

Definition 3.8 (Logically Self-consistent Language) A language is logically self-consistent if and only if

$$q(w|s) = \frac{p(s|w)p(w)}{p(s)}. \tag{3.4}$$

In this chapter, we do not impose any constraint on the logical self-consistency of the language; and hence, $q(w|s)$ does not need to satisfy (3.4) and can be any mapping. The advantage of such a formulation is that it encompasses a more practical semantic communication scenario with mismatched expression and interpretation.

Consider human communication as an example. The language of a human being is acquired from the world through interactions, which we refer to as the learning process. As illustrated in Fig. 3.4, the language of the world is supported by the knowledge of the world. Each person acquires both knowledge and language from the world, thereby developing a language system with its own message set and styles

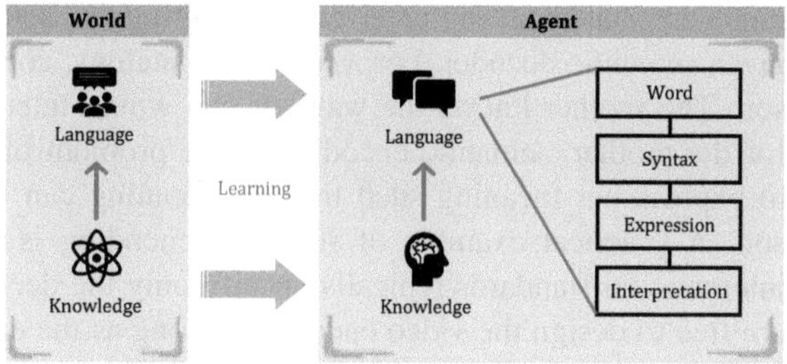

Fig. 3.4 A human being develops his/her own language by learning from the world

of expression and interpretation. In other words, the language of each agent can be treated as unique, i.e., a distinct pair of mappings $\boldsymbol{P}$ and $\boldsymbol{Q}$, which rely on the unique personal experience. The communication between two human beings, in this case, faces logical inconsistency, which is allowed in our formulation.

Next, we define the semantic channel.

Definition 3.9 (Semantic Channel) Consider the communication between a transmitter and a receiver. Suppose the transmitted message is $s \in \mathcal{S}$ and the received message is $\hat{s} \in \mathcal{S}$.[1] We characterize the technical channel by the transition probabilities from s to $\hat{s}$ and denote it by $\left\{c(\hat{s}|s) \in [0, 1] : s, \hat{s} \in \mathcal{S}, \sum_{\hat{s}} c(\hat{s}|s) = 1\right\}$, or the matrix form $\boldsymbol{C} \in \mathbb{R}^{M \times M}$. Given the process of expression and interpretation within the semantic language, the semantic channel can be written as

$$\Pr(\hat{w}|w) = \sum_{s,\hat{s}} p(s|w)c(\hat{s}|s)q(w|\hat{s}). \tag{3.5}$$

Our formulation abstracts technical communication as part of the semantic channel, the quality of which is determined by the manner technical communication is implemented (e.g., coding and modulation schemes, etc.).

As stated in Definition 3.1, semantic communication aims to minimize the misinterpretations of the receiver, leveraging the agreed semantic and technical languages between the transmitter and receiver. This can be achieved either from the transmitter's perspective (semantic encoding) or the receiver's perspective (semantic decoding). In the following, we formulate both problems based on the above definitions.

3.2.1.1 Semantic Encoding

From the transmitter's perspective, the semantic decoder at the receiver is dictated by the interpretation of the agreed language. The semantic encoding problem is thus how to encode the intended meaning using messages to minimize the

[1] More generally, one can define another message set $\hat{\mathcal{S}}$ at the receiver and define $\hat{s} \in \hat{\mathcal{S}}$.

misinterpretation at the receiver, considering the semantic channel as well as the given semantic decoder. For a figurative analogy, consider a mother talking to her son. The mother knows the way her son would interpret her message. Therefore, for the mother, semantic encoding is the problem of choosing the best messages to express her meaning such that the meaning can be precisely conveyed to the son. A practical example of semantic encoding is video compression, wherein international standards typically specify only the decoder [28]. The manufacturers are free to design the video encoders as long as the encoded video can be decoded by the standardized decoder. While we will assume that the stochastic semantic channel and interpretation mappings are known by the encoder in this work, a data-driven approach where this mapping can be learned through interactions is left as a potential future extension.

Definition 3.10 (Semantic Encoding Schemes) A semantic encoding scheme is a mapping from the meaning set $\mathcal{W}$ to the message set $\mathcal{S}$, denoted by

$$\left\{ u(s|w) \in [0, 1] : w \in \mathcal{W}, s \in \mathcal{S}, \sum_{s} u(s|w) = 1 \right\}, \tag{3.6}$$

or the matrix form $\boldsymbol{U} \in \mathbb{R}^{N \times M}$.

In principle, we can impose constraints on $u(s|w)$ such that the semantic encoding scheme complies with the semantic language. For example, if the semantic encoding scheme must comply with the interpretation, we can impose the constraint "$u(s|w) = 0$ if $p(s|w) = 0$"; if the transmitter is not allowed to deceive the receiver, we can impose the constraint "$u(s|w) = 0$ if $q(w|s) = 0$"; if the transmitter is not allowed to misdirect the receiver, we can impose the constraint "$u(s|w) \geq u(s'|w)$ if $q(w|s) \geq q(w|s')$". In this chapter, however, we consider a general setup and do not impose any constraints on $u(s|w)$.

To evaluate the performance of semantic encoding, we introduce two metrics below, one is semantic distortion and the other is semantic cost.

Definition 3.11 (Semantic Distortion) Let $w, \hat{w} \in \mathcal{W}$ be the transmitted and reconstructed meanings at the transmitter and receiver, respectively. We define a semantic distortion measure $d(w, \hat{w}) : \mathcal{W} \times \mathcal{W} \to \mathbb{R}^+$, where $\mathbb{R}^+$ is the set of non-negative real numbers. The average distortion achieved by a semantic encoding scheme $\boldsymbol{U}$ is given by

$$D_{U,Q} = \sum_{w,s,\hat{s},\hat{w}} p(w)u(s|w)c(\hat{s}|s)q(\hat{w}|\hat{s})d(w, \hat{w}). \tag{3.7}$$

Definition 3.12 (Semantic Cost) Each message is associated with a cost. Semantically, the cost of a message can be its conciseness, comprehensibility, or elegance. Technically, the cost of a message can be the length of the corresponding bit sequence. In general, we define a cost function $\ell(s) : \mathcal{S} \to \mathbb{R}^+$, $\forall s \in \mathcal{S}$. The

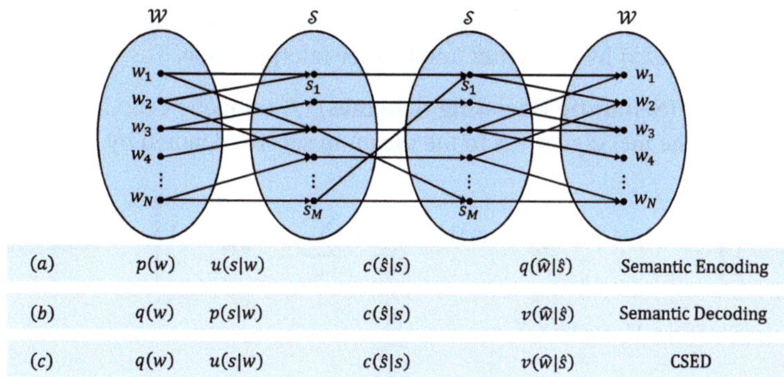

Fig. 3.5 A point-to-point semantic communication process with (**a**) semantic encoding at the transmitter, (**b**) semantic decoding at the receiver, or (**c**) combined semantic encoding and decoding at the transmitter and receiver, respectively. We have assumed that the receiver has the same message and meaning sets as the transmitter. More generally, we can define a new message set $\hat{S}$ and a new meaning set $\hat{W}$ for the receiver

average cost achieved by a semantic encoding scheme $\boldsymbol{U}$ is

$$L_U = \sum_{w,s} p(w)u(s|w)\ell(s). \tag{3.8}$$

Without loss of generality, we assume S is an ordered set such that for any two messages s_{m_1},s_{m_2}, $m_1, m_2 \in [M]$, we have $\ell(s_{m_1}) \leq \ell(s_{m_2})$ if $m_1 < m_2$.

Given the above definitions, we are ready to describe a point-to-point semantic communication process with semantic encoding at the transmitter. As shown in Fig. 3.5a, the transmitter first generates a meaning $w \sim p(w)$. To convey w to the receiver, the transmitter constructs a message $s \sim u(s|w)$ following its semantic encoding scheme. After passing through the semantic channel, the receiver observes $\hat{s} \sim c(\hat{s}|s)$ and reconstructs a meaning $\hat{w} \sim q(\hat{w}|\hat{s})$ following the interpretation of the language. The main problem of semantic encoding is to characterize the optimal encoding scheme that achieves the best trade-off between the semantic distortion and semantic cost.

3.2.1.2 Semantic Decoding

From the receiver's perspective, the semantic encoder at the transmitter is dictated by the expression of the agreed language. The semantic decoding problem is thus how to decode a received message to minimize semantic distortion, considering the semantic channel as well as the given semantic encoder at the transmitter. An analogy of semantic decoding is a son talking to his mother. The mother knows the way her son expresses his meaning and aims to find the best way to interpret the received messages. Another example of semantic decoding is reading a text. A reader tries to optimize his/her semantic decoder to minimize the distortion between the meaning intended by the writer and his/her interpretation.

Definition 3.13 (Semantic Decoding Schemes) A semantic decoding scheme is a mapping from the message set $\mathcal{S}$ to the meaning set $\mathcal{W}$, denoted by

$$\left\{ v(w|s) \in [0, 1] : w \in \mathcal{W}, s \in \mathcal{S}, \sum_{w} v(w|s) = 1 \right\}, \tag{3.9}$$

or the matrix form $\boldsymbol{V} \in \mathbb{R}^{M \times N}$.

Similar to the semantic encoding problem, we do not impose any constraints on $v(w|s)$. To evaluate the performance of a semantic decoding scheme, we define the semantic distortion below.

Definition 3.14 (Semantic Distortion of Semantic Decoding) Let $w, \hat{w} \in \mathcal{W}$ be the transmitted and reconstructed meanings at the transmitter and receiver, respectively. The average distortion achieved by a semantic decoding scheme $\boldsymbol{V}$ is

$$D_{\boldsymbol{P},\boldsymbol{V}} = \sum_{w,s,\hat{s},\hat{w}} p(w)p(s|w)c(\hat{s}|s)v(\hat{w}|\hat{s})d(w, \hat{w}). \tag{3.10}$$

The point-to-point semantic communication process with semantic decoding at the receiver is illustrated in Fig. 3.5b. Compared with semantic encoding, semantic decoding may consider a potentially inaccurate prior distribution $q(w)$ at the receiver and a fixed semantic encoder $\boldsymbol{P}$ dictated by the language. The goal is to find the optimal decoding scheme $\boldsymbol{V}^*$ such that the semantic distortion is minimized.

In the language utilization problem, the agreed language is the only common ground between the transmitter and receiver and there is no codebook negotiation. Therefore, semantic encoding and decoding are essentially two decoupled problems—jointly designing the semantic encoding and decoding falls into the language design problem.

Nevertheless, the transmitter and receiver can perform semantic encoding and decoding, separately, at the same time, which we refer to as the CSED, as illustrated in Fig. 3.5c. CSED will be defined rigorously later in Sect. 3.2.4.

3.2.2 *Semantic Encoding*

Based on the formulation in Sect. 3.2.1, this section focuses on the semantic encoding problem. For a given semantic language, our main goal in this section is to define and characterize the semantic distortion-cost region (and function) of semantic encoding.

3.2.2.1 Semantic Distortion-Cost Region and Function

A semantic encoding scheme can be evaluated from two angles: semantic distortion and semantic cost. We next define the semantic distortion-cost region for semantic encoding.

Definition 3.15 (The Distortion-Cost Region of Semantic Encoding) Consider a semantic language $(\mathcal{W}, \mathcal{S}, \boldsymbol{P}, \boldsymbol{Q})$, a semantic channel $\boldsymbol{C}$, a message cost function $\ell(s) : \mathcal{S} \to \mathbb{R}^+$, and a distortion measure $d(w, \hat{w}) : \mathcal{W} \times \mathcal{W} \to \mathbb{R}^+$. A distortion-cost pair (L, D) is achievable if there exists a semantic encoding scheme $\boldsymbol{U}$ such that $D_U = D$, $L_U = L$. The distortion-cost region R_{enc} is the set of all achievable distortion-cost pairs (L, D).

A trivial outer bound of the region is

$$\begin{cases} L_{\min} \leq L \leq L_{\max}, \\ D_{\min} \leq D \leq D_{\max}, \end{cases} \tag{3.11}$$

where

$$L_{\min} \triangleq \min_{(L,D)\in R_{\text{enc}}} L = \min_s \ell(s), \tag{3.12}$$

$$L_{\max} \triangleq \max_{(L,D)\in R_{\text{enc}}} L = \max_s \ell(s), \tag{3.13}$$

are the minimum and maximum costs, respectively, and

$$D_{\min} \triangleq \min_{(L,D)\in R_{\text{enc}}} D, \quad D_{\max} \triangleq \max_{(L,D)\in R_{\text{enc}}} D, \tag{3.14}$$

are the minimum and maximum achievable distortions with semantic encoding, respectively.

The exact distortion-cost region, on the other hand, can be more involved. For illustration purposes, we give a schematic plot of the region in Fig. 3.6.

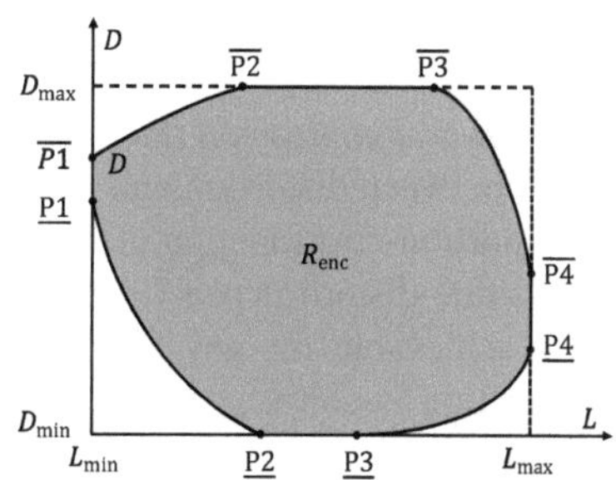

Fig. 3.6 An illustration of the distortion-cost region and function of semantic encoding

To characterize the region, we first define eight critical points as follows:

$$\underline{\text{P1}}: \left(L_{\min}, \min_{(L_{\min},D)\in R_{\text{enc}}} D\right), \ \underline{\text{P2}}: \left(\min_{(L,D_{\min})\in R_{\text{enc}}} L, D_{\min}\right),$$

$$\underline{\text{P3}}: \left(\max_{(L,D_{\min})\in R_{\text{enc}}} L, D_{\min}\right), \ \underline{\text{P4}}: \left(L_{\max}, \min_{(L_{\max},D)\in R_{\text{enc}}} D\right),$$

$$\overline{\text{P1}}: \left(L_{\min}, \max_{(L_{\min},D)\in R_{\text{enc}}} D\right), \ \overline{\text{P2}}: \left(\min_{(L,D_{\max})\in R_{\text{enc}}} L, D_{\max}\right),$$

$$\overline{\text{P3}}: \left(\max_{(L,D_{\max})\in R_{\text{enc}}} L, D_{\max}\right), \ \overline{\text{P4}}: \left(L_{\max}, \max_{(L_{\max},D)\in R_{\text{enc}}} D\right),$$

Each of these eight points corresponds to a semantic encoding scheme. For example, $\underline{\text{P1}}$ corresponds to the most "accurate" encoding scheme with minimum cost, while $\overline{\text{P1}}$ corresponds to the most "inaccurate" encoding scheme with minimum cost. On the other hand, $\underline{\text{P2}}$ corresponds to the most "cost-efficient" encoding scheme that achieves the minimum distortion, while $\underline{\text{P3}}$ corresponds to the most "cost-inefficient" encoding scheme that achieves the minimum distortion.

Since minimizing the distortion is the main concern of semantic communication, we are particularly interested in the lower-envelope of the region, i.e., the curve connecting $\underline{\text{P1}}$, $\underline{\text{P2}}$, $\underline{\text{P3}}$, $\underline{\text{P4}}$ in Fig. 3.6. As with Shannon's rate-distortion theory, we define the distortion-cost function for semantic encoding.

Definition 3.16 (The Distortion-Cost Function of Semantic Encoding) With semantic encoding, the minimum achievable semantic distortion under an average message cost L is called the distortion-cost function of semantic encoding and is given by

$$D^*_{U,Q}(L) = \min\{D : (L, D) \in R_{\text{enc}}\}. \tag{3.15}$$

As illustrated in Fig. 3.6, the distortion-cost function of semantic encoding consists of three segments in general, and we denote them by ($\underline{\text{P1}}$, $\underline{\text{P2}}$), ($\underline{\text{P2}}$, $\underline{\text{P3}}$), and ($\underline{\text{P3}}$, $\underline{\text{P4}}$), respectively. In ($\underline{\text{P1}}$, $\underline{\text{P2}}$), minimizing distortion and minimizing cost is a trade-off; in ($\underline{\text{P2}}$, $\underline{\text{P3}}$), the minimum distortion is achieved; in ($\underline{\text{P3}}$, $\underline{\text{P4}}$), minimizing distortion or cost boosts the other one.

Note that, in Shannon's rate-distortion theory [29],

1. A lower compression rate implies a larger distortion, and hence, there is a trade-off between distortion and rate. This corresponds to the segment ($\underline{\text{P1}}$, $\underline{\text{P2}}$) in the distortion-cost function of semantic encoding.
2. If a rate-distortion pair $(\mathfrak{R}, D)$ is achievable, then the rate-distortion pair $(\mathfrak{R}', D)$ is achievable for any $\mathfrak{R}' > \mathfrak{R}$. Thus, the rate-distortion (or distortion-rate) function is convex and monotonic – Shannon's rate-distortion function has no 'boosting' segment as in ($\underline{\text{P3}}$, $\underline{\text{P4}}$) in Fig. 3.6.

By contrast, the distortion-cost function of semantic encoding is determined by the semantic language as well as the exact forms of semantic cost and distortion. There can be an additional boost segment (P3, P4) because a larger semantic cost does not necessarily imply a smaller semantic distortion. For an intuitive example, consider the communication between a police officer and a suspect. The police officer wants to know the suspect's activities over a period. As the suspect's description increases, the police officer knows more and more about the suspect's activities, and the semantic distortion, in which case the uncertainty of the suspect's activities, decreases. However, when the suspect over-describes some details, the police may doubt the suspect's motive and the semantic distortion starts to increase. A more rigorous example will be given in Sect. 3.2.5.

3.2.2.2 Characterizing the Distortion-Cost Region

A semantic encoding scheme is a mapping from $\mathcal{W}$ to $\mathcal{S}$. To measure the expected distortion of mapping a meaning w to a message s, we define

$$\varphi(w, s) \triangleq \sum_{\hat{s},\hat{w}} c(\hat{s}|s)q(\hat{w}|\hat{s})d(w, \hat{w}). \tag{3.16}$$

The distortion of a semantic encoding scheme $\boldsymbol{U}$ can be written as

$$D_{U,Q} = \sum_{w,s} p(w)u(s|w)\varphi(w, s). \tag{3.17}$$

To characterize the distortion-cost region for semantic encoding, we introduce a special class of semantic encoding schemes: deterministic semantic encoding.

Definition 3.17 (Deterministic Semantic Encoding) A semantic encoding scheme $\boldsymbol{U}$ is said to be deterministic if $\boldsymbol{U}$ maps each meaning to a single message deterministically. Each row of a deterministic encoding scheme $\boldsymbol{U}$ has only one non-zero element. We denote a deterministic semantic code $\boldsymbol{U}$ by $\boldsymbol{\Delta}_{i_1,i_2,\ldots,i_N}$, where $i_1, i_2, \ldots, i_N \in [M]$ are the column indexes of the non-zero elements in the rows of $\boldsymbol{U}$. That is,

$$u(s_m|w_j) = \begin{cases} 1, & \text{if } m = i_j, \\ 0, & \text{otherwise.} \end{cases} \tag{3.18}$$

Proposition 3.1 *A semantic distortion-cost pair* (L, D) *can be achieved by a stochastic encoding scheme* $\boldsymbol{U}$ *if and only if* (L, D) *can be achieved by time sharing among a set of deterministic encoding schemes in*

$$\left\{\boldsymbol{\Delta}_{i_1,i_2,\ldots,i_N} : \forall i_1, i_2, \ldots, i_N \in [M]\right\}.$$

Proof See Appendix A.1. □

Proposition 3.1 suggests two important properties of the distortion-cost region:

- Any point in the distortion-cost region can be achieved by a set of deterministic semantic encoding schemes via time sharing.
- The distortion-cost region R_{enc} is a convex set. The distortion-cost function $D^*_{U,Q}(L)$ is a convex function.

As a result, we only need to focus on the class of deterministic semantic encoding schemes to characterize the distortion-cost region.

Definition 3.18 For a meaning $w \in \mathcal{W}$, we define six subsets of $\mathcal{S}$: $\underline{\mathcal{S}}'_w$, $\underline{\mathcal{S}}''_w$, $\underline{\mathcal{S}}_w$, $\overline{\mathcal{S}}'_w$, $\overline{\mathcal{S}}''_w$, $\overline{\mathcal{S}}_w \subseteq \mathcal{S}$:

1. $\underline{\mathcal{S}}'_w$ is the smallest subset of $\mathcal{S}$ such that $\forall s \in \mathcal{S}$, there exists an $s' \in \underline{\mathcal{S}}'_w$ that satisfies

$$\ell(s') \leq \ell(s), \quad \varphi(w, s') \leq \varphi(w, s).$$

2. $\underline{\mathcal{S}}''_w$ is the smallest subset of $\mathcal{S}$ such that $\forall s \in \mathcal{S}$, there exists an $s'' \in \underline{\mathcal{S}}''_w$ that satisfies

$$\ell(s'') \geq \ell(s), \quad \varphi(w, s'') \leq \varphi(w, s).$$

3. $\underline{\mathcal{S}}_w = \underline{\mathcal{S}}'_w \cup \underline{\mathcal{S}}''_w$.
4. $\overline{\mathcal{S}}'_w$ is the smallest subset of $\mathcal{S}$ such that $\forall s \in \mathcal{S}$, there exists an $s' \in \overline{\mathcal{S}}'_w$ that satisfies

$$\ell(s') \leq \ell(s), \quad \varphi(w, s') \geq \varphi(w, s).$$

5. $\overline{\mathcal{S}}''_w$ is the smallest subset of $\mathcal{S}$ such that $\forall s \in \mathcal{S}$, there exists an $s'' \in \overline{\mathcal{S}}''_w$ that satisfies

$$\ell(s'') \geq \ell(s), \quad \varphi(w, s'') \geq \varphi(w, s).$$

6. $\overline{\mathcal{S}}_w = \overline{\mathcal{S}}'_w \cup \overline{\mathcal{S}}''_w$.

For a $w \in \mathcal{W}$, the six subsets are defined based on the properties of s. Take the subset $\underline{\mathcal{S}}'_w$ for example. Each of s in $\underline{\mathcal{S}}'_w$ has the property that there exists no $s' \in \mathcal{S}\backslash\{s\}$ that is more accurate and cost-efficient than s at the same time. It will be shown later that, for each meaning w, we only need to consider the messages in $\underline{\mathcal{S}}_w$ and $\overline{\mathcal{S}}_w$ to construct the deterministic encoding schemes and characterize the distortion-cost region of semantic encoding. The procedures to discover the six sets for a w are described in Algorithm 3. The properties of the six subsets are summarized in Appendix A.2.

Algorithm 3 Discovering the six subsets of messages for a meaning w

Input: w; $\mathcal{S}$.
Output: $\underline{\mathcal{S}}_w$ and $\overline{\mathcal{S}}_w$;

$\underline{\mathcal{S}}'_w = \mathcal{S}, \underline{s}^l = s_1; \underline{\mathcal{S}}''_w = \mathcal{S}, \underline{s}^r = s_M;$
$\overline{\mathcal{S}}'_w = \mathcal{S}, \overline{s}^l = s_1; \overline{\mathcal{S}}''_w = \mathcal{S}, \overline{s}^r = s_M;$
for $m = 2, 3, \ldots, M$ **do**
    if $\varphi(w, \underline{s}^l) \leq \varphi(w, s_m)$ **then**
        $\underline{\mathcal{S}}'_w = \underline{\mathcal{S}}'_w \backslash s_m;$
    else if $\ell(\underline{s}^l) == \ell(s_m)$ **then**
        $\underline{\mathcal{S}}'_w = \underline{\mathcal{S}}'_w \backslash \underline{s}^l;$
        $\underline{s}^l = s_m;$
    else
        $\underline{s}^l = s_m;$
    if $\varphi(w, \underline{s}^r) \leq \varphi(w, s_{M-m+1})$ **then**
        $\underline{\mathcal{S}}''_w = \underline{\mathcal{S}}''_w \backslash s_{M-m+1};$
    else if $\ell(\underline{s}^r) == \ell(s_{M-m+1})$ **then**
        $\underline{\mathcal{S}}''_w = \underline{\mathcal{S}}''_w \backslash \underline{s}^r;$
        $\underline{s}^r = s_{M-m+1};$
    else
        $\underline{s}^r = s_{M-m+1};$
    if $\varphi(w, \overline{s}^l) \leq \varphi(w, s_m)$ **then**
        $\overline{\mathcal{S}}'_w = \overline{\mathcal{S}}'_w \backslash s_m;$
    else if $\ell(\overline{s}^l) == \ell(s_m)$ **then**
        $\overline{\mathcal{S}}'_w = \overline{\mathcal{S}}'_w \backslash \overline{s}^l;$
        $\overline{s}^l = s_m;$
    else
        $\overline{s}^l = s_m;$
    if $\varphi(w, \overline{s}^r) \leq \varphi(w, s_{M-m+1})$ **then**
        $\overline{\mathcal{S}}''_w = \overline{\mathcal{S}}''_w \backslash s_{M-m+1};$
    else if $\ell(\overline{s}^r) == \ell(s_{M-m+1})$ **then**
        $\overline{\mathcal{S}}''_w = \overline{\mathcal{S}}''_w \backslash \overline{s}^r;$
        $\overline{s}^r = s_{M-m+1};$
    else
        $\overline{s}^r = s_{M-m+1};$
$\underline{\mathcal{S}}_w = \underline{\mathcal{S}}'_w \cup \underline{\mathcal{S}}''_w, \overline{\mathcal{S}}_w = \overline{\mathcal{S}}'_w \cup \overline{\mathcal{S}}''_w.$

Definition 3.19 For a meaning $w \in \mathcal{W}$ and two messages $s, s' \in \mathcal{S}$, $\ell(s) \neq \ell(s')$, we define a function

$$G(w, s, s') = \frac{\varphi(w, s') - \varphi(w, s)}{\ell(s') - \ell(s)}. \tag{3.19}$$

The function G is important in that it guides us to discover a sequence of deterministic encoding schemes to characterize the boundary of the distortion-cost region. Theorem 3.1 below presents the main result of this section.

Theorem 3.1 (The Distortion-Cost Region of Semantic Encoding) *Consider a semantic language* $(\mathcal{W}, \mathcal{S}, \boldsymbol{P}, \boldsymbol{Q})$, *a semantic channel* $\boldsymbol{C}$, *a message cost function* $\ell(s) : \mathcal{S} \to \mathbb{R}^+$, *and a distortion measure* $d(w, \hat{w}) : \mathcal{W} \times \mathcal{W} \to \mathbb{R}^+$. *The boundary of the distortion-cost region of semantic encoding* R_{enc} *is the piecewise linear connection of* $\underline{T} + \overline{T} + 2$ *distortion-cost pairs* $\left(L_{\underline{\boldsymbol{U}}^{(t)}}, D_{\underline{\boldsymbol{U}}^{(t)}, \boldsymbol{Q}}\right)$, $t = 0, 1, 2, \ldots, \underline{T}$ *and* $\left(L_{\overline{\boldsymbol{U}}^{(t)}}, D_{\overline{\boldsymbol{U}}^{(t)}, \boldsymbol{Q}}\right)$, $t = 0, 1, 2, \ldots, \overline{T}$, *where* $\left\{\underline{\boldsymbol{U}}^{(t)} : t = 0, 1, 2, \ldots, \underline{T}\right\}$ *and* $\left\{\overline{\boldsymbol{U}}^{(t)} : t = 0, 1, 2, \ldots, \overline{T}\right\}$ *are deterministic semantic encoding schemes constructed in Algorithm 4.*

Proof See Appendix A.3. □

Theorem 3.1 states that the boundary of the distortion-cost region of semantic encoding is piecewise linear. The vertices of the boundary can be achieved by the deterministic encoding schemes constructed in Algorithm 4. Any non-vertex point on the boundary can be achieved by a stochastic encoding scheme or the time-sharing[2] of two deterministic encoding schemes associated with the two closest vertices. The distortion-cost function of semantic encoding $D^*_{\boldsymbol{U}, \boldsymbol{Q}}(L)$ is also piecewise linear and can be characterized by $\left\{\underline{\boldsymbol{U}}^{(t)} : t = 0, 1, 2, \ldots, \underline{T}\right\}$.

To conclude this section, we emphasize that the transmitter takes a proactive role in semantic communication, and hence, semantic encoding is often more efficient than semantic decoding as far as reducing distortion is concerned. In addition, the transmitter has the exact distribution of the intended meanings $p(w)$, while the receiver may only have an inaccurate prior $q(w)$. The semantic cost is fully determined by semantic encoding.

3.2.3 *Semantic Decoding*

In Sect. 3.2.2, we focused on semantic encoding from the transmitter's perspective. In this section, we study how to reduce semantic distortion from the receiver's perspective via semantic decoding.

In semantic decoding, the semantic encoder at the transmitter is dictated by the expression of the agreed language $\boldsymbol{P}$ and the receiver varies the mapping $\boldsymbol{V}$ to optimize the semantic distortion

$$D_{\boldsymbol{P},\boldsymbol{V}} = \sum_{\hat{s}, \hat{w}} v(\hat{w}|\hat{s}) \psi_{\boldsymbol{P}}(\hat{w}, \hat{s}), \tag{3.20}$$

[2] Time sharing means that the system alternates between two deterministic encoding schemes over separate time intervals within the same communication session.

Algorithm 4 Characterizing the distortion-cost region of semantic encoding R_{enc}

Input: $(\mathcal{W}, \mathcal{S}, \boldsymbol{P}, \boldsymbol{Q})$, $\ell(s)$, $d(w, \widehat{w})$.

Output: $\left\{\underline{\boldsymbol{U}}^{(t)}, \left(L_{\underline{\boldsymbol{U}}^{(t)}}, D_{\underline{\boldsymbol{U}}^{(t)}, \boldsymbol{Q}}\right) : t = 0, 1, 2, \ldots, \underline{T}\right\}$;

$\left\{\overline{\boldsymbol{U}}^{(t)}, \left(L_{\overline{\boldsymbol{U}}^{(t)}}, D_{\overline{\boldsymbol{U}}^{(t)}, \boldsymbol{Q}}\right) : t = 0, 1, 2, \ldots, \overline{T}\right\}$.

Step 1: Find $\underline{\mathcal{S}}_w$ and $\overline{\mathcal{S}}_w$ for each $w \in \mathcal{W}$.
for $n = 1, 2, \ldots, N$ **do**
  $\underline{\mathcal{S}}_{w_n}, \overline{\mathcal{S}}_{w_n} = \textbf{Algorithm1}(w_n, \mathcal{S})$.
Step 2: Construct a sequence of deterministic semantic source coding schemes $\underline{\boldsymbol{U}}^{(t)} = \boldsymbol{\Delta}_{\underline{i}_1^{(t)}, \underline{i}_2^{(t)}, \ldots, \underline{i}_N^{(t)}}$, $t = 0, 1, 2, \ldots, \underline{T}$.
$t = 0$;
Construct $\underline{\boldsymbol{U}}^{(0)} = \boldsymbol{\Delta}_{\underline{i}_1^{(0)}, \underline{i}_2^{(0)}, \ldots, \underline{i}_N^{(0)}}$, where $\underline{i}_n^{(0)} = \min\{m : s_m \in \underline{\mathcal{S}}_{w_n}\}$.
while $L_{\underline{\boldsymbol{U}}^{(t)}} < L_{\max}$ **do**
  $t = t + 1$;

$$\underline{n}^{(t)}, \underline{m}^{(t)} = \arg \min_{n, m : s_m \in \underline{\mathcal{S}}_{w_n}, m > \underline{i}_n^{(t-1)}} G\left(w_n, s_{\underline{i}_n^{(t-1)}}, s_m\right).$$

  Construct $\underline{\boldsymbol{U}}^{(t)} = \boldsymbol{\Delta}_{\underline{i}_1^{(t)}, \underline{i}_2^{(t)}, \ldots, \underline{i}_N^{(t)}}$ such that

$$\underline{i}_n^{(t)} = \begin{cases} \underline{i}_k^{(t-1)}, & \text{if } n \neq \underline{n}^{(t)}; \\ \underline{m}^{(t)}, & \text{if } n = \underline{n}^{(t)}. \end{cases}$$

$\underline{T} = t$.
Step 3: Construct a sequence of deterministic semantic source coding schemes $\overline{\boldsymbol{U}}^{(t)} = \boldsymbol{\Delta}_{\overline{i}_1^{(t)}, \overline{i}_2^{(t)}, \ldots, \overline{i}_N^{(t)}}$, $t = 0, 1, 2, \ldots, \overline{T}$.
$t = 0$;
Construct $\overline{\boldsymbol{U}}^{(0)} = \boldsymbol{\Delta}_{\overline{i}_1^{(0)}, \overline{i}_2^{(0)}, \ldots, \overline{i}_N^{(0)}}$, where $\overline{i}_n^{(0)} = \min\{m : s_m \in \overline{\mathcal{S}}_{w_n}\}$.
while $L_{\overline{\boldsymbol{U}}^{(t)}} < L_{\max}$ **do**
  $t = t + 1$;

$$\overline{n}^{(t)}, \overline{m}^{(t)} = \arg \max_{n, m : s_m \in \overline{\mathcal{S}}_{w_n}, m > \overline{i}_n^{(t-1)}} G\left(w_n, s_{\overline{i}_n^{(t-1)}}, s_m\right).$$

  Construct $\overline{\boldsymbol{U}}^{(t)} = \boldsymbol{\Delta}_{\overline{i}_1^{(t)}, \overline{i}_2^{(t)}, \ldots, \overline{i}_N^{(t)}}$ such that

$$\overline{i}_n^{(t)} = \begin{cases} \overline{i}_k^{(t-1)}, & \text{if } n \neq \overline{n}^{(t)}; \\ \overline{m}^{(t)}, & \text{if } n = \overline{n}^{(t)}. \end{cases}$$

$\overline{T} = t$.

where

$$\psi_P(\hat{w}, \hat{s}) \triangleq \sum_{w,s} p(w)p(s|w)c(\hat{s}|s)d(w, \hat{w}). \tag{3.21}$$

We first characterize the semantic distortion-cost region that can be achieved by semantic decoding.

Definition 3.20 (The Distortion-Cost Region of Semantic Decoding) A distortion-cost pair (L, D) is achievable if there exists a semantic decoding scheme $\boldsymbol{V}$ such that $D_{\boldsymbol{P},\boldsymbol{V}} = D$ and $L_{\boldsymbol{P}} = L$. The distortion-cost region R_{dec} is the set of all achievable distortion-cost pairs (L, D).

Since the semantic encoder $\boldsymbol{P}$ is fixed, the achievable cost of $\boldsymbol{V}$ is also fixed, i.e.,

$$L_{\boldsymbol{P}} = \sum_{w} p(w) \sum_{s} p(w|s)\ell(s). \tag{3.22}$$

Therefore, the distortion-cost region of semantic decoding is a vertical line in the distortion-cost space.

Definition 3.21 (Deterministic Semantic Decoding) A semantic decoding scheme $\boldsymbol{V} \in \mathbb{R}^{M\times N}$ is said to be deterministic if $\boldsymbol{V}$ maps each message to a single meaning deterministically. We denote a deterministic semantic decoding scheme $\boldsymbol{V}$ by $\widetilde{\boldsymbol{\Delta}}_{n_1,n_2,\ldots,n_M}$, where $n_1, n_2, \ldots, n_M \in [N]$ are the column indexes of the non-zero elements in the rows of $\boldsymbol{V}$.

Theorem 3.2 (The Distortion-Cost Region of Semantic Decoding) *Consider a semantic language* $(\mathcal{W}, \mathcal{S}, \boldsymbol{P}, \boldsymbol{Q})$, *a semantic channel* $\boldsymbol{C}$, *a message cost function* $\ell(s) : \mathcal{S} \to \mathbb{R}^+$, *and a distortion measure* $d(w, \hat{w}) : \mathcal{W} \times \mathcal{W} \to \mathbb{R}^+$. *The distortion-cost region of semantic decoding* R_{dec} *is the vertical line between*

$$\left(L_{\boldsymbol{P}}, D_{\boldsymbol{P},\widetilde{\boldsymbol{\Delta}}_{n'_1,n'_2,\ldots,n'_M}}\right) \text{ and } \left(L_{\boldsymbol{P}}, D_{\boldsymbol{P},\widetilde{\boldsymbol{\Delta}}_{n''_1,n''_2,\ldots,n''_M}}\right),$$

where

$$n'_m = \arg\min_{n} \psi_P(w_n, s_m),$$

$$n''_m = \arg\max_{n} \psi_P(w_n, s_m).$$

Proof Omitted. □

The distortion-cost region of semantic decoding gives all achievable distortion-cost pairs with semantic decoding. However, unlike semantic encoding, the receiver may not have accurate prior information on $p(w)$. As a consequence, the receiver may not be able to find the optimal semantic decoding scheme $\widetilde{\boldsymbol{\Delta}}_{n'_1,n'_2,\ldots,n'_M}$ that minimizes the distortion.

Under a potentially inaccurate prior distribution $q(w)$, we are interested in the semantic decoding scheme employed by the receiver and the conditions under which this scheme is optimal. In the following, we shall use the subscripts 'p' and 'q' to denote the statistics obtained under prior distributions $p(w)$ and $q(w)$, respectively. For example, we have

$$\psi_q(\hat{w}, \hat{s}) \triangleq \sum_{w,s} q(w)p(s|w)c(\hat{s}|s)d(w, \hat{w}), \tag{3.23}$$

under the prior $q(w)$.

Proposition 3.2 (Semantic Decoding Under Inaccurate Prior) *Consider the expression* $\boldsymbol{P}$ *at the transmitter, a semantic channel* $\boldsymbol{C}$*, and a prior distribution* $q(w)$ *at the receiver. The best semantic decoding scheme perceived by the receiver is* $\boldsymbol{V}_q^* = \widetilde{\boldsymbol{\Delta}}_{n_1^{(q)}, n_2^{(q)}, \ldots, n_M^{(q)}}$*, where*

$$n_m^{(q)} = \arg\min_n \psi_q(w_n, s_m). \tag{3.24}$$

To simplify notation, we denote by $\hat{w}_q(\hat{s})$ the meaning that a received message $\hat{s}$ will be mapped to under prior $q(w)$, i.e., $\hat{w}_q(\hat{s}) = w_{n_m^{(q)}}$. The achieved distortion of $\boldsymbol{V}_q^*$ can be written as

$$D_{\boldsymbol{P}, \boldsymbol{V}_q^*} = \sum_{\hat{s}} \psi_p(\hat{w}_q(\hat{s}), \hat{s}). \tag{3.25}$$

On the other hand, if the true prior $p(w)$ is available at the receiver, the optimal decoding scheme is $\boldsymbol{V}_p^* = \widetilde{\boldsymbol{\Delta}}_{n'_1, n'_2, \ldots, n'_M}$ and the minimum distortion is

$$D_{\boldsymbol{P}, \boldsymbol{V}_p^*} = \sum_{\hat{s}} \psi_p(\hat{w}_p(\hat{s}), \hat{s}). \tag{3.26}$$

The condition under which semantic decoding achieves the optimal distortion $D_{\boldsymbol{P}, \boldsymbol{V}_p^*}$ is non-trivial to establish for a general distortion function. In the following, we consider the Hamming distortion and study the condition under which $D_{\boldsymbol{P}, \boldsymbol{V}_q^*} = D_{\boldsymbol{P}, \boldsymbol{V}_p^*}$.

Definition 3.22 (Hamming Distortion) Hamming distortion measures the probability of semantic error and is given by

$$d(w, \hat{w}) = \begin{cases} 1, \text{ if } \hat{w} \neq w; \\ 0, \text{ if } \hat{w} = w. \end{cases} \tag{3.27}$$

Proposition 3.3 (Semantic Decoding with the Hamming Distortion) *Consider semantic decoding with a prior* $q(w)$ *and Hamming distortion. Semantic decoding achieves the optimal distortion, i.e.,* $D_{\boldsymbol{P},\mathbf{V}_q^*} = D_{\boldsymbol{P},\mathbf{V}_p^*}$, *if and only if*

$$\arg\max_{w} q(w)p(\hat{s}|w) \subseteq \arg\max_{w} p(w)p(\hat{s}|w), \ \forall \hat{s}, \tag{3.28}$$

where $p(\hat{s}|w) \triangleq \sum_s p(s|w)c(\hat{s}|s)$.

Proof See Appendix A.4. □

A geometric interpretation of (3.28) is given as follows. First, we normalize $q(w)p(\hat{s}|w)$ to a distribution $\boldsymbol{\alpha}_q(\hat{s})$ with each element being

$$\alpha_q(\hat{s}, w) = \frac{q(w)p(\hat{s}|w)}{\sum_w q(w)p(\hat{s}|w)}, \ \forall w \in \mathcal{W}. \tag{3.29}$$

The probability space in which $\boldsymbol{\alpha}_q(\hat{s})$ resides can be represented as a unit simplex. When $N = 3$, for example, the probability space is an equilateral triangle, as shown in Fig. 3.7a. Based on the semantic decoding rule, we partition the probability space into N regions, corresponding to the N meanings in $\mathcal{W}$. When $N = 3$, the equilateral triangle can be partitioned into three quadrilaterals, as shown in Fig. 3.7b. Under the prior $p(w)$ and $q(w)$, semantic decoding yields M points $\{\boldsymbol{\alpha}_p(\hat{s}) : \hat{s} \in \mathcal{S}\}$ and $\{\boldsymbol{\alpha}_q(\hat{s}) : \hat{s} \in \mathcal{S}\}$, respectively.

In the above context, the geometric interpretation of (3.28) is that $\boldsymbol{\alpha}_p(\hat{s})$ and $\boldsymbol{\alpha}_q(\hat{s})$ fall into the same region of the probability space, $\forall \hat{s} \in \mathcal{S}$. That is, under the Hamming distortion, the optimality of semantic decoding is governed by a "partition distance" between $\boldsymbol{\alpha}_p(\hat{s})$ and $\boldsymbol{\alpha}_q(\hat{s})$, i.e., whether they fall into the same region, but not the conventional distance measure between two probability distributions, such as the L1 or Kullback-Leibler (KL) distance.

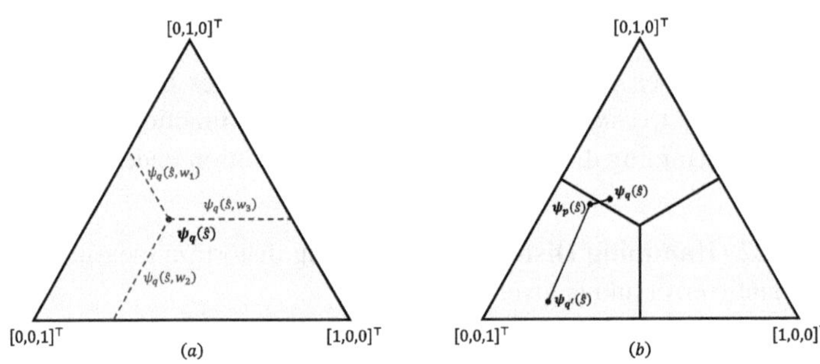

Fig. 3.7 (**a**) Probability space in which $\boldsymbol{\alpha}_q(\hat{s})$ resides is an equilateral triangle when $N = 3$. (**b**) We partition the equilateral triangle into three quadrilaterals. For a received message $\hat{s}$, the Hamming distortion of semantic decoding is determined by whether $\boldsymbol{\alpha}_p(\hat{s})$ and $\boldsymbol{\alpha}_q(\hat{s})$ fall into the same quadrilateral

Proposition 3.3 suggests that the condition that semantic decoding achieves the optimal distortion can be very demanding. In actuality, semantic decoding $\boldsymbol{V}$ can yield even larger distortion than the original interpretation of the language $\boldsymbol{Q}$. This can be easily understood as the inaccurate prior information $q(w)$ may mislead the receiver. A real-life example of this phenomenon is that when you talk to people with prejudice, they immediately sound harsh, even if they do not mean it that way.

Considering the Hamming distortion, the original interpretation of the language $\boldsymbol{Q}$ yields a distortion of

$$D_{\boldsymbol{P},\boldsymbol{Q}} = \sum_{\hat{s},\hat{w}} q(\hat{w}|\hat{s})\psi_p(\hat{w},\hat{s}) = 1 - \sum_{\hat{s},\hat{w}} p(\hat{w})p(\hat{s}|\hat{w})q(\hat{w}|\hat{s}).$$

The gap between $D_{\boldsymbol{P},\boldsymbol{Q}}$ and $D_{\boldsymbol{P},\boldsymbol{V}_q^*}$ can be written as

$$D_{\boldsymbol{P},\boldsymbol{Q}} - D_{\boldsymbol{P},\boldsymbol{V}_q^*} = \sum_{\hat{s}}\left(p\big(\hat{w}_q(\hat{s})\big)p\big(\hat{s}|\hat{w}_q(\hat{s})\big) - \sum_{\hat{w}} p(\hat{w})p(\hat{s}|\hat{w})q(\hat{w}|\hat{s})\right). \tag{3.30}$$

If the prior information is accurate, i.e., $p(w) = q(w)$, we have

$$D_{\boldsymbol{P},\boldsymbol{Q}} - D_{\boldsymbol{P},\boldsymbol{V}_q^*} = \sum_{\hat{s}}\left(\max_{w} p(w)p(\hat{s}|w) - \sum_{\hat{w}} p(\hat{w})p(\hat{s}|\hat{w})q(\hat{w}|\hat{s})\right) \geq 0. \tag{3.31}$$

and hence, $D_{\boldsymbol{P},\boldsymbol{V}_q^*}$ is strictly better than $D_{\boldsymbol{P},\boldsymbol{Q}}$. In general, however, $D_{\boldsymbol{P},\boldsymbol{Q}} - D_{\boldsymbol{P},\boldsymbol{V}_q^*}$ can be negative – the wrong prior leads to a semantic decoding scheme $\boldsymbol{V}$ worse than $\boldsymbol{Q}$ and it is better to stick to the original interpretation of the language. In this case, we propose two refinements on $\boldsymbol{Q}$ below to construct an improved random semantic decoding scheme $\boldsymbol{V}$.

Considering a received message $\hat{s}$, we define two subsets $\widehat{\mathcal{W}}(\hat{s}) \subseteq \mathcal{W}$ and $\mathcal{W}(\hat{s}) \subseteq \mathcal{W}$, where

$$\widehat{\mathcal{W}}(\hat{s}) \triangleq \big\{w : q(\hat{w}|\hat{s}) > 0\big\},$$
$$\mathcal{W}(\hat{s}) \triangleq \big\{w : p(s|w)c(\hat{s}|s) > 0\big\}.$$

Proposition 3.4 (Refining the Interpretation) *A random semantic decoding scheme* $\boldsymbol{V}$ *with* $D_{\boldsymbol{P},\boldsymbol{V}} \leq D_{\boldsymbol{P},\boldsymbol{Q}}$ *can be constructed by refining the interpretation* $\boldsymbol{Q}$ *as follows. First, let* $\boldsymbol{V} = \boldsymbol{Q}$.

Refinement 1: If $\exists\, w' \in \widehat{\mathcal{W}}(\hat{s})$, *s.t.,*

$$\min_{w\in\mathcal{W}(\hat{s})} d(w,w') \geq \max_{\substack{w\in\mathcal{W}(\hat{s}),\\ \widetilde{w}\in\widehat{\mathcal{W}}(\hat{s})\setminus\{w'\}}} d(w,\widetilde{w}), \tag{3.32}$$

we construct

$$v(w|\hat{s}) = \begin{cases} 0, & \textit{if } w = w', \\ \frac{q(w|\hat{s})}{1-q(w'|\hat{s})}, & \textit{if } w \neq w'. \end{cases} \tag{3.33}$$

Refinement 2: If $\exists\, w'' \in \widehat{\mathcal{W}}(\hat{s})$, *s.t.,*

$$\max_{w \in \mathcal{W}(\hat{s})} d(w, w'') \leq \min_{\substack{w \in \mathcal{W}(\hat{s}), \\ \widetilde{w} \in \widehat{\mathcal{W}}(\hat{s}) \setminus \{w''\}}} d(w, \widetilde{w}), \tag{3.34}$$

we construct

$$v(w|\hat{s}) = \begin{cases} 1, & \textit{if } w = w'', \\ 0, & \textit{if } w \neq w''. \end{cases} \tag{3.35}$$

In the first refinement (3.33), there exists a w' that is strictly worse than any other $\widetilde{w} \in \widehat{\mathcal{W}}(\hat{s})$. This is possible because the interpretation of the language can be suboptimal for some distortion functions. In this case, we remove the mapping between $\hat{s}$ and w' and assign the probability $q(w'|\hat{s})$ proportionally to other $\widetilde{w} \in \widehat{\mathcal{W}}(\hat{s})$. In the second refinement (3.35), there exists a w'' that is strictly better than any other $\widetilde{w} \in \widehat{\mathcal{W}}(\hat{s})$. As a result, we construct $v(w|\hat{s})$ such that $\hat{s}$ is deterministically decoded to w'' regardless of $q(w)$.

3.2.4 Combined Semantic Encoding and Decoding

In the problem of language utilization, semantic encoding and decoding are two decoupled processes since no negotiation is allowed between the transmitter and receiver. We have shown in Sects. 3.2.2 and 3.2.3 that the semantic distortion can be reduced by either semantic encoding or decoding. A natural question arises: what if the transmitter performs semantic encoding and the receiver performs semantic decoding simultaneously?

To answer this question, let us first define CSED formally.

Definition 3.23 (CSED) In CSED, the transmitter adopts a semantic encoding scheme

$$\boldsymbol{U} = \sum_{i=0}^{\underline{T}} \xi_i \underline{\boldsymbol{U}}^{(i)} + \sum_{i=\underline{T}+1}^{\underline{T}+\overline{T}+2} \xi_{i-\underline{T}-1} \overline{\boldsymbol{U}}^{(i-\underline{T}-1)}, \tag{3.36}$$

where $\boldsymbol{\xi} \triangleq \{\xi_i \geq 0 : \sum_{i=0}^{\underline{T}+\overline{T}+2} \xi_i = 1\}$ is a probability distribution; $\{\underline{\boldsymbol{U}}^{(i)} : i = 0, 1, 2, \ldots, \underline{T}\}$ and $\{\overline{\boldsymbol{U}}^{(i)} : i = 0, 1, 2, \ldots, \overline{T}\}$ are deterministic semantic encoding schemes derived from Theorem 3.1. The receiver performs semantic decoding and obtains $\boldsymbol{V}_q^* = \widetilde{\boldsymbol{\Delta}}_{n_1^{(q)}, n_2^{(q)}, \ldots, n_M^{(q)}}$ following Proposition 3.2. The semantic distortion and cost of CSED are given by

$$D_{\boldsymbol{U},\boldsymbol{V}_q^*} = \sum_{w,s,\hat{s},\hat{w}} p(w)u(s|w)c(\hat{s}|s)v(\hat{w}|\hat{s})d(w, \hat{w}). \tag{3.37}$$

$$L_{\boldsymbol{U}} = \sum_{w,s} p(w)u(s|w)\ell(s). \tag{3.38}$$

CSED leads to a new distortion-cost region.

Theorem 3.3 (The Distortion-Cost Region of CSED) *The distortion-cost region of CSED R_{csed}, defined as the set of all achievable distortion-cost pairs (L, D) via CSED, is a convex hull of the set of points:*

$$\left\{L_{\underline{\boldsymbol{U}}^{(i)}}, D_{\underline{\boldsymbol{U}}^{(i)},\boldsymbol{V}_q^*} : i = 0, 1, 2, \ldots, \underline{T}\right\},$$

$$\left\{L_{\overline{\boldsymbol{U}}^{(i)}}, D_{\overline{\boldsymbol{U}}^{(i)},\boldsymbol{V}_q^*} : i = 0, 1, 2, \ldots, \overline{T}\right\}.$$

Proof Theorem 3.3 follows from the time sharing argument that allows the system to interpolate among the performance characteristics of different deterministic encoding schemes, thereby enabling any point within the convex hull to be reachable. □

Correspondingly, we can define the distortion-cost function of CSED as the infimum of R_{csed}:

$$D^*_{\boldsymbol{U},\boldsymbol{V}_q^*}(L) \triangleq \inf\{D : (L, D) \in R_{\text{csed}}\}. \tag{3.39}$$

Given an average message cost L, the minimum distortion achieved by semantic encoding is $D^*_{\boldsymbol{U},\boldsymbol{Q}}(L)$. An interesting question is whether $D^*_{\boldsymbol{U},\boldsymbol{V}_q^*}(L)$ is smaller than $D^*_{\boldsymbol{U},\boldsymbol{Q}}(L)$? In addition, when $L = L_{\boldsymbol{P}}$, whether $D^*_{\boldsymbol{U},\boldsymbol{V}_q^*}(L_{\boldsymbol{P}})$ is smaller than $D_{\boldsymbol{P},\boldsymbol{V}_q^*}$ or $D^*_{\boldsymbol{U},\boldsymbol{Q}}(L_{\boldsymbol{P}})$? The answers to these questions are, unfortunately, negative. This can be easily understood via a counterexample.

Example 3.3 (The Node-Shake Example) Suppose the meaning set is $\mathcal{W} = \{\text{yes, no}\}$, the message set is $\mathcal{S} = \{\text{head nod, head shake}\}$, and the costs of the two messages are the same. The expression and interpretation of the semantic language

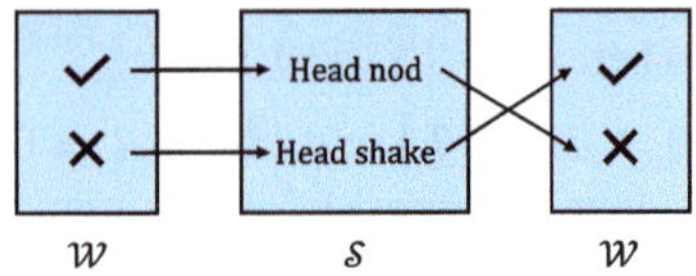

Fig. 3.8 The nod-shake example. Both semantic encoding and semantic decoding is better than CSED

are quite opposite, as shown in Fig. 3.8. At the transmitter, "head nod" stands for "yes" and "head shake" stands for "no", while at the receiver, "head nod" stands for "no" and "head shake" stands for "yes". Suppose the prior distribution $p(w)$ is known at the receiver and we consider the Hamming distortion.

In this example, the Hamming distortion can be reduced to 0 with either semantic encoding or semantic decoding. However, when both semantic encoding and decoding are performed, the Hamming distortion is 1. This suggests that CSED does not necessarily improve upon separate semantic encoding or decoding.

In what follows, we consider an error-free semantic channel and establish sufficient conditions under which CSED is better than semantic encoding and decoding.

Theorem 3.4 (CSED with an Error-Free Semantic Channel) *Consider a semantic language $(\mathcal{W}, \mathcal{S}, \boldsymbol{P}, \boldsymbol{Q})$, an error-free semantic channel $\boldsymbol{C}$, and a symmetric distortion measure $d(w, \hat{w}) = d(\hat{w}, w)$.*

Under the conditions that

1. *The receiver has perfect prior information, i.e., $p(w) = q(w)$.*
2. *The language is logically self-consistent, i.e., $q(w|s) = \frac{p(s|w)p(w)}{p(s)}$.*
3. *$\forall w, w' \in \mathcal{W}$,*

$$\arg\min_{s} \varphi(w, s) \cap \arg\min_{s} \varphi(w', s) = \emptyset. \tag{3.40}$$

4. *$\forall s, s' \in \{s : u(s|w) > 0, s \in \mathcal{S}, w \in \mathcal{W}\}$,*

$$\arg\min_{w} \psi_q(w, s) \cap \arg\min_{w} \psi_q(w, s') = \emptyset. \tag{3.41}$$

We have

$$D^*_{\boldsymbol{U},\boldsymbol{V}^*_q}(L) = \min_{\boldsymbol{U}',\boldsymbol{V}'} D_{\boldsymbol{U}',\boldsymbol{V}'}(L), \tag{3.42}$$

where $\boldsymbol{U}', \boldsymbol{V}'$ can be any semantic encoding and decoding schemes. In this case, CSED is better than both semantic encoding and semantic decoding, i.e., $D^*_{\boldsymbol{U},\boldsymbol{V}^*_q}(L) \leq D^*_{\boldsymbol{U},\boldsymbol{Q}}(L)$ and $D^*_{\boldsymbol{U},\boldsymbol{V}^*_q}(L_{\boldsymbol{P}}) \leq D_{\boldsymbol{P},\boldsymbol{V}^*_q}$.

Proof See Appendix A.5. □

More About CSED

The CSED scheme discussed in Theorem 3.23 consists of the semantic encoding scheme $\boldsymbol{U}$ and decoding scheme $\boldsymbol{V}$. In particular, $\boldsymbol{U}$ is optimized based on the interpretation $\boldsymbol{Q}$ of the language, and $\boldsymbol{V}$ is optimized based on the expression $\boldsymbol{P}$ of the language.

We have shown in the node-shake example that this can lead to a poorer performance than semantic encoding or decoding. However, what if the transmitter is clever enough and predicts that the receiver would improve its interpretation from $\boldsymbol{Q}$ to $\boldsymbol{V}$? In this case, the transmitter can optimize $\boldsymbol{P}$ to $\boldsymbol{U}'$ based on $\boldsymbol{V}$, as opposed to $\boldsymbol{Q}$. In the node-shake example, $\boldsymbol{U}'$ and $\boldsymbol{V}$ would lead to zero Hamming distortion.

Following this idea, the receiver can further predict that the transmitter would improve its expression from $\boldsymbol{P}$ to $\boldsymbol{U}'$ and optimize $\boldsymbol{Q}$ based on $\boldsymbol{U}'$ instead of $\boldsymbol{Q}$, and so on. The resulting Hamming distortion would alternate between 0 and 1, depending on how many times the transmitter and receiver have thought about each other's encoding and decoding schemes. This phenomenon resembles human communication—two smart people may not be able to communicate efficiently as they would keep inferring each other's expression and interpretation and changing theirs continuously.

3.2.5 *Language Utilization for IoT Communications: An Example*

This section provides a tangible example to illustrate the principles of language utilization as outlined in this chapter. By utilizing the established semantic and technical languages, we demonstrate how semantic encoding, semantic decoding, and CSED can be applied to minimize semantic distortion within IoT applications.

3.2.5.1 The Grid World and Languages

Example 3.4 (Grid World) We consider a world of square grids, as shown in Fig. 3.9.

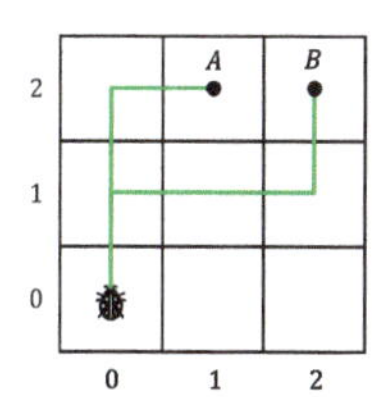

Fig. 3.9 An IoT device walking in the grid world

- The grids are indexed by $(i, j), i, j \in \{0, 1, 2\}$.
- An IoT device sets out from (0, 0) and walks towards one of the two destinations A: (1, 2) and B: (2, 2).
- At each grid, the IoT device can only go up or right.
- There is a transmitter and a receiver. Only the transmitter observes the trajectory of the IoT device.

The transmitter and receiver have agreed on semantic language and technical language.

The Message and Meaning Sets

There are two words '$\mathcal{U}$' and '$\mathcal{R}$' in the semantic language, representing the actions of the IoT device (i.e., "up" and "right"). A message is composed of a sequence of words, representing the sequential actions (i.e., a trajectory) of the IoT device starting from (0, 0). The message '$\mathcal{URR}$', for example, means that the IoT device goes up, right, and right from (0, 0). The maximal number of moves in the grid world is 4; hence, the number of possible messages is $\sum_{i=0}^{4} 2^i = 31$. The message set can be written as

$$\begin{aligned} \mathcal{S}_{\text{all}} = \{&\emptyset, \mathcal{U}, \mathcal{R}, \mathcal{UU}, \mathcal{UR}, \mathcal{RU}, \mathcal{RR}, \mathcal{UUU}, \mathcal{UUR}, \mathcal{URU}, \\ &\mathcal{URR}, \mathcal{RUU}, \mathcal{RUR}, \mathcal{RRU}, \mathcal{RRR}, \mathcal{UUUU}, \mathcal{UUUR}, \mathcal{UURU}, \\ &\mathcal{UURR}, \mathcal{URUU}, \mathcal{URUR}, \mathcal{URRU}, \mathcal{URRR}, \mathcal{RUUU}, \mathcal{RUUR}, \\ &\mathcal{RURU}, \mathcal{RURR}, \mathcal{RRUU}, \mathcal{RRUR}, \mathcal{RRRU}, \mathcal{RRRR}\} \end{aligned}$$

where $\emptyset$ means no message is transmitted.

Given the message set, the transmitter can convey various meanings such as the trajectory and the destination of the IoT device. In this example, we assume the receiver is only interested in the destination of the IoT device and define the meaning set as $\mathcal{W} = \{A, B\}$ and $p(A) = 1/3$, $p(B) = 2/3$. We assume the receiver has a prior distribution $q(A) = 1/2$ and $q(B) = 1/2$.

Error-Free Technical Communication

For a given message, the technical language maps the words '$\mathcal{U}$' and '$\mathcal{R}$' to bit sequences '0' and '10', respectively. The set of channel symbols can be constructed by substituting '$\mathcal{U}$' and '$\mathcal{R}$' in $\mathcal{S}_{\text{all}}$ with '0' and '10', respectively.

The mappings between the messages and channel symbols are one-to-one. Throughout this section, we assume that the technical channel is error-free, and hence, the semantic channel is also error-free—a transmitted message can be perfectly received by the receiver. The generalization to an error-prone technical channel is straightforward.

We consider the Hamming distortion. The cost function $\ell(s)$ is defined as the length of the bit sequence associated with s.

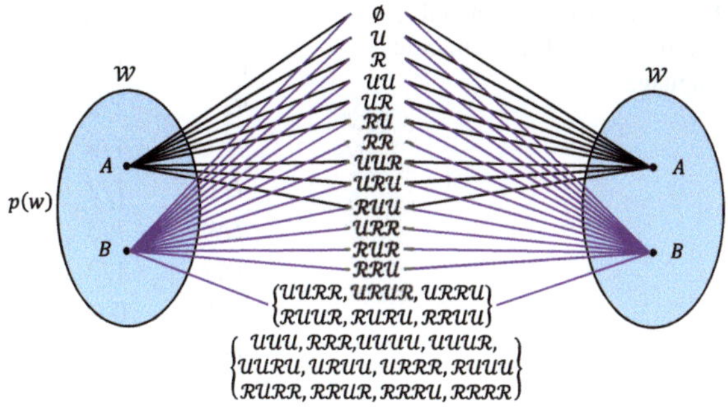

Fig. 3.10 The expression and interpretation of the semantic language

Expression and Interpretation

Given the message and meaning sets, we illustrate the expression and interpretation of the semantic language in Fig. 3.10. As can be seen, to convey meanings in $\mathcal{W}$, the class of messages

$$\mathcal{S}_{\text{null}} = \{\mathcal{UUU}, \mathcal{RRR}, \mathcal{UUUU}, \mathcal{UUUR}, \mathcal{UURU}, \mathcal{URUU}, \mathcal{URRR}, \mathcal{RUUU}, \mathcal{RURR}, \mathcal{RRUR}, \mathcal{RRRU}, \mathcal{RRRR}\}$$

are illegitimate, and the class of messages

$$\{\mathcal{UURR}, \mathcal{URUR}, \mathcal{URRU}, \mathcal{RUUR}, \mathcal{RURU}, \mathcal{RRUU}\}$$

are equivalent as they have the same cost and lead to the same distortion. We can use a single message $\mathcal{UURR}$ to represent them in order to simplify notation. This gives us a refined message set

$$\mathcal{S} = \mathcal{S}_{\text{all}} \backslash \mathcal{S}_{\text{null}} = \{\emptyset, \mathcal{U}, \mathcal{R}, \mathcal{UU}, \mathcal{UR}, \mathcal{RU}, \mathcal{RR}, \mathcal{UUR}, \mathcal{URU}, \mathcal{RUU}, \mathcal{URR}, \mathcal{RUR}, \mathcal{RRU}, \mathcal{UURR}\},$$

as shown in Fig. 3.10. Note that we have arranged them in ascending order according to the message cost. Next, we define the interpretation and expression, respectively.

Denoting by ϕ_A and ϕ_B the sets of trajectories that lead to A and B, respectively, we have

$$\phi_A = \{\mathcal{UUR}, \mathcal{URU}, \mathcal{RUU}\},$$

$$\phi_B = \{\mathcal{UURR}, \mathcal{URUR}, \mathcal{URRU}, \mathcal{RRUU}, \mathcal{RURU}, \mathcal{RUUR}\}.$$

The interpretation of the semantic language is defined as

$$q(\hat{w} = w|\hat{s}) = \frac{\sum_{s \in \phi_w} \hat{s} \sqsubseteq s}{\sum_{w \in \mathcal{W}} \sum_{s \in \phi_w} \hat{s} \sqsubseteq s} \tag{3.43}$$

Table 3.1 Expression and interpretation of the semantic language

s	$q(A\|s)$	$q(B\|s)$	$p(s\|A)$	$p(s\|B)$	$\ell(s)$
$\emptyset$	1/3	2/3	1/9	1/19	0
$\mathcal{U}$	2/5	3/5	1/9	1/19	1
$\mathcal{R}$	1/4	3/4	1/9	1/19	2
$\mathcal{UU}$	1/2	1/2	1/9	1/19	2
$\mathcal{UR}$	1/3	2/3	1/9	1/19	3
$\mathcal{RU}$	1/3	2/3	1/9	1/19	3
$\mathcal{RR}$	0	1	0	1/19	4
$\mathcal{UUR}$	1/2	1/2	1/9	1/19	4
$\mathcal{URU}$	1/2	1/2	1/9	1/19	4
$\mathcal{RUU}$	1/2	1/2	1/9	1/19	4
$\mathcal{URR}$	0	1	0	1/19	5
$\mathcal{RUR}$	0	1	0	1/19	5
$\mathcal{RRU}$	0	1	0	1/19	5
$\mathcal{UURR}$	0	1	0	6/19	6

where $\hat{s} \sqsubseteq s = 1$ if $\hat{s}$ is a prefix of s, and $\hat{s} \sqsubseteq s = 0$, otherwise. For example, suppose the received message $\hat{s} = \mathcal{U}$, we have $q(\hat{w} = A|\mathcal{U}) = 2/5$ because $\mathcal{U}$ is a prefix of two messages in ϕ_A and three messages in ϕ_B. A complete list of $q(w|s)$ is given in Table 3.1.

Correspondingly, for a meaning w, we define the expression of the language $p(s|w)$ as the uniform distribution over $\{s : q(w|s) > 0\}$. Table 3.1 lists $p(s|w)$, $\forall w, s$.

Semantic Encoding

This subsection studies how to reduce semantic distortion from the transmitter's perspective via semantic encoding. We shall follow Theorem 3.1 and Algorithm 4 to characterize the semantic distortion-cost region with semantic encoding.

For an error-free semantic channel and Hamming distortion measure, the expected distortion of mapping a meaning w to a message s can be written as

$$\varphi(w, s) = \sum_{\hat{s}} c(\hat{s}|s) \sum_{\hat{w}} q(\hat{w}|\hat{s}) d(w, \hat{w}) = 1 - q(\hat{w} = w|s).$$

A list of $\varphi(w, s)$ is given in Table 3.2

To characterize the semantic distortion-cost region, the first step is to construct the six subsets for $w \in \mathcal{W}$. According to Algorithm 3, the six subsets are given by

$$\underline{S}'(A) = \{\emptyset, \mathcal{U}, \mathcal{UU}\}, \quad \underline{S}''(A) = \{\mathcal{UUR}, \mathcal{UURR}\},$$
$$\underline{S}(A) = \{\emptyset, \mathcal{U}, \mathcal{UU}, \mathcal{UUR}, \mathcal{UURR}\},$$
$$\overline{S}'(A) = \{\emptyset, \mathcal{RR}\}, \quad \overline{S}''(A) = \{\mathcal{UURR}\},$$
$$\overline{S}(A) = \{\emptyset, \mathcal{RR}, \mathcal{UURR}\},$$

Table 3.2 $\varphi(w, s)$ and $\alpha_q(w, s)$

s	$\varphi(A, s)$	$\varphi(B, s)$	$\alpha_q(A, s)$	$\alpha_q(B, s)$
$\emptyset$	2/3	1/3	19/28	9/28
$\mathcal{U}$	3/5	2/5	19/28	9/28
$\mathcal{R}$	3/4	1/4	19/28	9/28
$\mathcal{UU}$	1/2	1/2	19/28	9/28
$\mathcal{UR}$	2/3	1/3	19/28	9/28
$\mathcal{RU}$	2/3	1/3	19/28	9/28
$\mathcal{RR}$	1	0	0	1
$\mathcal{UUR}$	1/2	1/2	19/28	9/28
$\mathcal{URU}$	1/2	1/2	19/28	9/28
$\mathcal{RUU}$	1/2	1/2	19/28	9/28
$\mathcal{URR}$	1	0	0	1
$\mathcal{RUR}$	1	0	0	1
$\mathcal{RRU}$	1	0	0	1
$\mathcal{UURR}$	1	0	0	1

$$
\begin{aligned}
&\underline{S}'(B) = \{\emptyset, \mathcal{RR}\}, \quad \underline{S}''(B) = \{\mathcal{UURR}\},\\
&\underline{S}(B) = \{\emptyset, \mathcal{RR}, \mathcal{UURR}\},\\
&\overline{S}'(B) = \{\emptyset, \mathcal{U}, \mathcal{UU}\}, \quad \overline{S}''(B) = \{\mathcal{UUR}, \mathcal{UURR}\},\\
&\overline{S}(B) = \{\emptyset, \mathcal{U}, \mathcal{UU}, \mathcal{UUR}, \mathcal{UURR}\},
\end{aligned}
$$

The second step constructs a sequence of deterministic encoding schemes that characterize the distortion-cost function. To simplify notation, we denote a deterministic encoding scheme by a set with two elements in this example, where the two elements correspond to the messages that A and B are mapped to. First, we construct $\underline{U}^{(0)} = \{\emptyset, \emptyset\}$, where $D_{\underline{U}^{(0)}, Q} = p(A)\varphi(A, \emptyset) + p(B)\varphi(B, \emptyset) = \frac{4}{9}$ and $L_{\underline{U}^{(0)}} = p(A)\ell(\emptyset) + p(B)\ell(\emptyset) = 0$.

To construct $\underline{U}^{(1)}$, we first compute the G function and choose $\underline{w}^{(1)}$ and $\underline{s}^{(1)}$ that minimizes the G function. In this case, we have $G(w = A, \emptyset, \mathcal{U}) = -\frac{1}{15}$, $G(w = A, \emptyset, \mathcal{UU}) = -\frac{1}{12}$, and $G(w = B, \emptyset, \mathcal{RR}) = -\frac{1}{12}$.

Therefore, we choose

$$
\underline{w}^{(1)} = A, \quad \underline{s}^{(1)} = \mathcal{UU}, \quad \underline{U}^{(1)} = \{\mathcal{UU}, \emptyset\}, \tag{3.44}
$$

and $D_{\underline{U}^{(1)}, Q} = p(A)\varphi(A, \mathcal{UU}) + p(B)\varphi(B, \emptyset) = \frac{7}{18}$ and $L_{\underline{U}^{(1)}} = p(A)\ell(\mathcal{UU}) + p(B)\ell(\emptyset) = \frac{2}{3}$.

Likewise, we can derive all the deterministic encoding schemes and the achieved distortion and cost. Here, we omit all the derivations and list the resulting encoding schemes in Table 3.3, based on which the distortion-cost region of semantic encoding R_{enc} is characterized in Fig. 3.11a.

Table 3.3 The discovered encoding schemes

Encoding schemes	D	L
$\underline{U}^{(0)} = \{\emptyset, \emptyset\}$	4/9	0
$\underline{U}^{(1)} = \{\mathcal{UU}, \emptyset\}$	7/18	2/3
$\underline{U}^{(2)} = \{\mathcal{UU}, \mathcal{RR}\}$	1/6	10/3
$\underline{U}^{(3)} = \{\mathcal{UUR}, \mathcal{RR}\}$	1/6	4
$\underline{U}^{(4)} = \{\mathcal{UUR}, \mathcal{UURR}\}$	1/6	16/3
$\underline{U}^{(5)} = \{\mathcal{UURR}, \mathcal{UURR}\}$	1/3	6
$\overline{U}^{(0)} = \{\emptyset, \emptyset\}$	4/9	0
$\overline{U}^{(1)} = \{\mathcal{RR}, \emptyset\}$	5/9	4/3
$\overline{U}^{(2)} = \{\mathcal{RR}, \mathcal{UU}\}$	2/3	8/3
$\overline{U}^{(3)} = \{\mathcal{UURR}, \mathcal{UU}\}$	2/3	10/3
$\overline{U}^{(4)} = \{\mathcal{UURR}, \mathcal{UUR}\}$	2/3	14/3
$\overline{U}^{(5)} = \{\mathcal{UURR}, \mathcal{UURR}\}$	1/3	6

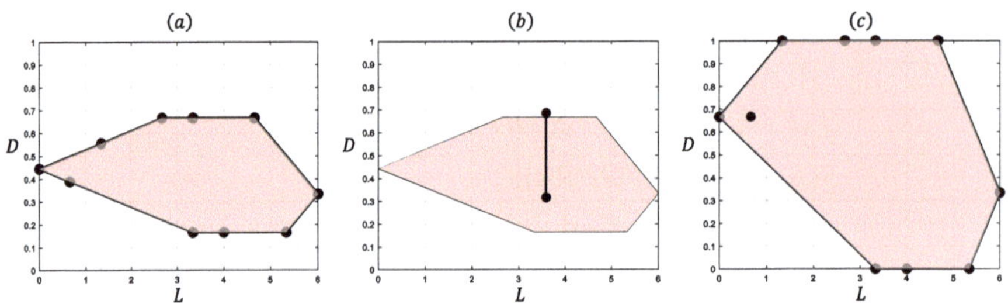

Fig. 3.11 The distortion-cost regions of (**a**) semantic encoding R_{enc}, (**b**) semantic decoding R_{dec}, and (**c**) CSED R_{csed}

Semantic Decoding and CSED

Next, we study how to reduce the distortion from the receiver's perspective via semantic decoding. We first characterize the distortion-cost region of semantic decoding.

Under the Hamming distortion, we have $\psi_P(\hat{w}, s) = p(s) - p(\hat{w})p(s|\hat{w})$. The semantic decoding schemes $\widetilde{\mathbf{\Delta}}_{n'_1,n'_2,\ldots,n'_M}$ and $\widetilde{\mathbf{\Delta}}_{n''_1,n''_2,\ldots,n''_M}$ can be derived from Theorem 3.2, as listed in Table 3.4, and

$$D_{\widetilde{\mathbf{\Delta}}_{n'_1,n'_2,\ldots,n'_M}} \approx 0.3158, \quad D_{\widetilde{\mathbf{\Delta}}_{n''_1,n''_2,\ldots,n''_M}} \approx 0.6842,$$
$$L_{\boldsymbol{P}} \approx 3.5887.$$

The distortion-cost region of semantic decoding R_{dec} is plotted in Fig. 3.11b. In this example, the receiver has an inaccurate prior distribution. According to Proposition 3.2, we derive the optimal decoding scheme perceived by the receiver $\boldsymbol{V}_q^*$ in Table 3.4. As can be seen, $\boldsymbol{V}_q^*$ matches $\widetilde{\mathbf{\Delta}}_{n'_1,n'_2,\ldots,n'_M}$. This is because Proposition 3.3 is satisfied. In this case, semantic decoding with the inaccurate prior achieves the optimal distortion, i.e., $D_{\boldsymbol{P},\boldsymbol{V}_q^*} \approx 0.3158$, as shown in Fig. 3.11b. It

Table 3.4 Semantic decoding schemes

s	$\widetilde{\boldsymbol{\Delta}}_{n'_1,n'_2,\dots,n'_M}$	$\widetilde{\boldsymbol{\Delta}}_{n''_1,n''_2,\dots,n''_M}$	$\boldsymbol{V}_q^*$
$\emptyset$	A	B	A
$\mathcal{U}$	A	B	A
$\mathcal{R}$	A	B	A
$\mathcal{UU}$	A	B	A
$\mathcal{UR}$	A	B	A
$\mathcal{RU}$	A	B	A
$\mathcal{RR}$	B	A	B
$\mathcal{UUR}$	A	B	A
$\mathcal{URU}$	A	B	A
$\mathcal{RUU}$	A	B	A
$\mathcal{URR}$	B	A	B
$\mathcal{RUR}$	B	A	B
$\mathcal{RRU}$	B	A	B
$\mathcal{UURR}$	B	A	B

is worth noting that, in this example, the original interpretation of the language $\boldsymbol{Q}$ yields a distortion of $D_{\boldsymbol{P},\boldsymbol{Q}} = 1 - \sum_{s,w} p(w)p(s|\hat{w})q(w|s) \approx 0.3262$.

Finally, when combining both semantic encoding and decoding, the new distortion-cost region of CSED R_{csed} can be obtained from Theorem 3.3, and we plot it in Fig. 3.11c.

Impact of Encoding Scheme Selection on the Distortion-Cost Region of CSED

When characterizing the distortion-cost function of semantic encoding, Algorithm 4 sequentially constructs $\underline{T}+1$ encoding schemes $\{\underline{\boldsymbol{U}}^{(t)}\}$. At a step t, it is possible to find several pairs of $(\underline{n}^{(t)}, \underline{m}^{(t)})$ that are equally optimal as they yield the minimum G (see line 12 of Algorithm 4). Likewise, when generating $\overline{\boldsymbol{U}}^{(t)}$, it is possible to find several pairs of $(\overline{n}^{(t)}, \overline{m}^{(t)})$ that are equally optimal as they yield the maximum G (see line 20 of Algorithm 4). Although these equally optimal pairs correspond to different encoding schemes, they finally yield the same distortion-cost region for semantic encoding. Therefore, we can randomly sample one from them in Algorithm 4. However, the selection does have an impact on the distortion-cost region of CSED. According to Definition 3.23 and Theorem 3.3, CSED and its distortion-cost region are determined by the chosen semantic encoding schemes. Selecting different semantic encoding schemes in Algorithm 4 leads to different distortion-cost regions for CSED. For example, if we choose

$$\underline{w}^{(1)} = B,\ \ \underline{s}^{(1)} = \mathcal{RR},\ \ \underline{U}^{(1)} = \{\emptyset, \mathcal{RR}\}, \tag{3.45}$$

in (3.44), the distortion-cost pair $\left(\frac{2}{3}, \frac{2}{3}\right)$ will change to $\left(\frac{2}{3}, 0\right)$, leading to reduced semantic distortion.

3.3 Language Design via Deep Joint Source-Channel Coding

In the preceding section, we explored the language utilization problem, focusing on how to effectively leverage fixed or partially fixed semantic and technical languages to minimize misinterpretations and enhance communication efficiency within IoT systems. In this section, we shift focus to the language design problem, where both the semantic and technical languages are fully designable. This flexibility allows for the integration of advanced AI techniques to develop deep joint source-channel coding (DeepJSCC) encoders and decoders [8]. By tailoring these models to specific types of data in IoT networks, we can significantly reduce the end-to-end distortion, thereby optimizing the transmission quality and efficiency in a variety of IoT applications.

3.3.1 The DeepJSCC Framework

3.3.1.1 From JSCC to DeepJSCC

JSCC is a communication technique designed to integrate source coding (compression) and channel coding (error correction) into a unified coding operation [30]. Traditionally, these two processes are distinctly separate: source coding primarily focuses on reducing data redundancy to minimize the bandwidth requirements for transmission, while channel coding strategically introduces redundancy to safeguard the data against noise and other errors prevalent in the transmission medium. This traditional separation, though systematic, often leads to suboptimal performance when dealing with constraints on bandwidth and error sensitivity, particularly in environments where channel conditions are unpredictable and highly variable.

JSCC seeks to transcend these limitations by harmonizing source and channel coding into a single, cohesive strategy. This approach leverages the interdependencies between the compression of data and the protection against errors, aiming to optimize the overall coding process. By doing so, JSCC dynamically adjusts its coding parameters to align closely with the specific characteristics of the source data and the prevailing conditions of the communication channel. This dynamic adaptation allows JSCC to effectively balance the trade-offs between data fidelity, robustness to errors, and efficient use of bandwidth.

A notable limitation of traditional JSCC is its analytical complexity, especially when seeking optimal design solutions. Historically, the success of JSCC designs has been limited to a few specific scenarios where both the source and the channel could be well-characterized, such as Gaussian sources transmitted over AWGN channels [31]. These conditions, while useful in theoretical exploration, often do not translate directly into practical applications, particularly in diverse and dynamic environments like those encountered in IoT communications.

To address these challenges and extend the utility of JSCC beyond these traditional confines, data-driven and AI techniques have been increasingly adopted [32–35]. By leveraging the capabilities of DL, researchers and engineers have developed a new paradigm known as DeepJSCC [8, 36]. This approach redefines JSCC by employing DNNs to automate and optimize the coding process, moving away from the dependency on predefined algorithms and models that characterize traditional JSCC.

Furthermore, DeepJSCC not only simplifies the implementation of JSCC by reducing the need for explicit mathematical modeling of the source and channel but also enhances performance by allowing for continuous improvement and optimization through training. As the network encounters new data or channel conditions, it can learn and adjust in real-time [34], offering a significant advantage over static, conventional JSCC methods. This dynamic and flexible approach makes DeepJSCC a highly promising technology for managing the complex and evolving demands of modern IoT communication systems.

3.3.1.2 Key Elements of DeepJSCC

The key ingredients of DeepJSCC encompass an encoder, a channel, a decoder, and end-to-end training.

Encoder
The encoder in a DeepJSCC system is a crucial component, designed to handle two primary tasks: data compression and encoding for transmission. This encoder leverages a DNN to map raw input data into a lower-dimensional, compact representation that retains critical information while being robust against potential channel impairments. This mapping is learned during training, where the network is exposed to a variety of data samples and channel conditions, allowing it to identify and maintain the most essential features of the data that need to be preserved.

The design of the encoder often involves different DNN architectures tailored to the data type being processed. The encoder is also trained to anticipate and mitigate specific types of noise and interference typical of the communication channels expected in deployment.

Channel
The channel component represents the physical pathways through which the encoded signals are transmitted. In IoT contexts, these can range from unstable wireless connections subject to fading and interference to more stable wired connections. Each type of channel has its characteristics, such as noise levels, bandwidth limitations, and error rates, which can significantly affect the performance of DeepJSCC systems.

During the training of DeepJSCC systems, realistic channel models are often used to simulate the conditions under which the communication will occur. These models help in fine-tuning the encoder and decoder by providing them with scenarios of varying complexity and difficulty, such as high noise environments or low bandwidth situations commonly encountered in IoT deployments.

Decoder
The decoder, like the encoder, is a DNN but functions in reverse. It receives the encoded data, which may have been corrupted by the channel, and works to reconstruct the original data as accurately as possible. This process is challenging, particularly in noisy environments, and the effectiveness of the decoder is critical to the overall performance of the system.

Techniques such as denoising, statistical inference, and even reinforcement learning can be integrated into the decoder to enhance its ability to correct errors and improve data recovery. The decoder's architecture is typically a mirror of the encoder, utilizing similar layers and structures to effectively reverse the encoding process.

End-to-End Training
The hallmark of the DeepJSCC strategy is its end-to-end training approach, where the encoder and decoder are trained simultaneously as a single system. This approach ensures that both components learn to complement each other's functionality, optimizing the overall system performance.

The training of both encoder and decoder involves minimizing a loss function that quantitatively measures the quality of the reconstructed output against the original input. This loss function can be a standard mean squared error (MSE) for tasks where exact reconstruction is necessary, or it can be more complex, incorporating aspects of perceptual quality for data like images and video.

This comprehensive framework positions DeepJSCC as a potent solution for the demands of modern IoT communications, where the efficiency, reliability, and adaptability of data transmission are paramount.

3.3.2 IoT Image Transmission with DeepJSCC

Building on the foundational concepts introduced in the DeepJSCC framework, we now turn our attention to a practical application within the IoT that exemplifies the cutting-edge capabilities of this approach [34, 37, 38]. This section focuses specifically on a learning-based, channel-adaptive joint source and channel coding (CA-JSCC) scheme [37] designed for the efficient transmission of wireless images over multipath fading channels.

This innovative method employs an end-to-end autoencoder architecture enhanced by a dual-attention mechanism and utilizes OFDM for transmission. This approach is particularly distinguished by its adaptability to the dynamic conditions of wireless channels, which is critical for IoT applications where environments can vary dramatically and unpredictably. By integrating a dual-attention mechanism that leverages estimated channel state information (CSI), the DeepJSCC encoder dynamically maps features and allocates transmission power across subcarriers, optimizing resource use based on real-time channel conditions. This section will explore how the CA-JSCC scheme not only meets but exceeds current benchmarks, providing robust performance and efficient utilization of limited channel resources under varying conditions.

3.3.2.1 System Model

We consider OFDM-based JSCC of images over a multipath fading channel with L_t paths. We transmit each input image using N_s OFDM symbols accompanied with N_p pilots for channel estimation of L_f OFDM subcarriers. As shown in Fig. 3.12, an encoding function $E_{\boldsymbol{\theta}}$ first maps the input image $\boldsymbol{x} \in \mathbb{R}^{c \times h \times w}$ into a complex matrix $\boldsymbol{Y} \in \mathbb{C}^{N_s \times L_f}$, where c, h and w denote the color, height and width. The generated channel input can be denoted by $\boldsymbol{Y} = E_{\boldsymbol{\theta}}(\boldsymbol{x}, \hat{\boldsymbol{h}})$, where $\hat{\boldsymbol{h}}$ is the estimated CSI vector available at the transmitter.

Channel input $\boldsymbol{Y}$ is subject to an average power constraint P_s: $\frac{1}{N_s L_f}\mathbb{E}\big[\|\boldsymbol{Y}\|_{\mathrm{F}}^2\big] \leq P_s$, where the expectation is taken over the input images, and $\|\cdot\|_{\mathrm{F}}$ denotes the Frobenius norm. Without loss of generality, we set $P_s = 1$.

Each OFDM symbol is passed through the inverse discrete Fourier transform (IFFT) module, appended with the cyclic prefix (CP), and transmitted to the receiver over the multipath channel. The transfer function of the multipath fading channel with L_t paths is defined as: $\hat{\boldsymbol{y}} = h_c(\boldsymbol{y}) = \boldsymbol{h}_t * \boldsymbol{y} + \boldsymbol{w}$, where $\boldsymbol{y}$ and $\hat{\boldsymbol{y}}$ denote the input and output vectors, respectively; $*$ is the linear convolution operation, $\boldsymbol{h}_t \in \mathbb{C}^L$ is the channel impulse response, and $\boldsymbol{w}$ is the AWGN term.

The receiver first demodulates $\hat{\boldsymbol{y}}$ by removing the CP and applying fast Fourier transform (FFT). The equivalent frequency-domain transfer function of h_c can be written as:

$$\hat{Y}[i,k] = H[k]\bar{Y}[i,k] + W[i,k], \tag{3.46}$$

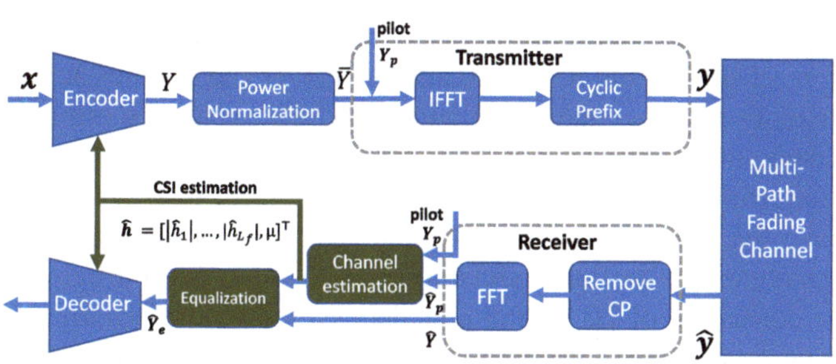

Fig. 3.12 Our proposed channel-adaptive CA-JSCC scheme

where the frequency-domain channel matrix $\boldsymbol{H} \in \mathbb{C}^{(L_f, L_f)}$ is a diagonal matrix with the k-th diagonal element being $H[k]$ (frequency-domain channel response of the k-th subcarrier). Let $\bar{\boldsymbol{Y}} \in \mathbb{C}^{(N_s, L_f)}$ denote the output of the power normalization module at the transmitter (i.e., the inputs to the IFFT module), where $\bar{Y}[i,k]$ denotes the symbol at the k-th subcarrier of the i-th OFDM symbol. $\boldsymbol{W} \in \mathbb{C}^{(N_s, L_f)}$ is the frequency domain noise matrix, where $W[i,k] \sim \mathcal{CN}(0, \sigma^2)$ and independent of each other.

Given the FFT output of the pilots at the receiver, we use minimum mean square error (MMSE) or least square (LS) channel estimator to estimate the CSI ($H[k]$) in frequency-domain (Eq. (3.46)). The estimated CSI vector $\hat{\boldsymbol{h}}$ is then used to equalize the data (MMSE equalizer). We have

$$\hat{\boldsymbol{h}} \triangleq \left[|\hat{h}_1|, |\hat{h}_2|, \cdots, |\hat{h}_{L_f}|, \mu\right]^\top, \tag{3.47}$$

where $\hat{h}_i, i = 1, \ldots, L_f$, is the estimated channel gain of the i-th subcarrier; while μ is the average SNR defined as $\mu = 10\log_{10}\frac{P_s}{\sigma^2}$ dB for transmit power P_s and noise power σ^2.

Based on the equalized data $\hat{\boldsymbol{Y}}_{\boldsymbol{e}}$ and the estimated CSI vector $\hat{\boldsymbol{h}}$, the decoder $D_{\boldsymbol{\phi}}$ reconstructs the transmitted image as $\hat{\boldsymbol{x}}$, i.e., $\hat{\boldsymbol{x}} = D_{\boldsymbol{\phi}}(\hat{\boldsymbol{Y}}_{\boldsymbol{e}}, \hat{\boldsymbol{h}})$. The performance indicator is the peak signal-to-noise ratio (PSNR), defined as $\text{PSNR} = 10\log_{10}\frac{(\max \boldsymbol{x})^2}{\text{MSE}(\boldsymbol{x}, \hat{\boldsymbol{x}})}$ (dB), where $\max \boldsymbol{x}$ denotes the maximum possible value of the input signal $\boldsymbol{x}$, $\text{MSE}(\boldsymbol{x}, \hat{\boldsymbol{x}}) \triangleq E[\|\boldsymbol{x} - \hat{\boldsymbol{x}}\|_2^2]$ and the expectation is taken over all pixels [39].

We then jointly train the encoder and decoder to minimize the loss function $\mathcal{L}(\boldsymbol{\theta}, \boldsymbol{\phi}) = \mathbb{E}\left[\text{PSNR}(\boldsymbol{x}, \hat{\boldsymbol{x}})\right]$, where the expectation is taken over the randomness both in the source and channel distributions.

3.3.2.2 Dual-Attention Mechanism

In OFDM systems, different subcarriers face different channel gains, and a judicious transmission scheme should be able to allocate power and features across appropriate subcarriers to adapt to channel variations. To fully exploit the estimated CSI, we propose a dual-attention based CA-JSCC scheme.

The architecture of our method is shown in Fig. 3.13, both the encoder and the decoder have five feature learning (FL) and four channel learning (CL) modules in a comb structure, allowing modulating each feature with a different scale. The FL module consisting of 2D convolution/deconvolution, batch normalization, and PReLU layers, is designed to learn an intermediate representation of the input. The dual-attention based CL module is designed to learn an attention mask to map the intermediate features to appropriate subcarriers based on the estimated CSI and input features. The CL module consists of a *channel-wise* attention module and a *spatial* attention module. Its operation is presented in Algorithm 5 in detail.

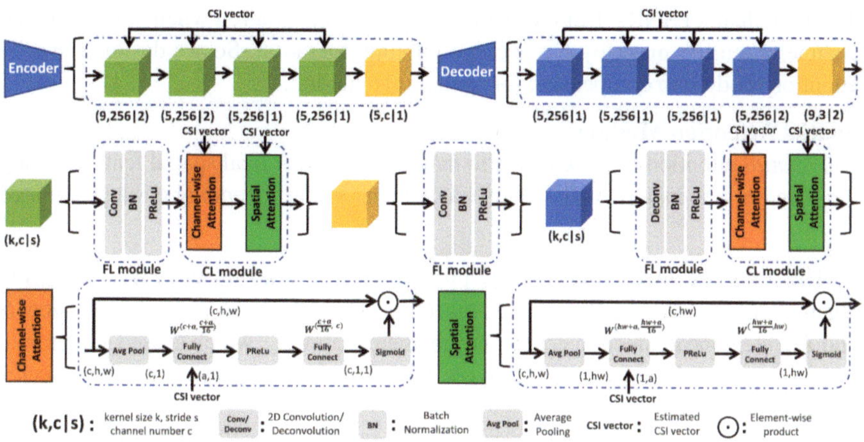

Fig. 3.13 Basic blocks of our dual-attention encoder and decoder architectures

---

Algorithm 5 Dual-attention based CL

Input: $\boldsymbol{F_{in}} \in \mathbb{R}^{c \times h \times w}$, $\hat{\boldsymbol{h}} \in \mathbb{R}^{L_f+1}$
Output: $\boldsymbol{F_{out}} \in \mathbb{R}^{c \times h \times w}$
Stage 1: Channel-wise attention
$\boldsymbol{F_{ca}} = Ave_c(\boldsymbol{F_{in}}) \in \mathbb{R}^{(c)}$
$\boldsymbol{i_c} = concatenate(\boldsymbol{F_{ca}}, \hat{\boldsymbol{h}}) \in \mathbb{R}^{(c+L_f+1)}$
$\boldsymbol{S_c} = f_c(\boldsymbol{i_c}) \in \mathbb{R}^{(c,1,1)}$
for i= 0:1:c **do**
    $\boldsymbol{F_{cout}}[i, h, w] = \boldsymbol{S_c}[i] \odot \boldsymbol{F_{in}}[i, h, w] \in \mathbb{R}^{(c,h,w)}$
Stage 2: Spatial attention
$\boldsymbol{F_{sa}} = Ave_s(\boldsymbol{F_{cout}}) \in \mathbb{R}^{(1,h,w)} \Rightarrow \mathbb{R}^{(hw)}$
$\boldsymbol{i_s} = concatenate(\boldsymbol{F_{sa}}, \hat{\boldsymbol{h}}) \in \mathbb{R}^{(hw+L_f+1)}$
$\boldsymbol{S_s} = f_s(\boldsymbol{i_s}) \in \mathbb{R}^{(hw)} \Rightarrow \mathbb{R}^{(h,w)}$
for j= 0:1:h **do**
    for k= 0:1:w **do**
        $\boldsymbol{F_{out}}[c, j, k] = \boldsymbol{S_s}[j, k] \odot \boldsymbol{F_{cout}}[c, j, k] \in \mathbb{R}^{(c,h,w)}$

Channel-Wise Attention Module

Our *channel-wise* attention module is inspired by [40], which adapts to a single SNR value in an AWGN channel model. Instead, CA-JSCC learns an attention mask to allocate and map features based on the estimated CSI of all N_s subcarriers.

We first apply average pooling $Ave_c(\boldsymbol{F_{in}})$ on input features $\boldsymbol{F_{in}}$ along the spatial direction to get vector $\boldsymbol{F_{ca}}$, where $Ave_c(\boldsymbol{F_{in}}) \triangleq \frac{1}{hw}\sum_{j=1}^{h}\sum_{k=1}^{w} F_{in}[c, h, w]$. $\boldsymbol{F_{ca}}$ is then concatenated with $\hat{\boldsymbol{h}}$ to get the intermediate vector $\boldsymbol{i_c}$ to compute the *channel-wise* attention mask $\boldsymbol{S_c}$ by several fully connected (FC) layers: $S_c = f_c(\boldsymbol{i_c})$, where f_c represents the FC layers followed by PReLU functions. Finally, we get the output of our *channel-wise* attention module as $\boldsymbol{F_{cout}} = \boldsymbol{S_c} \odot \boldsymbol{F_{in}}$.

Our *channel-wise* attention module learns to map features from the input to the subcarriers based on the estimated CSI, allowing JSCC to dynamically adjust to different channel SNRs. But the spatial information is ignored when computing the channel attention mask by the average pooling operation. So we design a *spatial* attention module to compensate for spatial information.

Spatial Attention Module
Our *spatial* attention module learns to match the more critical spatial features along the h and w dimensions with better channel conditions depending on the estimated CSI.

This time we first apply average pooling $Ave_s(\boldsymbol{F_{cout}})$ on the $\boldsymbol{F_{cout}}$ along the channel direction to get $\boldsymbol{F_{sa}}$, where $Ave_s(\boldsymbol{F_{cout}}) \triangleq \frac{1}{c}\sum_{j=1}^{c} F_{cout}[c, h, w]$. Then, $\hat{\boldsymbol{h}}$ and $\boldsymbol{F_{sa}}$ are concatenated to get the intermediate feature $\boldsymbol{i_s}$, which is used to compute the *spatial* mask $\boldsymbol{S_s}$ by several FC layers. We compute the final output feature vector as $\boldsymbol{F_{out}} = \boldsymbol{S_s} \odot \boldsymbol{F_{cout}}$.

The *spatial* attention module can further improve the PSNR performance by exploiting the spatial information and helping JSCC encoder to do more adaptive power allocation, which matches critical features with better channels.

3.3.2.3 Training and Evaluation

This section presents numerical experiment results to evaluate the performance of our CA-JSCC scheme. The Exp-JSCC scheme introduced in [41] is the most related work to the current study in the literature. We use Exp-JSCC trained on different channel conditions as a benchmark to compare with the proposed CA-JSCC scheme.

Experimental Setup
If not specified otherwise, all experiments were performed on the CIFAR-10 dataset [42] with PyTorch. Models were trained until the performance on a validation set (selected separately from the training dataset) stops improving. The Adam optimizer is used to perform backpropagation.

We set the number of subcarriers to $L_f = 64$. The Zadoff-Chu (ZC) sequence [43], denoted by $\boldsymbol{Y_p} \in \mathbb{C}^{(2,64)}$, is used as the pilot. The values of channel gains $\{H[k] : k = 1, 2, \ldots, L_f\}$ are sampled from a complex Gaussian distribution $\mathcal{CN}(0, 1)$. Unless specified otherwise, the frequency-domain channel responses in the experiments are estimated by an MMSE estimator. We also sort the channels based on their estimated CSI to make training process easier; that is, we have $|H[1]|^2 \geq \cdots \geq |H[L]|^2$. Following [44], we define the bandwidth ratio (i.e., bandwidth usage to source symbol ratio) as $R \triangleq \frac{N_s L_f}{c \times h \times w}$, where $N_s L_f$ is the number of symbols transmitted per image.

Channel-Gain Adaptability
We first verify the adaptability of the CA-JSCC scheme to channel-gain variations. Specifically, under a fixed bandwidth ratio and a given SNR, we want to see if our dual-attention mechanism can instruct the transmitter to exploit better channels and allocate power to different subcarriers judiciously.

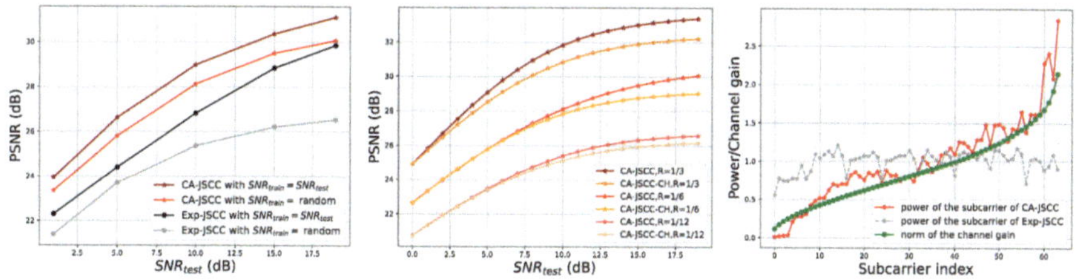

(a) (b) (c)

Fig. 3.14 (**a**) Comparison of our CA-JSCC scheme with Exp-JSCC when the values of the train and test SNRs are the same. (**b**) Ablation experiments for the attention strategy. (**c**) Visualization of the power allocation for the CA-JSCC and the Exp-JSCC

The experimental results are shown in Fig. 3.14a, where we set the number of OFDM symbols to $N_s = 8$ and the bandwidth ratio to $R = 1/6$. The Exp-JSCC and CA-JSCC models are both trained at a fixed SNR, and tested at the same SNR. As can be seen, by feeding the estimated CSI to the transmitter, CA-JSCC can exploit the channel-gain information and adaptively allocate power to different subcarriers. We can see a significant gain compared to Exp-JSCC at all SNRs. This can be attributed to the advantage of our method in better exploiting the channel gains and allocating power to different subcarriers.

SNR Adaptability

Next, we evaluate the SNR adaptability of our scheme. If not specified otherwise, we train the CA-JSCC model at random SNR values of each training episode chosen uniformly from [0, 20] dB and test the well-trained model at different SNRs.

Compared with the CA-JSCC scheme trained at specific SNRs in Fig. 3.14a, we observe that there is a slight performance degradation when it is trained at random SNR values. We conclude that, while CA-JSCC can learn to adapt to different channel SNRs, this flexibility comes at the expense of some loss in the PSNR (up to 1 dB). However, this CA-JSCC model trained at random SNR values still outperforms the Exp-JSCC models trained at specific SNR values.

We also compare the performance of dual-attention based architecture CA-JSCC with an alternative using only channel-wise attention, called (CA-JSCC-CH), as an ablation study. As shown in Fig. 3.14b, for three different bandwidth ratios, CA-JSCC architecture outperforms CA-JSCC-CH at all SNR values, which shows that the *spatial* attention mechanism is essential to achieve the improved performance provided by CA-JSCC. We also observe larger gains by our dual-attention method at higher bandwidth ratios and SNR_{test} values, where more spatial information and better CSI adaptability benefit both feature mapping and power allocation. To visualize the power allocation executed by CA-JSCC and Exp-JSCC, we plot the average channel gain and the average power allocated to each subcarrier in Fig. 3.14c, where we set SNR $= 1$ dB and R=1/6. The channel gains are ordered in an increasing manner in the plot. Compared with Exp-JSCC, Fig. 3.14c shows

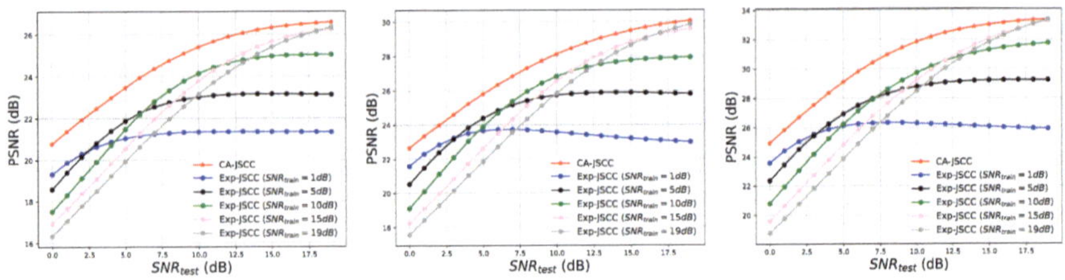

(a) $R = 1/12$ (b) $R = 1/6$ (c) $R = 1/3$

Fig. 3.15 Performance of our CA-JSCC model compared with the Exp-JSCC model of different bandwidth ratios. (**a**) $R = 1/12$. (**b**) $R = 1/6$. (**c**) $R = 1/3$

that CA-JSCC generally allocates more power for the subcarrier with better channel conditions, as one would desire.

It is worth noting that the Exp-JSCC scheme is not SNR-adaptive, which means the training and test SNRs of Exp-JSCC must match to achieve a sound performance, as shown in Fig. 3.14a. Figure 3.15 presents the PSNR versus test SNR results for bandwidth ratios of $R = 1/12, 1/6, 1/3$ (we set $L_f = 64$ and vary N_s to attain different R values). As stated above, our CA-JSCC scheme can be trained with random SNRs, and yields a single model for each bandwidth ratio to be verified on a range of test SNRs. The Exp-JSCC scheme, on the other hand, is trained at five different SNRs, yielding five different models under each bandwidth ratio. The Exp-JSCC scheme performs the best when the training and test SNRs match. However, our CA-JSCC scheme is SNR-adaptive and consistently outperforms Exp-JSCC at all SNRs and bandwidth ratios by a considerable margin.

Additional experiments by training over the ImageNet dataset is shown in Fig. 3.16a. We train the models with randomly cropped 64×64 patches from ImageNet, and evaluate the models on the Kodak dataset. Results show that training on a sufficiently large dataset (ImageNet) can allow our CA-JSCC model to perform well on a never-seen dataset (Kodak). CA-JSCC can still achieve state-of-the-art performance with the additional capability of channel adaptability.

Impact of CSI Estimation Errors

In the above experiments, we have assumed MMSE estimated CSI. In this subsection, we look into the effect of channel estimation errors on the performance of CA-JSCC. We repeat the experiment in Fig. 3.15b with three types of CSI: (i) perfect CSI, H_{per}; (ii) MMSE estimated CSI, H_{mmse}; and (iii) LS estimated CSI, H_{ls}. We remark that H_{mmse} provides a more accurate estimate than H_{ls}.

The experimental results are shown in Fig. 3.16b, where we train our CA-JSCC models with H_{mmse} and H_{per}, respectively, and evaluate these two models with H_{per}, H_{mmse} and H_{ls}, respectively. As expected, the model trained and tested with H_{per} achieves the best performance. On the other hand, the model trained with perfect CSI is not robust to CSI errors during test time. Its performance gets worse as the quality of channel estimation degrades. Instead, we see that models trained

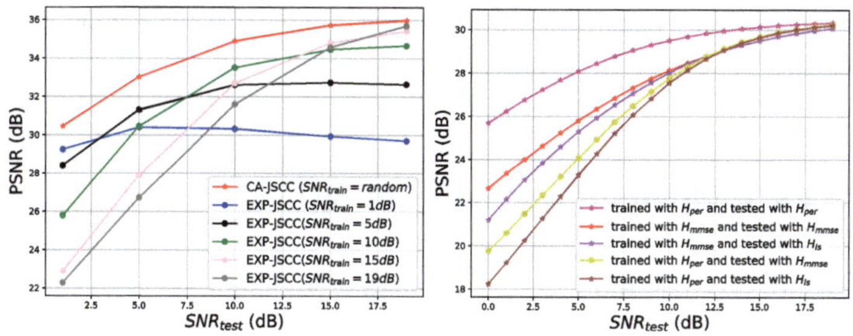

(a) (b)

Fig. 3.16 (**a**) Additional experiments with training on the ImageNet dataset tested on the Kodak dataset. (**b**) Comparison of CA-JSCC with different channel estimation methods

with H_{mmse} perform better, since during training they learn to compensate for CSI estimation errors. We conclude from these results that a more accurate CSI during testing is generally beneficial, and the performance improves if training is done with the same type of CSI.

3.3.2.4 Summary

In this section, we have explored the transformative potential of DeepJSCC in the context of IoT communications, focusing on the design and implementation of "language" systems that integrate source and channel coding into a unified framework. By leveraging advanced DL techniques, we presented the CA-JSCC scheme, a model that exemplifies the practical application and benefits of DeepJSCC in handling complex wireless transmission scenarios over multipath fading channels.

The CA-JSCC scheme, with its dual-attention mechanism, adapts dynamically to the fluctuations in channel conditions, utilizing the available Channel State Information (CSI) to optimize the allocation of both data features and power resources across OFDM subcarriers. This method not only enhances the efficiency and reliability of data transmission but also ensures that crucial data features are matched with the most reliable transmission paths, thereby maximizing the quality of service in varied SNR and bandwidth environments.

Through rigorous numerical experiments, we demonstrated that the CA-JSCC model achieves state-of-the-art performance, surpassing traditional approaches in both efficiency and adaptability. This is especially significant in the IoT space, where devices often operate under stringent power constraints and in dynamically changing environments.

Overall, the section underscores the importance of innovative language design in IoT communications and establishes DeepJSCC as a cornerstone technology for future developments in this field. By integrating deep learning into the core of JSCC, we pave the way for more intelligent, robust, and scalable communication systems that can keep pace with the increasing complexity and demands of the IoT ecosystem.

3.4 PAPR Reduction: A Practical Concern

An important ingredient of DeepJSCC systems is discrete-time analog transmission (DTAT) [1, 8, 16, 35], because DNN-based encoder and decoder are capable of exploiting discrete-time continuous amplitude signals, yielding more freedom than discrete constellations. The practical use of DTAT, however, faces an important challenge due to the peak-to-average power ratio (PAPR) [45], especially when used in conjunction with the prevailing OFDM transceiver, which is now adopted in most IEEE standards. PAPR stems from the saturation effect of the power amplifier: a low PAPR of the transmitted signal is desired as the power amplifier can operate more efficiently and the coverage of the transmission is larger. Said in another way, for a given communication range, low-PAPR signals save the transmission power. Thus, PAPR is more critical in the uplink transmission of a mobile communication system since the battery power of mobile users is limited.

In traditional digital communications, the OFDM signal exhibits a large PAPR since independent QAM-modulated waveforms are linearly combined. To address this problem, various methods, such as clipping, coding, scrambling, and linear precoding, have been proposed in the literature [45]. In addition to traditional schemes, DL-based approaches to reducing the PAPR of OFDM signals are proposed in [46–48]. Specifically, these works focus on the transmission of independent and identically distributed (i.i.d.) constellations. The OFDM system is modeled and trained in an end-to-end fashion, and the main idea is to incorporate the PAPR into the loss function in addition to the original decoding loss, e.g., bit error rate (BER). In so doing, the DL-based transceiver can learn to generate low-PAPR signals as training progresses. In DeepJSCC-enabled semantic communications, the coded symbols have continuous amplitude. The PAPR of the transmitted signal after OFDM modulation is even more severe than that in digital communications because both DTAT and OFDM modulation contribute to the high PAPR.

In this section, we study the PAPR performance of DeepJSCC-enabled semantic communications considering the uplink transmission from mobile users to the base station in cellular networks. The cellular network is operated with orthogonal frequency-division multiple access (OFDMA), where the radio channel is divided into multiple subchannels/subcarriers and each user is allocated with a subset of subcarriers. We develop a passband transceiver for the OFDMA system and evaluate the PAPR at the passband—unlike above-mentioned works that consider only baseband signals.

We show that there is a trade-off between achieving low-PAPR transmission and high-quality reconstruction. Focusing on the application of semantic image transmission, we investigate three PAPR reduction techniques, i.e., linearly precoded OFDMA, clipping, and PAPR loss, and characterize the trade-off between

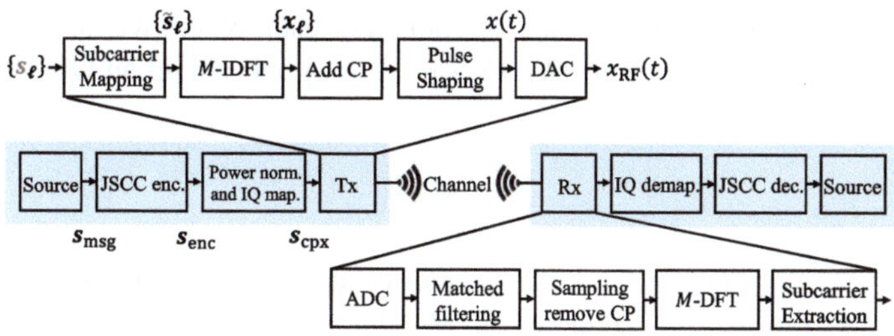

Fig. 3.17 DeepJSCC-enabled semantic communications with a passband transceiver

PAPR and PSNR of the reconstructed image by means of the relative operating characteristic (ROC) curve.

3.4.1 The Passband OFDMA Transceiver

We consider the uplink transmission from a mobile user to the base station in an OFDMA cellular network. The radio channel is divided into M subcarriers indexed by $\mathcal{M} = \{1, 2, 3, \ldots, M\}$ and the mobile user is allocated with N ($N \leq M$) subcarriers indexed by $\{k_n : k_n \in \mathcal{M}, n = 1, 2, \ldots, N\}$. The goal is to deliver a source message, which can be a bitstream, an image, a text, a video, etc., through the physical wireless channel. The DeepJSCC-enabled semantic communication system is illustrated in Fig. 3.17.

At the transmitter, we denote the source message by a vector s_{msg} of length L_s; the elements of s_{msg} can be continuous or discrete. After DeepJSCC encoding, the source message is transformed to a vector of real coded symbols $s_{\text{enc}} \in \mathcal{R}^{L_e \times 1}$. To align with prior works, we define the system bandwidth ratio as $R \triangleq L_e/2L_s$. Note that when the source message is a sequence of i.i.d. bits, the DeepJSCC encoder reduces to a deep channel encoder.

The encoded symbols are subject to an average power constraint P. Thus, we normalize s_{enc} such that the average power of the coded symbols generated from a source message is P, yielding

$$s_{\text{norm}} = \sqrt{P} \frac{s_{\text{enc}} - \mathbb{E}[s_{\text{enc}}]}{\|s_{\text{enc}} - \mathbb{E}[s_{\text{enc}}]\|_2}, \tag{3.48}$$

where $\mathbb{E}[s_{\text{enc}}]$ is the mean of s_{enc} and $\| * \|_2$ denotes the L2 norm of a vector. Given the power normalized real vector $s_{\text{norm}} \in \mathcal{R}^{L_e \times 1}$, we construct a complex vector $s_{\text{cpx}} \in \mathcal{C}^{L_e/2 \times 1}$, where the real and imaginary components of s_{cpx} are the odd and even elements of s_{norm}.

Since the transmitter is allocated with N subcarriers, we partition the complex vector s_{cpx} into $L = \lceil L_e/2N \rceil$ blocks, denoted by $\{s_\ell : \ell = 0, 1, 2, \ldots, L-1\}$, and each block s_ℓ consists of N complex symbols. For each block, we map $s_\ell \in C^{N\times 1}$ onto the N allocated subcarriers $\{k_1, k_2, k_3, \ldots, k_N\}$ and obtain an OFDM symbol by

$$x_\ell = F_M^H \widetilde{s}_\ell \tag{3.49}$$

where $x_\ell \in C^{N\times 1}$ is the time-domain samples of the OFDM symbol generated from the ℓ-th block; $F_M \in C^{M\times M}$ denotes the M-dimensional discrete Fourier transform (DFT) matrix, hence F_M^H is the inverse DFT (IDFT) matrix; $\widetilde{s}_\ell \in C^{M\times 1}$ is a vector of M complex symbols given by

$$\widetilde{s}_\ell[m] = \begin{cases} s_\ell[n], & \text{if } m = k_n, \\ 0, & \text{otherwise.} \end{cases} \tag{3.50}$$

Note that the average power of s_ℓ is $2P$ and the average power of $\widetilde{s}_\ell$ and x_ℓ is $\frac{2N}{M}P$.

Next, we add cyclic prefix (CP) to the OFDM symbols, yielding $\widetilde{x}_\ell \in C^{(M+L_{\text{cp}})\times 1}$, where L_{cp} is the CP length. After pulse shaping, the baseband continuous-time signal can be written as

$$x(t) = \sum_{\ell=0}^{L-1} \sum_{k=0}^{M+L_{\text{cp}}-1} \widetilde{x}_\ell[k] p(t - kT - \ell T_{\text{OFDM}}), \tag{3.51}$$

in which T is the baseband sampling period (i.e., the baseband baud rate is $1/T$); $T_{\text{OFDM}} = (M + L_{\text{CP}})T$ is the OFDM symbol duration; and $p(t)$ is the root-raised-cosine (RRC) pulse with a roll-off factor β.

Finally, the passband signal can be constructed by

$$x_{\text{RF}}(t) = \text{Re}\left\{x(t)e^{j2\pi f_c t}\right\}, \tag{3.52}$$

where Re$\{*\}$ denotes the real component of a given signal and f_c is the carrier frequency.

After passing through the wireless channel, the received signal is given by

$$r_{\text{RF}}(t) = h(t) \otimes x_{\text{RF}}(t) + w(t), \tag{3.53}$$

where $h(t)$ is the real channel response function; $\otimes$ denotes the linear convolution operation; and $w(t)$ is additive white Gaussian noise (AWGN) with a double-sided power spectral density of N_0. The transmit signal-to-noise ratio (SNR) is defined as $\eta \triangleq P/N_0$. To study the PAPR of DL-enabled semantic communications, this section focuses on the AWGN channel case with $h(t) = 1$. The extensions to fading channels are straightforward.

As shown in Fig. 3.17, the receiver down converts $r_{\text{RF}}(t)$ to the baseband, matched filters and samples the baseband signal; performs OFDM demodulation,

IQ demapping, and DeepJSCC decoding (or deep channel decoding when the source message is a sequence of i.i.d. bits) to recover the transmitted message. Denote by $\widehat{s}_{\text{msg}}$ the reconstructed source message. The communication system is trained end-to-end to minimize a loss function $\mathcal{L}$ between s_{msg} and $\widehat{s}_{\text{msg}}$. Depending on whether s_{msg} is continuous or discrete, typical loss functions are the mean-squared error (MSE) and the cross-entropy loss.

3.4.2 *PAPR of DTAT*

Achieving high reconstruction performance and low PAPR is a trade-off, because a low-PAPR transmitted signal offers a limited degree of freedom for message reconstruction.

DeepJSCC-enabled semantic communications often exhibit much better message reconstruction performance than digital communications, but the PAPR of the transmitted signal is also much larger. A natural question that arises is whether the reconstruction gains of DeepJSCC come from the higher-PAPR signal. Said in another way, can the excellent reconstruction performance be retained if the PAPR of the transmitted signal is bounded to levels permissible for practical implementations? Our main objective in this section is to answer the above questions and characterize the trade-off between the reconstruction performance and the PAPR performance of DeepJSCC with DTAT.

We measure the PAPR performance of a communication system by the PAPR of the passband signal $x_{\text{RF}}(t)$:

$$\rho = \frac{\max |x_{\text{RF}}(t)|^2}{\mathbb{E}\left[|x_{\text{RF}}(t)|^2\right]}. \tag{3.54}$$

The complementary cumulative distribution function (CCDF) of ρ is defined as $\Pr(\rho > \Gamma)$ for a threshold Γ (i.e., the tail distribution). As a PAPR performance indicator, we shall use the 99.9-percentile PAPR, denoted by Γ_{-3}, where $\Pr(\rho > \Gamma_{-3}) = 10^{-3}$.

3.4.2.1 PAPR Analysis

For a well-trained DeepJSCC-based communication system, the PAPR of the passband signal is non-trivial to derive in general, because it depends on the source message, the channel distribution, and the learning process, which are elusive to characterize analytically. A simple analysis is given below by assuming that the DeepJSCC coded symbols s_{enc} follow i.i.d. Gaussian distributions [16, 49].

Under this assumption, s_{enc} is a multivariate Gaussian random vector. After power normalization, the vector s_{norm} follows $s_{\text{norm}} \sim \mathcal{N}(\mathbf{0}, P\boldsymbol{I})$, where $\mathbf{0}$ is an all-zero vector and $\boldsymbol{I}$ is the identity matrix. After IQ mapping, partition, and subcarrier mapping, each block of complex symbols $\boldsymbol{s}_\ell \sim \mathcal{CN}(\mathbf{0}, 2P\boldsymbol{I})$ and $\widetilde{\boldsymbol{s}_\ell} \sim \mathcal{CN}(\mathbf{0}, 2P\boldsymbol{\Lambda})$, where $\boldsymbol{\Lambda}$ is an M-dimensional diagonal matrix. In particular, the m-th diagonal element of $\boldsymbol{\Lambda}$ is 1 if $m \in \{k_1, k_2, k_3, \ldots, k_N\}$ and 0 otherwise. After IDFT, the time-domain OFDM samples follow $\boldsymbol{x}_\ell \sim \mathcal{CN}(\mathbf{0}, 2P\boldsymbol{F}^H\boldsymbol{\Lambda}\boldsymbol{F})$. In particular, each element of $\boldsymbol{x}_\ell$ follows $\boldsymbol{x}_\ell[m] \sim \mathcal{CN}(\mathbf{0}, \frac{2N}{M}P)$ and $\|\boldsymbol{x}_\ell[m]\|_2^2 \sim \exp(\frac{M}{2NP})$, $\forall m$. The average power of $\boldsymbol{x}_\ell$ is $\frac{2N}{M}P$, hence, the PAPR of $\boldsymbol{x}_\ell$ is given by

$$\rho(\boldsymbol{x}_\ell) = \frac{M}{2NP} \max_m \|\boldsymbol{x}_\ell[m]\|_2^2. \tag{3.55}$$

The CCDF of $\rho(\boldsymbol{x}_\ell)$ can be approximated by

$$\begin{aligned}
\Pr(\rho(\boldsymbol{x}_\ell) > \Gamma) &= \Pr\left(\max_m \|\boldsymbol{x}_\ell[m]\|_2^2 > \frac{2NP}{M}\Gamma\right) \\
&\approx 1 - \prod_{m=0}^{M-1} \Pr\left(\|\boldsymbol{x}_\ell[m]\|_2^2 < \frac{2NP}{\kappa_1 M}\Gamma\right) \\
&= 1 - \left(1 - e^{-\frac{\Gamma}{\kappa_1}}\right)^M.
\end{aligned} \tag{3.56}$$

where the approximation comes from the assumption that the elements of $\boldsymbol{x}_\ell$ are independent, while they are in fact correlated since the covariance of $\boldsymbol{x}_\ell$ is $2P\boldsymbol{F}^H\boldsymbol{\Lambda}\boldsymbol{F}$. To compensate for the loss of this independent assumption, a hyperparameter κ_1 is introduced. It is worth noting that the independent assumption is valid when $N = M$, since $\boldsymbol{x}_\ell \sim \mathcal{CN}(0, 2P\boldsymbol{I})$, in which case we set $\kappa_1 = 1$.

For the passband signal $x_{\mathrm{RF}}(t)$, the CCDF of ρ in (3.54) can be approximated by that of oversampled baseband signal $\boldsymbol{x}_\ell$ [50], yielding,

$$\rho \approx 1 - \left(1 - e^{-\frac{\Gamma}{\kappa_1}}\right)^{\kappa_2 M}, \tag{3.57}$$

where κ_2 is another hyperparameter. Both κ_1 and κ_2 can be found by parametric fitting.

3.4.2.2 PAPR Reduction

A low PAPR is essential for achieving high power efficiency at the transmitter. This section explores three PAPR reduction techniques and characterizes the trade-off between message reconstruction and PAPR with these techniques.

Linearly Precoded OFDMA

In digital communications, an efficient scheme to reduce the PAPR of the transmitted signal without compromising the BER performance is linearly precoded

OFDMA (a.k.a., single carrier FDMA) [51, 52], which is standardized in both 4G LTE and 5G NR for uplink low-PAPR transmission.

Linearly-precoded OFDMA generates low-PAPR waveforms by changing the signal modulation and subcarrier allocation schemes. Specifically, with OFDMA, the N allocated subcarriers are randomly distributed over $\mathcal{M}$ and a block of complex symbols $\boldsymbol{s}_\ell \in \mathcal{C}^{N\times 1}$ is directly mapped onto the N subcarriers. With linearly-precoded OFDMA, on the other hand, we first DFT-precode $\boldsymbol{s}_\ell$, yielding

$$\boldsymbol{s}'_\ell = \boldsymbol{F}_N \boldsymbol{s}_\ell, \tag{3.58}$$

and then map the precoded symbols $\boldsymbol{s}'_\ell$ onto the subcarriers. In particular, the N allocated subcarriers are localized (contiguous) or interleaved (evenly-spaced) over the spectrum. With localized and interleaved subcarrier allocations, the corresponding linear precoded OFDMA systems are called LFDMA and IFDMA, respectively.

In digital communications, the BER performances of OFDMA, LFDMA, and IFDMA are exactly the same, but the IFDMA waveform exhibits the lowest PAPR [53, 54]. The reason is as follows. Consider a block of discrete constellations $\boldsymbol{s}_\ell$. After N-DFT precoding, interleaved subcarrier mapping, and M-IDFT modulation, it can be shown that the resulting signal is given by

$$\boldsymbol{x}_\ell[m] = \frac{N}{M} e^{j\frac{2\pi m}{M}} \boldsymbol{s}_\ell[m \bmod N], \tag{3.59}$$

for $m = 0, 1, 2, \ldots, M-1$. That is, $\boldsymbol{x}_\ell$ can be constructed directly from $\boldsymbol{s}_\ell$ by simple repetition and frequency upshifting. This implies that the IFDMA waveform exhibits nearly the same PAPR properties as the discrete constellations $\boldsymbol{s}_\ell$, and hence, has much lower PAPR than that of LFDMA and OFDMA.

A visual illustration is given in Fig. 3.18a, where we simulate a digital communication system with QPSK and 16QAM modulations. As can be seen, with 16QAM, the 99.9-percentile PAPR Γ_{-3} of OFDMA is improved by 2 dB and 9 dB

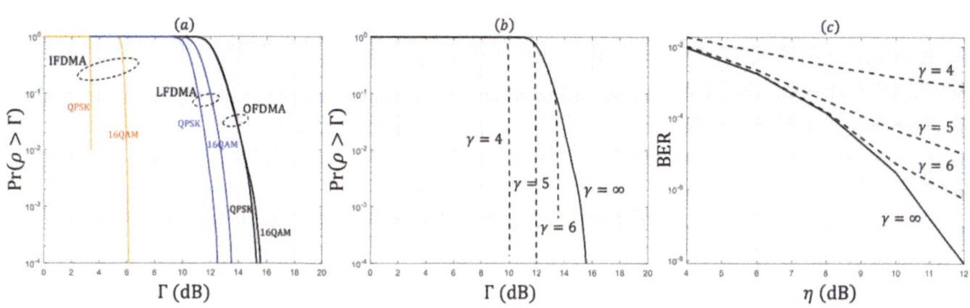

Fig. 3.18 PAPR reduction in digital communications. The system configurations are given in Table 3.5. (**a**) PAPR performance of linearly precoded OFDMA; (**b**) PAPR performance with clipping, where the modulation is 16QAM, no channel code is considered, and $\gamma = \infty$ means no clipping. (**c**) Impact of clipping on the BER performance

with LFDMA and IFDMA, respectively. The performance gains are even larger when lower-order modulations are used. For QPSK, the Γ_{-3} gains of LFDMA and IFDMA over OFDMA are up to 3 dB and 11.5 dB, respectively.

Efficient Transceiver Design and Subcarrier Allocation for IFDMA

While IFDMA offers significantly lower PAPR compared to LFDMA and OFDMA, it is often perceived to impose rigid constraints on subcarrier allocation due to its interleaved structure. This perception has led to the belief that IFDMA requires more complex transceiver designs and subcarrier allocation algorithms. However, we show in Appendix A that these concerns are largely misconceptions. The seemingly rigid interleaved constraint actually reveals an elegant mathematical structure that can be leveraged to simplify, rather than complicate, the transceiver design and subcarrier allocation processes.

Clipping

Clipping the large amplitude of $x_{\mathrm{RF}}(t)$ is a straightforward PAPR reduction scheme [55]. Specifically, let $\bar{x}_{\mathrm{RF}} = \mathbb{E}\left(|x_{\mathrm{RF}}(t)|\right)$ be the average amplitude of $x_{\mathrm{RF}}(t)$. We clip $x_{\mathrm{RF}}(t)$ such that the magnitude of the clipped signal does not exceed a threshold $\gamma\bar{x}_{\mathrm{RF}}$, where γ is a clipping ratio measuring the severity of clipping. The clipped signal can be written as

$$x_{\mathrm{clip}}(t) = \begin{cases} x_{\mathrm{RF}}(t), & \text{if } |x_{\mathrm{RF}}(t)| \le \gamma\bar{x}_{\mathrm{RF}}; \\ \gamma\bar{x}_{\mathrm{RF}}, & \text{if } |x_{\mathrm{RF}}(t)| > \gamma\bar{x}_{\mathrm{RF}}. \end{cases} \tag{3.60}$$

Figure 3.18b presents the PAPR reduction performance of clipping in digital OFDMA communication systems, where the modulation is 16QAM and various clipping ratios are considered. As shown, increasingly lower PAPR can be obtained as we decrease γ. When $\gamma = 5$, for example, the 99.9-percentile PAPR is improved by 3.1 dB. Clipping, however, causes both out-of-band radiation and in-band inter-carrier interference (ICI). The out-of-band radiation can be addressed by filtering according to the spectrum mask. The in-band ICI, on the other hand, leads to an inevitable BER loss since the orthogonality among subcarriers is destroyed. Fig. 3.18c shows the impact of clipping on the BER. When $\gamma = 5$, the BER performance deteriorates for 2.7 dB to achieve a BER of 10^{-5}.

PAPR Loss

Since our goal is to minimize both the reconstruction error and the PAPR of $x_{\mathrm{RF}}(t)$, a natural idea is to add a PAPR loss to the original reconstruction loss [46], and minimizing

$$\mathcal{L}' = \mathcal{L} + \lambda\mathbb{E}[\rho], \tag{3.61}$$

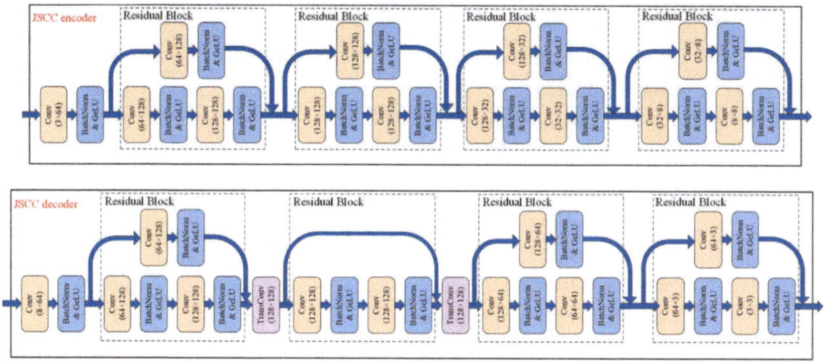

Fig. 3.19 The DeepJSCC encoder and decoder architectures for semantic image transmission

where λ is a hyperparameter. With this new loss function, the communication system learns to minimize both metrics simultaneously, where λ determines the operating points on the trade-off curve between MSE and PAPR.

3.4.3 Simulation Results

3.4.3.1 Transmission of CIFAR10 Images

This section analyzes the PAPR performance of DL-enable semantic communications considering the wireless transmission of CIFAR10 images. In this application, the source messages are assumed to be colored images from the CIFAR10 dataset. To retain the spatial feature of images, we denote a source message by a three-dimensional matrix $\boldsymbol{S}_{\text{msg}} \in \mathcal{R}^{32\times32\times3}$. If reshaped to a vector $\boldsymbol{s}_{\text{msg}} \in \mathcal{R}^{L_s\times1}$, we have $L_s = 32 \times 32 \times 3$. The DeepJSCC encoder is designed to be a residual convolutional neural network (CNN) with convolutional layers, batch normalization (BN) layers, and Gaussian error linear unit (GeLU) activation functions. The detailed network architecture is depicted in Fig. 3.19. For a given compression ratio R, the source image is transformed to a feature matrix $\boldsymbol{S}_{\text{enc}} \in \mathcal{R}^{8\times8\times96R}$ and then reshaped to a vector $\boldsymbol{s}_{\text{enc}} \in \mathcal{R}^{L_e\times1}$ with $L_e = 8 \times 8 \times 96R$.

Following the signal flow defined in the passband transceiver, we process $\boldsymbol{s}_{\text{enc}}$ by power normalization, IQ mapping, and Tx modulation at the transmitter; Rx demodulation and IQ demapping at the receiver, after which we obtain the noisy feature matrix $\widehat{\boldsymbol{S}}_{\text{enc}} \in \mathcal{R}^{8\times8\times96R}$. Then, we feed $\widehat{\boldsymbol{S}}_{\text{enc}}$ into the DeepJSCC decoder to reconstruct the source image $\widehat{\boldsymbol{S}}_{\text{msg}} \in \mathcal{R}^{32\times32\times3}$. The DeepJSCC decoder is designed to be a residual network with transposed convolutional layers, BN layers, and GeLU activation functions, as depicted in Fig. 3.19.

Table 3.5 Hyperparameter settings

	Hyper parameters	Symbols	Values
System	Bandwidth ratio	R	$1/12$
	Total number of subcarriers	M	128
	Number of allocated subcarriers	N	64
	Length of CP	L_{cp}	16
	Roll-off factor of RRC	β	0.5
	Carrier frequency	f_c	25 MHz
	Baseband baud rate	$1/T$	1 MHz
	Baseband oversampling rate		10 MHz
Learning	Number of training epochs		100
	Batch size		256
	Learning rate		10^{-3}
	Weight decay		5×10^{-3}
	Optimizer		adamW

The communication goal is image reconstruction and we aim to maximize the PSNR of the reconstructed image:

$$\text{PSNR} = 10 \log \frac{L_s}{\mathbb{E}\|\widehat{\boldsymbol{S}}_{\text{msg}} - \boldsymbol{S}_{\text{msg}}\|_F^2}, \tag{3.62}$$

where $\| * \|_F$ denotes the Frobenius norm of a matrix. To this end, the loss function is chosen to be the MSE loss

$$\mathcal{L} = \mathbb{E}\left[\frac{1}{L_s}\|\widehat{\boldsymbol{S}}_{\text{msg}} - \boldsymbol{S}_{\text{msg}}\|_F^2\right]. \tag{3.63}$$

As a baseline, we train the semantic image transmission system with OFDMA in an end-to-end fashion to minimize (3.63) under various transmit SNRs η from 0 dB to 20 dB. The system and learning configurations are summarized in Table 3.5 unless otherwise specified. The PSNR versus η performance is presented in Fig. 3.20a, where the benchmarks are the traditional digital communication systems with better portable graphics (BPG) image compression, low-density parity-check code (LDPC) with different rates, and various modulation schemes. As shown, the PSNR of DeepJSCC outperforms that of digital communication. At an SNR of 10 dB, DeepJSCC achieves a PSNR of 27.83, while the best PSNR achieved by digital communication is 22.26.

Let us further evaluate the PAPR performance of both systems. For digital communication, the best PSNR at $\eta = 10$ dB is achieved by BPG, 16QAM, and LDPC with a $1/2$ rate, the 99.9-percentile PAPR of which is about 15 dB with OFDMA. On the other hand, the 99.9-percentile PAPR of DeepJSCC is up to 21 dB with OFDMA. This indicates that, despite the excellent PSNR performance, PAPR is a limitation for the practical implementation of DeepJSCC-enabled semantic communication. PAPR reduction techniques must be introduced.

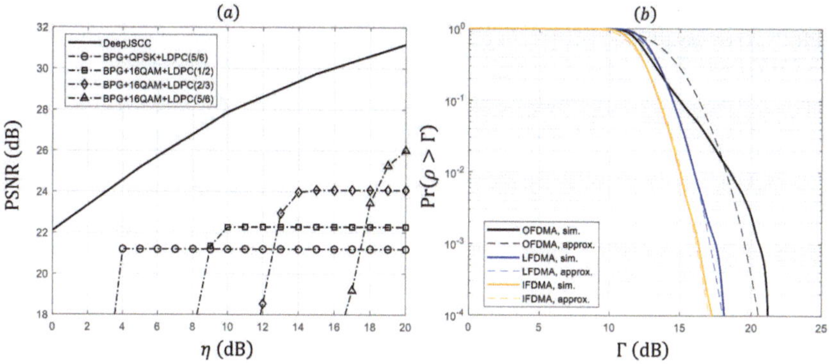

(a) (b)

Fig. 3.20 PSNR and PAPR performances of DeepJSCC. (**a**) PSNR under various SNR η; (**b**) CCDF of PAPR for the model trained at $\eta = 10$ dB. The dashed curves are plotted using (3.57)

3.4.3.2 DeepJSCC with PAPR Reduction

To reduce the PAPR of the continuous-amplitude signal, this section applies the three PAPR reduction techniques discussed in Sect. 3.4.2.2 and studies the PSNR versus PAPR trade-off for DeepJSCC-enabled semantic communications.

Linearly-Precoded OFDMA

We first study the PAPR reduction capability of linearly precoded OFDMA. At an SNR of 10 dB, the CCDF of PAPR is presented in Fig. 3.20b. We have two main observations,

- With linearly-precoded OFDMA, the PAPR reduction in DeepJSCC is not as significant as that in digital communication: the 99.9-percentile PAPR of OFDMA is only reduced by 3.5 dB and 4.5 dB with LFDMA and IFDMA, respectively. This observation is not surprising, because DeepJSCC outputs continuous amplitude signals as opposed to discrete constellations.
- With IFDMA, both DeepJSCC and digital communications exhibit the lowest PAPR. The 99.9-percentile PAPR of DeepJSCC is 10.5 dB larger than that of digital communication with 16QAM modulation.

PAPR Loss

To further reduce the PAPR, we introduce an additional PAPR loss to the PSNR loss. The new loss function is given by

$$\mathcal{L}' = \mathbb{E}\left[\frac{1}{L_s}\|\widehat{\boldsymbol{S}}_{\text{msg}} - \boldsymbol{S}_{\text{msg}}\|_F^2\right] + \lambda\mathbb{E}[\rho]. \tag{3.64}$$

With different λ, DeepJSCC can balance PSNR and PAPR. To illustrate the trade-off, we plot the ROC curve in Fig. 3.21. The ROC space is defined by

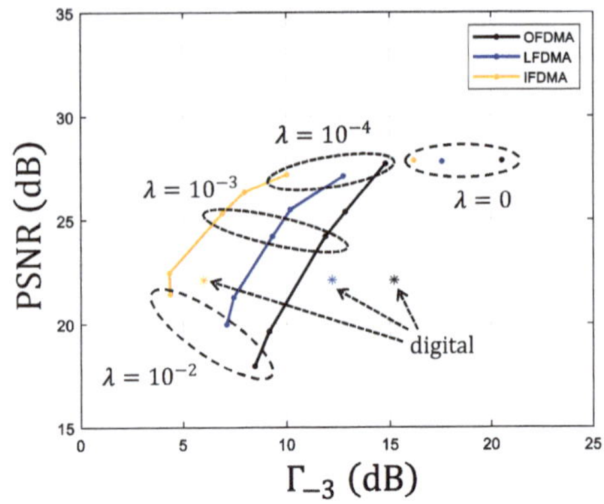

Fig. 3.21 PSNR versus PAPR (Γ_{-3}) of DeepJSCC with PAPR loss

the 99.9-percentile PAPR Γ_{-3} (in dB) and PSNR (in dB) as the x and y axis. The performance of a DeepJSCC-based semantic communication system can be represented by one point in the ROC space and the best system yields a point in the upper left corner of the space, in which case the PSNR is maximized and the PAPR is minimized.

We have two main observations from Fig. 3.21.

- With the increase in λ, PAPR weighs more than MSE in (3.64). As a result, the well-trained DeepJSCC model exhibits lower PAPR but also lower PSNR.
- IFDMA exhibits a better trade-off than LFDMA and OFDMA, thanks to its simpler signal structure at the transmitter. That is, with IFDMA, it is easier to learn a communication system with low PAPR and high PSNR by minimizing (3.64). When $\lambda = 10^{-4}$, for example, IFDMA, LFDMA, and OFDMA yield almost the same PSNR performance, but the 99.9-percentile PAPR of IFDMA is about 2.5 dB and 4.9 dB less than that of LFDMA and OFDMA, respectively.

Clipping

Finally, we evaluate the performance of clipping when applied to DeepJSCC-based image transmission. Figure 3.22a characterizes the PSNR and PAPR trade-off when we clip $x_{\rm RF}(t)$ of a well-trained DeepJSCC encoder according to (3.60). Note that the signal after clipping has to be re-normalized such that the power of $\widetilde{x}_{\rm clip}(t)$ is the same as that of $x_{\rm RF}(t)$, that is,

$$\widetilde{x}_{\rm clip}(t) = \sqrt{\frac{x_{\rm RF}^2(t)}{x_{\rm clip}^2(t)}}\, x_{\rm clip}(t). \tag{3.65}$$

As shown in Fig. 3.22a, clipping can reduce the PAPR by a large margin without compromising too much PSNR performance. When $\gamma = 3$, the 99.9-percentile PAPR of OFDMA is reduced by 13.5 dB, while the PSNR only deteriorates by 0.45 dB. When $\gamma = 2$, the 99.9-percentile PAPR of OFDMA is reduced by 16 dB, while the PSNR only deteriorates by 1.58 dB. In other words, the DeepJSCC

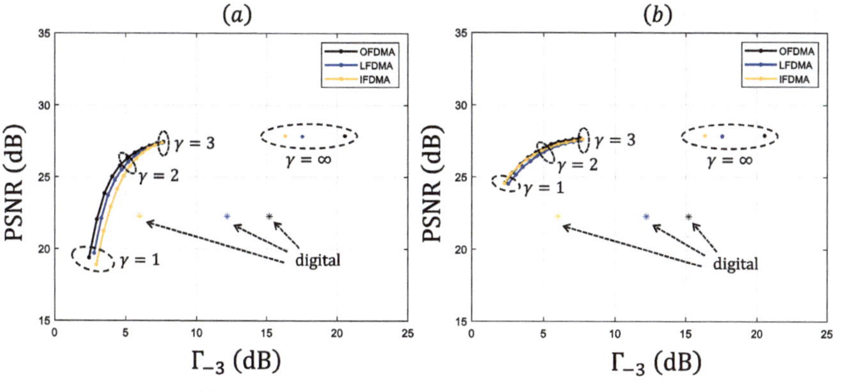

(a) (b)

Fig. 3.22 PSNR versus PAPR (Γ_{-3}) of DeepJSCC with clipping: (**a**) clipping without retraining; (**b**) clipping with retraining

decoder is robust to the severe ICI caused by clipping, and a good trade-off between PSNR and PAPR can be achieved.

Moreover, we can incorporate the clipping operation into the training of the communication system and further enhance the robustness of the trained system to clipping. To this end, we implement a trainable clipping operation by

$$\widetilde{x}_{\text{clip}}(t) = x_{\text{RF}}(t)\left(1 - \frac{\text{ReLU}\left(|x_{\text{RF}}(t)| - \gamma \bar{x}_{\text{RF}}\right)}{|x_{\text{RF}}(t)| + \varepsilon}\right), \tag{3.66}$$

where $\varepsilon = 10^{-8}$ is a small constant for numerical stability. Figure 3.22b characterizes the PSNR and PAPR trade-off when clipping is incorporated into the training phase. As shown,

- Compared with Fig. 3.22a, DeepJSCC is even more robust to small clipping ratios. When $\gamma = 1$, the 99.9-percentile PAPR of OFDMA is reduced by 18.5 dB, while the PSNR loss is only 3.27 dB.
- There is no remarkable difference between OFDMA and linearly-precoded OFDMA after clipping. Thus, the additional DFT (IDFT) operation at the transmitter (receiver) is unnecessary.
- Suppose the 99.9-percentile PAPR of 16QAM and IFDMA in digital communications is an acceptable PAPR performance, i.e., $\Gamma_{-3} = 6$ dB. With clipping, the PSNR is up to 27.42 when the target 99.9-percentile PAPR is 6 dB.

To confirm the above results, we further extend our experiments to ImageNet and evaluate the impact of clipping with retraining on the PAPR and PSNR performances. ImageNet consists of 1.2 million high-resolution images. In the training phase, we randomly sample mini-batches of size 32 from ImageNet and crop each image to size 128 × 128. As the experiments on CIFAR10, the bandwidth

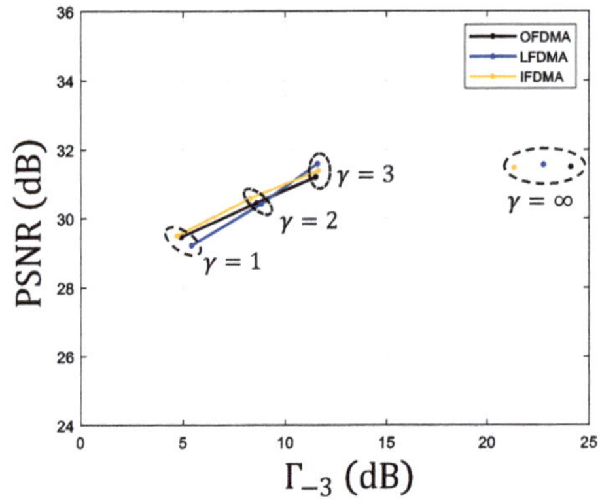

Fig. 3.23 PSNR versus PAPR (Γ_{-3}) of DeepJSCC with clipping (with retraining) on ImageNet

ratio is fixed to $1/12$ and each transmission packet consists of 256 complex symbols (4 OFDM symbols). The number of training epochs is fixed to 5. In the test phase, we evaluate the trained DeepJSCC model on the Kodak dataset, which consists of 24 images of size 768×512. The experimental results are shown in Fig. 3.23. As can be seen, setting $\gamma = 1$ satisfies the target 99.9-percentile PAPR of 6 dB, while the PSNR degradation is only 2 dB compared with the no clipping case.

3.5 Concluding Remarks

AI-empowered semantic communication marks a fundamental shift in how IoT networks exchange and interpret information. Traditional communication paradigms, which focus solely on data accuracy, are inadequate for modern IoT systems that require context-aware, efficient, and scalable information exchange mechanisms. This chapter has introduced the conceptual foundations of semantic communication, emphasizing how AI can transform IoT networks by enabling intelligent meaning extraction, adaptive communication strategies, and optimized resource allocation.

Through a structured exploration of semantic encoding, decoding, and DeepJSCC techniques, we have demonstrated how AI can enhance the efficiency of IoT networks by reducing transmission overhead and improving interpretability. The discussion on DeepJSCC and AI-driven compression methods has illustrated how deep learning can bridge the gap between Shannon's theory and real-world communication needs, ensuring that messages are not only transmitted accurately but also understood effectively.

Looking ahead, semantic communication will be pivotal in next-generation IoT architectures, enhancing the ability of smart environments, autonomous agents, and industrial networks to collaborate with minimal human intervention. The integration of AI and semantic communication will not only enhance system intelligence but also redefine the very nature of connectivity, paving the way for truly intelligent, goal-driven IoT ecosystems.

References

1. H. Xie, Z. Qin, G.Y. Li, B.-H. Juang, Deep learning enabled semantic communication systems. IEEE Trans. Signal Process. **69**, 2663–2675 (2021)
2. Y. Shao, Q. Cao, D. Gündüz, A theory of semantic communication. IEEE Trans. Mob. Comput. **23**(12), 12211–12228 (2024)
3. Q. Cao, Y. Shao, S. Ge, On the zero-error capacity of semantic channels with input and output memories. IEEE Wirel. Commun. Lett. **14**(3), 896–900 (2025)
4. H. Cui, Y. Du, Q. Yang, Y. Shao, S.C. Liew, LLMind: orchestrating AI and IoT with LLMs for complex task execution. IEEE Commun. Mag. **63**(4), 214–220 (2024)
5. W. Weaver, Recent contributions to the mathematical theory of communication. ETC: Rev. Gen. Semant. **10**(4), 261–281 (1953)
6. C.E. Shannon, A mathematical theory of communication. Bell Syst. Tech. J. **27**(3), 379–423 (1948)
7. B. Juba, A.T. Kalai, S. Khanna, M. Sudan, Compression without a common prior: an information-theoretic justification for ambiguity in language, in *Proceedings of the Innovations in Computer Science* (2011)
8. E. Bourtsoulatze, D.B. Kurka, D. Gündüz, Deep joint source-channel coding for wireless image transmission. IEEE Trans. Cogn. Commun. Netw. **5**(3), 567–579 (2019)
9. T.-Y. Tung, D. Gündüz, Deepwive: deep-learning-aided wireless video transmission. IEEE J. Sel. Areas Commun. **40**(9), 2570–2583 (2022)
10. Y. Shao, C. Bian, L. Yang, Q. Yang, Z. Zhang, D. Gunduz, Point cloud in the air (2024). arXiv:2401.00658
11. Y. Shao, D. Gündüz, S.C. Liew, Federated edge learning with misaligned over-the-air computation. IEEE Trans. Wirel. Commun. **21**(6), 3951–3964 (2021)
12. Y. Blau, T. Michaeli, Rethinking lossy compression: the rate-distortion-perception tradeoff, in *International Conference on Machine Learning*, PMLR (2019), pp. 675–685
13. E. Beck, C. Bockelmann, A. Dekorsy, Semantic communication: an information bottleneck view (2022). arXiv:2204.13366
14. Y. Shao, S.C. Liew, D. Gündüz, Denoising noisy neural networks: a bayesian approach with compensation. IEEE Trans. Signal Process. **71**, 2460–2474 (2023)
15. J. Xu, T.-Y. Tung, B. Ai, W. Chen, Y. Sun, D.D. Gündüz, Deep joint source-channel coding for semantic communications. IEEE Commun. Mag. **61**(11), 42–48 (2023)
16. Y. Shao, D. Gunduz, Semantic communications with discrete-time analog transmission: a PAPR perspective. IEEE Wirel. Commun. Lett. **12**(3), 510–514 (2022)
17. R. Carnap, Y. Bar-Hillel, et al., *An Outline of a Theory of Semantic Information* (Research Laboratory of Electronics, Massachusetts Institute of Technology, Cambridge, 1952)
18. J. Bao, P. Basu, M. Dean, C. Partridge, A. Swami, W. Leland, J.A. Hendler, Towards a theory of semantic communication, in *2011 IEEE Network Science Workshop* (IEEE, 2011), pp. 110–117
19. B. Juba, M. Sudan, Universal semantic communication i, in *Proceedings of the Annual ACM Symposium on Theory of Computing* (2008), pp. 123–132
20. Y. Zhong, A theory of semantic information. China Commun. **14**(1), 1–17 (2017)
21. M.K. Farshbafan, W. Saad, M. Debbah, Common language for goal-oriented semantic communications: a curriculum learning framework, in *ICC 2022-IEEE International Conference on Communications* (IEEE, 2022), pp. 1710–1715
22. J. Dai, S. Wang, K. Tan, Z. Si, X. Qin, K. Niu, P. Zhang, Nonlinear transform source-channel coding for semantic communications. IEEE J. Sel. Areas Commun. **40**(8), 2300–2316 (2022)
23. D. Gunduz, Z. Qin, I.E. Aguerri, H.S. Dhillon, Z. Yang, A. Yener, K.K. Wong, C.-B. Chae, Beyond transmitting bits: context, semantics, and task-oriented communications (2022). arXiv:2207.09353
24. Z. Qin, X. Tao, J. Lu, G.Y. Li, Semantic communications: principles and challenges (2021). arXiv:2201.01389

25. Q. Lan, D. Wen, Z. Zhang, Q. Zeng, X. Chen, P. Popovski, K. Huang, What is semantic communication? a view on conveying meaning in the era of machine intelligence. J. Commun. Inf. Netw. **6**(4), 336–371 (2021)
26. B. Güler, A. Yener, A. Swami, The semantic communication game. IEEE Trans. Cogn. Commun. Netw. **4**(4), 787–802 (2018)
27. J. Liu, S. Shao, W. Zhang, H.V. Poor, An indirect rate-distortion characterization for semantic sources: general model and the case of gaussian observation (2022). arXiv preprint arXiv:2201.12477
28. D. Le Gall, MPEG: a video compression standard for multimedia applications. Commun. ACM **34**(4), 46–58 (1991)
29. T. Berger, Rate-distortion theory, in *Wiley Encyclopedia of Telecommunications* (Wiley, New York, 2003)
30. M. Fresia, F. Peréz-Cruz, H.V. Poor, S. Verdú, Joint source and channel coding. IEEE Signal Process. Mag. **27**(6), 104–113 (2010)
31. Z. Reznic, R. Zamir, M. Feder, Joint source-channel coding of a Gaussian mixture source over the Gaussian broadcast channel. IEEE Trans. Inf. Theory **48**(3), 776–781 (2002)
32. H. Wu, Y. Shao, C. Bian, K. Mikolajczyk, D. Gündüz, Deep joint source-channel coding for adaptive image transmission over MIMO channels. IEEE Trans. Wirel. Commun. **23**(10), 15002–15017 (2024)
33. C. Bian, Y. Shao, H. Wu, E. Ozfatura, D. Gündüz, Process-and-forward: deep joint source-channel coding over cooperative relay networks. IEEE J. Sel. Areas Commun. **43**(4), 1118–1134 (2025)
34. H. Wu, Y. Shao, E. Ozfatura, K. Mikolajczyk, D. Gündüz, Transformer-aided wireless image transmission with channel feedback. IEEE Trans. Wirel. Commun. **23**(9), 11904–11919 (2024)
35. C. Bian, Y. Shao, D. Gündüz, A deep joint source-channel coding scheme for hybrid mobile multi-hop networks. IEEE J. Sel. Areas Commun. **43**(7), 2543–2559 (2025)
36. N. Farsad, M. Rao, A. Goldsmith, Deep learning for joint source-channel coding of text, in *IEEE International Conference on Acoustics, Speech and Signal Processing (ICASSP)* (IEEE, 2018), pp. 2326–2330
37. H. Wu, Y. Shao, K. Mikolajczyk, D. Gündüz, Channel-adaptive wireless image transmission with OFDM. IEEE Wirel. Commun. Lett. **11**(11), 2400–2404 (2022)
38. C. Bian, Y. Shao, D. Gündüz, Wireless point cloud transmission, in *IEEE International Workshop on Signal Processing Advances in Wireless Communications (SPAWC)* (IEEE, 2024), pp. 851–855
39. H. Wu, Y. Shao, C. Bian, K. Mikolajczyk, D. Gündüz, Vision transformer for adaptive image transmission over MIMO channels, in *IEEE International Conference on Communications* (IEEE, 2023), pp. 3702–3707
40. J. Xu, B. Ai, W. Chen, A. Yang, P. Sun, M. Rodrigues, Wireless image transmission using deep source channel coding with attention modules. IEEE Trans. Circuits Syst. Video Technol. **32**(4), 2315–2328 (2022)
41. M. Yang, C. Bian, H.-S. Kim, OFDM-guided deep joint source channel coding for wireless multipath fading channels. IEEE Trans. Cogn. Commun. Netw. **8**(2), 584–599 (2022)
42. A. Krizhevsky, G. Hinton, et al., Learning multiple layers of features from tiny images. University of Toronto, Tech. Rep. (2009)
43. D. Chu, Polyphase codes with good periodic correlation properties (corresp.). IEEE Trans. Inf. Theory **18**(4), 531–532 (1972)
44. D.B. Kurka, D. Gündüz, Deepjscc-f: deep joint source-channel coding of images with feedback. IEEE J. Sel. Areas Inf. Theory **1**(1), 178–193 (2020)
45. Y. Rahmatallah, S. Mohan, Peak-to-average power ratio reduction in OFDM systems: a survey and taxonomy. IEEE Commun. Surv. Tutor. **15**(4), 1567–1592 (2013)
46. M. Kim, W. Lee, D.-H. Cho, A novel PAPR reduction scheme for OFDM system based on deep learning. IEEE Commun. Lett. **22**(3), 510–513 (2017)

47. L. Li, C. Tellambura, X. Tang, Improved tone reservation method based on deep learning for PAPR reduction in OFDM system, in *IEEE Int. Conf. Wireless Commun. and Signal Proc.* (2019)
48. M. Goutay, F.A. Aoudia, J. Hoydis, J.-M. Gorce, Learning OFDM waveforms with PAPR and ACLR constraints (2021). arXiv:2110.10987
49. Y. Cao, Y. Shao, R.Q. Hu, Fractional fourier domain PAPR reduction (2024). arXiv preprint arXiv:2411.08473
50. R. Van Nee, A. De Wild, Reducing the peak-to-average power ratio of OFDM, in *IEEE Veh. Tech. Conf.*, vol. 3 (1998), pp. 2072–2076
51. H.G. Myung, J. Lim, D.J. Goodman, Single carrier FDMA for uplink wireless transmission. IEEE Veh. Tech. Mag. **1**(3), 30–38 (2006)
52. Y. Shao, S.C. Liew, Flexible subcarrier allocation for interleaved frequency division multiple access. IEEE Trans. Wirel. Commun. **19**(11), 7139–7152 (2020)
53. S.C. Liew, Y. Shao, New transceiver designs for interleaved frequency-division multiple access. IEEE Trans. Wirel. Commun. **19**(12), 7765–7778 (2020)
54. Y. Du, S.C. Liew, Y. Shao, Efficient FFT computation in IFDMA transceivers. IEEE Trans. Wirel. Commun. **22**(10), 6594–6607 (2023)
55. Y.-C. Wang, Z.-Q. Luo, Optimized iterative clipping and filtering for papr reduction of OFDM signals. IEEE Trans. Commun. **59**(1), 33–37 (2010)

Chapter 4
AI-Driven Multiple Access for IoT

Where many speak, only intelligence ensures they are heard.

4.1 Overview

In wireless communication systems, multiple access refers to the methods and protocols that enable multiple users or devices to share the same communication medium or spectrum resources efficiently and with minimal interference [1]. The primary goal of multiple access techniques is to facilitate simultaneous data transmission from numerous devices while optimizing the use of available bandwidth and minimizing collisions or data loss.

Over the years, various multiple access schemes have been developed, each employing different strategies to allocate resources among users. These schemes can be broadly categorized into:

- **Scheduled (Deterministic) Multiple Access**: Resources are allocated to users in a predetermined or centrally controlled manner, reducing the likelihood of collisions. This category includes:
  - Time Division Multiple Access (TDMA). Allocates distinct time slots to different users, so that each user transmits in their assigned time slot.
  - Frequency Division Multiple Access (FDMA). Assigns separate frequency bands to each user, allowing simultaneous transmissions over different frequency channels.
  - Code Division Multiple Access (CDMA). Uses unique spreading codes for each user, enabling multiple users to share the same time and frequency resources while remaining separable at the receiver due to code orthogonality.
  - Orthogonal Frequency Division Multiple Access (OFDMA). Combines orthogonal frequency division multiplexing with multiple access by assigning subsets of subcarriers to individual users.

Y. Shao, *AI-Empowered IoT Communications*, Wireless Networks,
https://doi.org/10.1007/978-3-031-98816-5_4

- **Random (Contention-Based) Multiple Access**: Users compete for channel access, which can lead to collisions that must be managed or avoided. This category includes:
  - ALOHA and Slotted ALOHA. Users transmit whenever they have data, leading to possible collisions; slotted ALOHA reduces collision probability by synchronizing transmissions into time slots.
  - Carrier Sense Multiple Access with Collision Avoidance (CSMA/CA). Devices sense the channel before transmitting and use random backoff times to avoid collisions.
  - Carrier Sense Multiple Access with Collision Detection (CSMA/CD). Similar to CSMA/CA but with collision detection, used in wired networks like Ethernet.

These multiple access techniques have been fundamental in enabling communication in various wireless systems, including cellular networks, Wi-Fi, satellite communications, and IoT networks. They provide the necessary frameworks to manage access to the communication medium, ensuring that users can transmit data reliably and efficiently.

4.1.1 Multiple Access in IoT Networks

IoT networks exhibit several unique characteristics that impact the design and implementation of multiple access schemes:

1. **Massive Device Density**. IoT networks may consist of thousands or even millions of devices within a limited geographical area or network segment, leading to high contention for communication resources.
2. **Heterogeneous Traffic Patterns**. IoT devices generate data with varying frequency, size, and latency requirements, ranging from sporadic sensor readings to continuous video streams. This diversity necessitates flexible access methods that can accommodate different types of traffic.
3. **Resource Constraints**. Many IoT devices are limited in terms of power, processing capabilities, and memory. Energy-efficient communication protocols are essential to prolong battery life and ensure sustainable operation.
4. **Dynamic Topologies**. Devices may be mobile or subject to changing network conditions due to environmental factors or device behavior. Adaptive communication strategies are required to maintain reliable connections under varying circumstances.

While traditional multiple access techniques have been effective in various wireless communication systems [2], they face significant challenges when applied to IoT networks.

One of the primary issues is scalability. Traditional schemes often cannot scale efficiently to support the massive number of devices present in IoT environments. As the number of connected devices increases, congestion becomes more likely, leading to degraded performance and higher packet loss rates. Additionally, many multiple access protocols require control messages and coordination mechanisms that grow with the number of devices. This increased protocol overhead consumes valuable bandwidth, reduces effective throughput, and elevates latency.

Another significant limitation is inefficient resource utilization. Techniques like TDMA, FDMA, and OFDMA allocate resources in a fixed manner. In IoT networks characterized by sporadic or bursty traffic patterns, this static allocation can lead to substantial underutilization of resources when devices are idle. Conversely, contention-based protocols such as CSMA/CA rely on devices competing for the channel. As the number of devices grows, the likelihood of collisions increases dramatically, resulting in more retransmissions and higher delays.

Energy consumption is a critical concern in IoT networks. Some protocols require devices to continuously monitor the channel to detect transmission opportunities or avoid collisions. This constant vigilance consumes significant energy, which is problematic for battery-powered IoT devices that need to operate for extended periods without maintenance. Moreover, collisions and interference necessitate retransmissions, further draining device batteries and reducing overall network efficiency.

Latency constraints also pose a substantial challenge. Increased contention and extended backoff times in crowded networks lead to higher latency. For time-sensitive IoT applications, such as industrial automation, real-time monitoring, and autonomous vehicles, these delays are unacceptable and can compromise system performance or safety.

The complexity and cost associated with implementing advanced multiple access schemes add another layer of difficulty. Such schemes may require sophisticated hardware and processing capabilities, increasing the cost and complexity of IoT devices. For large-scale deployments involving low-cost devices, this becomes impractical and economically unfeasible, hindering widespread adoption.

Finally, interference management becomes more challenging in dense environments, such as smart factories or urban IoT deployments. In settings with a high density of devices, traditional protocols may not provide adequate mechanisms to mitigate interference. This inadequacy affects the reliability and quality of communication, leading to degraded network performance and user experience.

These challenges highlight the need for new multiple access solutions tailored to the specific needs of IoT networks. Traditional approaches may not suffice due to their inherent limitations in scalability, efficiency, and adaptability. Innovative techniques that can intelligently manage the communication medium, adapt to dynamic conditions, and optimize resource utilization are essential for the successful deployment and operation of large-scale IoT systems.

4.1.2 Conceptual Shifts in Multiple Access Protocol Design

The advent of AI and ML has introduced significant opportunities to revolutionize multiple-access protocol design in IoT networks [3–5]. Traditional protocols, with their static rules and limited adaptability, are increasingly inadequate for the dynamic and complex environments of modern IoT systems. AI-driven approaches offer the potential to create intelligent, autonomous, and efficient communication strategies that address the unique challenges of IoT.

One of the primary ways AI enhances multiple access in IoT networks is through dynamic resource allocation. AI algorithms can analyze historical and real-time data to predict traffic patterns, device mobility, and network congestion. This predictive modeling enables dynamic allocation of spectrum resources, optimizing utilization and reducing the likelihood of collisions. Machine learning models adjust resource allocation in real-time based on evolving network conditions and specific device requirements. This adaptability ensures that devices with critical data or stringent latency needs receive priority access when necessary.

Another significant contribution of AI is in autonomous decision-making. Devices employ reinforcement learning to make independent decisions about when and how to access the communication medium. By learning from interactions with the environment, devices optimize their transmission strategies, reducing reliance on predefined protocols like CSMA/CA and improving overall efficiency. In networks with numerous devices, multi-agent reinforcement learning allows devices to learn cooperative behaviors. This collective intelligence enhances network performance by minimizing interference and efficiently sharing the communication medium.

AI also plays a crucial role in interference mitigation. Advanced signal processing techniques powered by AI enhance signal detection and decoding in the presence of noise and interference. Deep learning models can better distinguish between desired signals and background noise, improving communication reliability. Additionally, machine learning models monitor network traffic to identify and mitigate sources of interference or anomalous behavior. Early detection enables prompt corrective actions, maintaining network integrity and performance.

Energy efficiency is another area where AI makes a significant impact. AI algorithms determine optimal transmission timing and power levels to minimize energy consumption while maintaining communication quality. This is crucial for battery-powered IoT devices deployed in remote or inaccessible locations. Furthermore, devices use AI to schedule sleep and wake times effectively, conserving battery life without compromising communication needs. Balancing activity and rest periods extends device longevity and reduces maintenance costs.

Lastly, AI enhances scalability and adaptability in IoT networks. AI-driven approaches scale efficiently by learning from data and adapting to new patterns as the network grows. Continuous learning ensures that performance remains robust despite increasing device numbers. Processing data at the network edge reduces latency and bandwidth usage. Deploying AI models on edge devices enables real-time decision-making and rapid adaptation to local conditions, further improving network responsiveness.

Overall, the integration of AI and ML into multiple-access protocols represents a significant conceptual shift from static, rule-based systems to dynamic, learning-driven approaches. This shift addresses the inherent limitations of traditional methods and paves the way for more resilient, efficient, and intelligent IoT networks capable of meeting the demands of an increasingly connected world.

4.1.3 Chapter Organization

In this chapter, we explore how AI-driven approaches can revolutionize multiple access in IoT networks by addressing the specific challenges previously outlined. We focus on two critical areas that both embody AI-driven multiple access but from different perspectives: using AI to design new multiple access techniques, and designing multiple access techniques specifically for AI applications.

First, we examine how AI can be utilized to create new multiple access methods [5]. Specifically, we delve into how multi-agent deep reinforcement learning (MADRL) [6] can replace traditional protocols like CSMA/CA. By employing MADRL, IoT devices autonomously learn optimal access strategies in a decentralized manner. This approach addresses the limitations of CSMA/CA in dense IoT environments, such as scalability issues, increased collisions, and energy inefficiency. We discuss the fundamentals of MADRL, explaining how agents (IoT devices) learn through interactions with the environment to make optimal decisions on when and how to access the communication medium. This results in improved network performance, better resource utilization, reduced latency, and energy savings. Readers will gain insights into how AI techniques can be harnessed to design innovative multiple access protocols that enhance the efficiency and scalability of IoT networks.

Second, we explore how multiple access techniques can be specially designed to facilitate AI applications, particularly focusing on federated learning (FL) in IoT edge computing [7]. Implementing FL across numerous IoT devices introduces significant communication overhead and multiple-access challenges, often referred to as the multiple-access bottleneck. To address this, we investigate the use of Over-the-Air Computation (AirComp) as a multiple access technique tailored for AI applications. AirComp exploits the superposition property of wireless channels to enable simultaneous transmission and aggregation of data, significantly reducing communication rounds, saving energy, and improving scalability. By integrating AirComp with FL, we demonstrate how this specialized multiple access technique enhances the performance of AI applications in IoT networks. Readers will learn how designing multiple access methods specifically for AI tasks can address communication bottlenecks and enable efficient, privacy-preserving collaborative model training in IoT edge environments.

By focusing on these two areas, the chapter provides a comprehensive understanding of AI-driven multiple access in IoT networks from both angles: using AI to innovate multiple access techniques, and crafting multiple access solutions to support AI applications. Readers will gain insights into the transformative potential of AI in managing communication resources among vast numbers of IoT devices, practical implementations of MADRL and AirComp in real-world scenarios, and future directions for research and development in AI-driven multiple access for IoT. Ultimately, the chapter demonstrates how AI can both revolutionize multiple access methods and benefit from specialized multiple access designs, paving the way for advanced IoT applications and services.

Roadmap

The remainder of this chapter is articulated in the following manner. Section 4.2 presents AI-driven multiple access protocols in realizing autonomous channel access. Section 4.3 presents a DRL-based MAC design for feedback-enhanced IoT communications, aiming to assess the potential enhancements in lifespan for an IoT cell comprising multiple devices. Section 4.4 presents federated learning-driven multiple access designs with over-the-air computation techniques. Section 4.5 concludes this chapter.

4.2 Autonomous Channel Access

Autonomous channel access (AutoCA) is a class of multiple access control (MAC) problems, where a group of IoT terminals communicate with an access point (AP) on a shared channel in a decentralized fashion [3, 4, 8–12]. Due to the multiple-access interference, at any time only one terminal can communicate with the AP; when two or more terminals transmit simultaneously, a collision happens and the transmissions fail. In AutoCA, the goal of the terminals is to discover a set of transmission policies autonomously such that the system throughput is maximized and the fairness among terminals is guaranteed. AutoCA is essential to communication networks that lack well-governed infrastructures, such as IoT communications systems operated on the unlicensed band [13–15], to provide high bandwidth efficiency and low-latency services.

The classical solution to the AutoCA problem is human-crafted random access policies. A typical example is the contention-based IEEE 802.11 carrier-sense multiple access with collision avoidance (CSMA/CA) [8, 13], which is now widely used in Wi-Fi networks. When operated with CSMA/CA, distributed terminals compete with each other for channel access opportunities: each terminal carrier senses the channel before transmission (i.e., the listen-before-talk (LBT) protocol); and retransmits after a binary exponential back-off (BEB) in the case of transmission failures, i.e., the average back-off time is doubled after a transmission failure to reduce the collision probability.

Recent advances in deep reinforcement learning multiple access (DRLMA) open up a new perspective for solving the AutoCA problem [3, 4, 9–11, 16–22]. With DRLMA, distributed terminals endowed with intelligence learn to cooperate, as opposed to competing, to share the spectrum harmonically. In particular, the group of terminals are expected to adapt to each other's transmission policies to avoid collisions and balance channel-access opportunities. To this end, the AutoCA problem is formulated as either a Markov decision process (MDP) [3, 9, 11, 16, 22, 23] or a partially observable MDP (POMDP) [4, 10, 17–19, 21], depending on whether the environment is fully observable to the terminals or not. The MDP or POMDP is then solved by DRL or multi-agent DRL (MADRL), wherein terminals progressively learn a good set of policies from their past transmission history, carrier sensed behaviors of other terminals, and additional information fed back from the AP/channel. Compared with legacy random access policies, DRLMA often exhibits much better performance in terms of throughput, delay, and jitter, demonstrating the great potential to solve the AutoCA problem in a variety of network scenarios.

A notorious problem in wireless networks is the hidden terminal problem [24, 25]. Hidden terminals are terminals that are out of each other's communication range. Hidden terminals cannot detect each other's transmission behaviors via carrier sensing, and hence, their transmissions to the common receiver can collide, resulting in a waste of spectrum. In Wi-Fi networks, for example, the LBT protocol breaks down when there are hidden terminals, due to the inaccurate carrier-sensed channel states; and the throughput and delay performances deteriorate significantly [25]. To alleviate this problem, the IEEE 802.11 standard introduced the request to send/clear to send (RTS/CTS) mechanism. However, RTS/CTS falls short of an ideal protocol as it incurs additional handshake overhead and new problems such as hidden receiver and exposed receiver [26, 27] – as a consequence, it can even deteriorate the performance of CSMA/CA [27]. This is even more pronounced in machine-type communications with a small payload size and the latest IEEE 802.11 standards, where the bandwidth becomes larger and larger, yielding packets with increasingly shorter duration. On the other hand, prior works on DRLMA focused exclusively on the ideal case, where all the terminals are audible to each other and the hidden terminal problem is completely omitted. In the practical setup with hidden terminals, a terminal can observe inaccurate channel states; its transmissions can collide with the transmissions of a hidden terminal that it has never seen. Worse still, the transmission policy of the hidden terminal varies across time, yielding non-stationary hidden dynamics to the terminal. In this context, designing a DRLMA algorithm to tackle the impact of hidden terminals is an important and challenging task.

This section explores a learning-based approach to solve the hidden terminal problem. We present MADRL-HT, a partially observable MADRL paradigm tailored for AutoCA in the presence of hidden terminals. Three innovations that underpin the design of MADRL-HT are terminal clustering, the look-back mechanism, and a window-based reward function.

1. **Terminal clustering**. For individual terminal, we put forth a clustering method to partition all other terminals according to their number of communication hops to the considered terminal. In so doing, the environment faced by each terminal can be summarized as the dynamics of two consolidated groups of terminals, thereby simplifying the observation space by a large margin.
2. **Look-back mechanism**. To compensate for the partial observability and alleviate the hidden non-stationarity, we proposed a look-back mechanism to revise the carrier-sensed channel states and infer the hidden terminals' behaviors via the acknowledgment (ACK) of the AP.
3. **Window-based reward**. Unlike prior works where the reward is often designed to be event-triggered, we put forth a new reward function based on the relative number of successful transmissions in a look-back window of time. Our reward motivates the agents to avoid collision and maximize the network throughput, while balancing the successful transmissions among terminals to ensure fairness.

Considering short-packet machine-type communications with various network topologies, we perform extensive numerical experiments to verify the superior performance of the MADRL-HT solution. Compared with the legacy CSMA/CA protocol, the MADRL-HT solution exhibits remarkable performance gains: the α-fairness is improved by up to 46%; the average packet collision rate (PCR) is reduced by up to 37%; the average packet delay is reduced by 99.2%, and the delay jitter is reduced by 99.9%.

Related Work

The standard solution in IEEE 802.11 to the AutoCA problem is CSMA/CA with binary exponential back-off (BEB) [8, 13]. To be more specific, the back-off time of CSMA/CA is randomly sampled from $[0, \mathrm{CW}]$, where CW is the length of a contention window. In the case of a collision, CW will be doubled to reduce the probability of contention. The setting of CW has a significant impact on the efficiency of CSMA/CA. As BEB is a heuristic scheme, several recent works [11, 16] employed DRL to learn better policies for setting CW under different network conditions, whereby the system throughput is optimized.

As opposed to optimizing the parameters on top of CSMA/CA, a more promising and interesting method that exploits the full potential of DRL is learning AutoCA protocols from scratch [3, 4, 9, 17, 18]. In this line of work, early studies [3, 9, 17] focused on the heterogeneous setup where an intelligent terminal (or a cluster of terminals with a central controller) learns to coexist with other terminals operated with legacy MAC protocols, such as Aloha, TDMA, CSMA/CA, and LTE. These works formulated single-agent RL problems, in which an intelligent agent learns to adapt to other non-intelligent agents' unknown policies from its observations. Compared with MARL, single-agent RL is less challenging as the environment faced by the agent is stationary.

Motivated by these research efforts, two works [4, 18] made further advances on the AutoML problem. In [18], the authors generalized [17] to a multi-agent setup, where multiple intelligent terminals make independent transmission decisions in the heterogeneous network. The unreliable physical channel is also taken into account in [18]. Ref. [4], on the other hand, considered a homogeneous setup where all terminals are intelligent and adapted the well-known MARL algorithm, Qmix [28], to solve the AutoCA problem. In [4, 17, 18], the intelligent terminals are endowed with carrier sensing capabilities, and they can utilize the feedback from the AP to make transmission decisions. Another line of work [29, 30] studied the joint learning of MAC signaling and access protocols with MADRL.

A shortcoming of the above-mentioned works is that the hidden terminal problem is completely omitted. By carrier sensing, the channel occupation is fully observable to the agents. In a more practical setup with hidden terminals, a terminal can observe inaccurate channel states; its transmissions can collide with the transmissions of a hidden terminal that it has never seen. Worse still, the transmission policy of the hidden terminal varies across time, yielding non-stationary hidden dynamics to each terminal. In this context, designing a DRLMA algorithm to tackle the impact of hidden terminals is a challenging task.

It is worth noting that another line of work that addresses the multiple-access problems with DRL is dynamic spectrum access (DSA) in cognitive radios [10, 19–22]. Under their framework, there are two classes of users in the network: primary users (PUs) and secondary users (SUs). The spectrum is licensed to the PUs, and hence, they have the highest channel access priorities. In an effort to maximize the spectrum efficiency, the SUs are allowed to access the unused spectrum provided that they do not harm the channel usages of PUs. To this end, the SUs learn to identify the unused spectrum, predict the channel-access patterns of the PUs, and cooperate harmonically with each other. As can be seen, DSA with cognitive radios can be viewed as an AutoCA problem under the heterogeneous setup with a class of prioritized PUs.

4.2.1 Problem Statement

We consider a typical IoT basic service set (BSS) with one AP and N IoT terminals formed in a star topology, as shown in Fig. 4.1a. The IoT terminals are indexed by $n = 1, 2, \ldots, N$. We focus on the uplink transmissions from the terminals to the AP. All terminals share the same uplink frequency spectrum, which is referred to as the common channel. Time is divided into slots. We consider the saturated traffic setup, i.e., the transmission buffers of the terminals are nonempty in any time slot, and a terminal has a packet to transmit whenever there is an idle slot. Throughout the section, we assume that a transmission failure can only be caused by collisions, but not the quality of the underlying physical channel.

Not all terminals are within each other's communication range. In Fig. 4.1a, for example, terminals A and B are within each other's communication range. Therefore, terminal A can sense terminal B's behaviors (i.e., transmit or not) when A is idle, and vice versa. On the other hand, terminals A and C are out of each other's communication range, and hence, they cannot infer each other's transmissions from carrier sense—they are "hidden terminals" to each other.

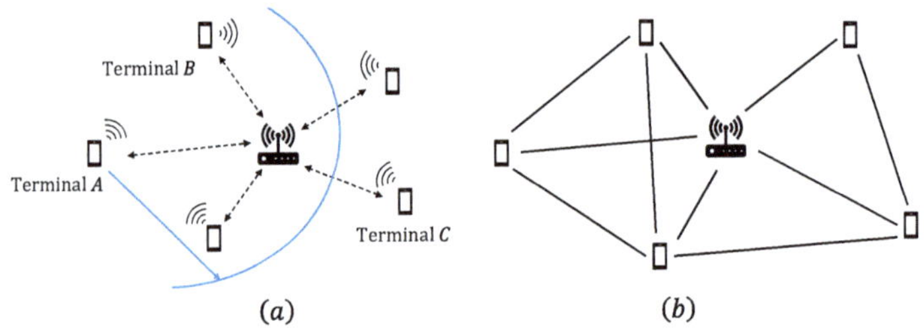

Fig. 4.1 An IoT BSS with one AP and N terminals sharing a common uplink channel: (**a**) the physical network; (**b**) the graph whose vertices represent the terminals and the edges connect terminals that are within each other's communication range

In the presence of hidden terminals, this section aims to address the AutoCA problem, i.e., how a group of agents can reach a multiple-access protocol autonomously to maximize the network throughput while guaranteeing fairness among each other. More rigorously, we formulate the AutoCA problem using the α-fairness measure in the following.

Let us denote the set of channel-access policies of the N terminals by $\boldsymbol{\pi} = \{\pi_n : n = 1, 2, \ldots, N\}$, where π_n is the policy of the n-th terminal. When operated with $\boldsymbol{\pi}$, we denote by $\mathcal{T}_n(\boldsymbol{\pi})$ the throughput achieved by the n-th terminal. In particular, $\mathcal{T}_n(\boldsymbol{\pi})$ can be written as

$$\mathcal{T}_n(\boldsymbol{\pi}) \triangleq \frac{M_n(\boldsymbol{\pi}, T)D}{T}, \tag{4.1}$$

where D is the packet length measured in terms of number of slots and we assume all terminals have the same packet length; and $M_n(\boldsymbol{\pi}, T)$ is the number of packets successfully transmitted by the n-th terminal in a period of T slots under the set of policies $\boldsymbol{\pi}$. The α-fairness value of the n-th terminal is then given by [31]

$$f\big(\mathcal{T}_n(\boldsymbol{\pi})\big) = \begin{cases} \log\big(\mathcal{T}_n(\boldsymbol{\pi})\big), & \alpha = 1, \\ \frac{\left(\mathcal{T}_n(\boldsymbol{\pi})\right)^{1-\alpha}}{(1-\alpha)}, & \alpha \neq 1. \end{cases} \tag{4.2}$$

In AutoCA, the goal of the terminals is to find the optimal set of channel-access policies $\boldsymbol{\pi}^*$ in a decentralized manner to maximize the α-fairness function $F(\boldsymbol{\pi})$, which is the sum of all α-fairness values. Specifically, we have

$$\boldsymbol{\pi}^* = \arg\max_{\boldsymbol{\pi}} \sum_{n=1}^{N} f\big(\mathcal{T}_n(\boldsymbol{\pi})\big) \triangleq \arg\max_{\boldsymbol{\pi}} F(\boldsymbol{\pi}). \tag{4.3}$$

With the α-fairness measure, the optimal set of channel-access policies $\boldsymbol{\pi}^*$ is the policy set that maximizes the total throughput of the IoT BSS while balancing the throughput of individual terminals equally.

This section explores an MADRL solution to the AutoCA problem, where individual IoT terminals in the BSS learn to optimize their transmission policies based on their sensing and perception of the environment. In particular, under the IEEE 802.11 standards, two pieces of information that a terminal can exploit to learn the optimal policy are:

1. The carrier-sensed channel states. When it is idle, a terminal can actively sense the occupancy of the common channel within its communication range, and hence, infer and adapt to other terminals' transmission behaviors. There are two possible carrier-sensed channel states: "idle" and "busy".
2. Feedback from the AP. The AP broadcasts an ACK/NACK signal (NACK means that a collision happens and no ACK is received at the end of transmission) following every transmission, indicating whether a packet has been received successfully or not. We assume that this 1-bit feedback from the AP is received by all the terminals reliably.

It is worth noting that packet length is conventionally defined in terms of the number of bytes, such as the maximum transmission unit (MTU). Here, we define the packet length in the manner of transmit opportunity (TXOP) [32]: the time duration for which a terminal can transmit after it has gained the channel. That is, when a terminal decides to transmit, it occupies the channel for D slots until the whole packet is transmitted.

4.2.2 *Learn the Multiple Access Protocol*

This section presents a learning-based approach for the IoT BSS to autonomously learn a MAC protocol from the interactions of IoT terminals.

4.2.2.1 Terminal Clustering

For any IoT terminal in the BSS, the environment it faces consists of the other $N-1$ terminals. To develop a sound transmission policy that coexists well with others, the first step is to infer each of the other terminals' transmission policies. The partial observability of our problem, however, makes the inference an elusive task since it is impossible to extrapolate each of the other $N-1$ terminals' transmission histories from the limited carrier-sensed channel states and AP's feedback.

To address the above challenge, this section puts forth a terminal clustering method: for each terminal in the BSS, we will show how to partition the other $N-1$ terminals such that their transmission histories can be summarized as the histories of two consolidated terminals, thereby yielding a compact observation space for each terminal in the BSS. Along with the discussion, we will also define some basic terminologies to be used in the rest of this chapter.

To start with, we study the topology of an IoT BSS by treating it as an undirected graph—its vertices correspond to the terminals and the AP; and there is an edge connecting two vertices if and only if the corresponding terminals are within each other's communication range. As an example, the IoT BSS in Fig. 4.1a can be transformed to the graph shown in Fig. 4.1b.

Definition 4.1 (n-Hop Neighbor) Consider two vertices A and B on a graph. Vertex B is an ℓ-hop neighbor of vertex A if and only if vertex A can reach vertex B in at least ℓ hops.

Given Definition 4.1, we have two immediate results from Fig. 4.1b, thanks to the star topology of the BSS.

- In the IoT BSS, all terminals are one-hop (OH) neighbors to the AP.
- For a terminal A in the BSS, any other terminal is either a OH or a two-hop (TH) neighbor of terminal A.

In light of the above, for the n-th terminal in the IoT BSS, we can partition the other $N - 1$ terminals into two sets: one consists of all OH neighbors of the n-th terminal, denoted by OH(n), and the other consists of all TH neighbors of the n-th terminal, denoted by TH(n). An example is given in Fig. 4.2, in which any other terminal in the BSS is either an OH or a TH neighbor to terminal A or C. We emphasize that different terminals have different sets of OH and TH neighbors.

Another key observation is that, for the n-th terminal, all the terminals in OH(n) can be viewed as a single consolidated terminal because (1) the n-th terminal senses them in the same way; and (2) their actions (transmit or not) have the same effects on the n-th terminal. Likewise, all the terminals in TH(n) can also be viewed as a single consolidated terminal. Therefore, from the perspective of the n-th terminal, the transmission actions of all other terminals can be summarized as the transmission actions of two consolidated terminals: OH(n) and TH(n).

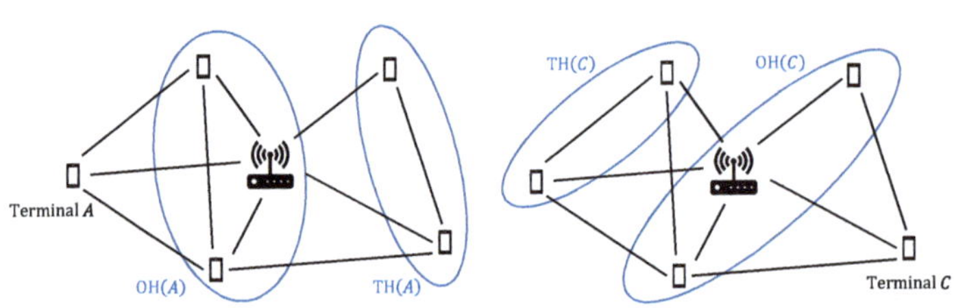

Fig. 4.2 Any terminal in the IoT BSS has only OH and TH neighbors. Different terminals observe different sets of OH and TH neighbors

Definition 4.2 (Actions of OH and TH Neighbors) Suppose there are N terminals in an IoT BSS. The action of the n-th terminal in slot t, denoted by $a_n^t \in \{0, 1\}$, indicates whether the n-terminal is transmitting ($a_n^t = 1$) or idle ($a_n^t = 0$) in slot t.

Suppose the n-th terminal has $M - 1$ OH neighbors and $N - M$ TH neighbors. Let $\mathrm{OH}(n) \triangleq \{i_1, i_2, \ldots, i_{M-1}\}$ and $\mathrm{TH}(n) \triangleq \{j_1, j_2, \ldots, j_{N-M}\}$. We define

$$a_{\mathrm{OH}(n)}^t \triangleq a_{i_1}^t \vee a_{i_2}^t \vee \cdots \vee a_{i_{M-1}}^t,$$

$$a_{\mathrm{TH}(n)}^t \triangleq a_{j_1}^t \vee a_{j_2}^t \vee \cdots \vee a_{j_{N-M}}^t,$$

to indicate, respectively, whether an OH and a TH neighbor of the n-th terminal transmits in slot t or not, where $\vee$ stands for the OR operation.

Listen-Before-Talk

To be compatible with legacy IEEE 802.11 MAC standards, we impose the listen-before-talk (LBT) constraint on terminals' actions. Specifically, a terminal tries to access the channel only after it sensed that the channel has been idle for at least a distributed coordination function interframe spacing (DIFS). As a result, there will be at least DIFS idle slots between consecutive transmissions of a single terminal or two terminals that are within OH.

The n-th terminal carrier senses the channel as long as $a_n^t = 0$. Two outputs of carrier sensing, "idle" and "busy", correspond to $a_{\mathrm{OH}(n)}^t = 0$ and $a_{\mathrm{OH}(n)}^t = 1$, respectively.

4.2.2.2 An MADRL Solution

Based on the previous discussions, this section formulates AutoCA as a multi-agent reinforcement learning (MARL) problem and presents MADRL-HT, an MADRL solution tailored for AutoCA in the presence of hidden terminals. In MADRL-HT, terminals interact and negotiate autonomously with each other to reach a multiple-access protocol. During the interactions, each terminal infers other terminals' behaviors (summarized as two consolidated terminals) from the carrier-sensed channel states and AP's feedback. Transmission decisions are then made to adapt to other terminals' policies.

A general MARL problem [28, 33] can be defined as a tuple $(\mathcal{N}, \mathcal{S}, \{O_n\}_{n\in\mathcal{N}}, \mathcal{A}, \{r_n\}_{n\in\mathcal{N}})$, where $\mathcal{N} = \{1, 2, \ldots, N\}$ denotes the set of N agents; $\mathcal{S}$ denotes the state space of the environment that the agents reside in; O_n denotes the observation space of the n-th agent; $\mathcal{A}$ denotes the action space of each agent; $r_n : \mathcal{S} \times \mathcal{A} \times \mathcal{S} \to \mathcal{R}$

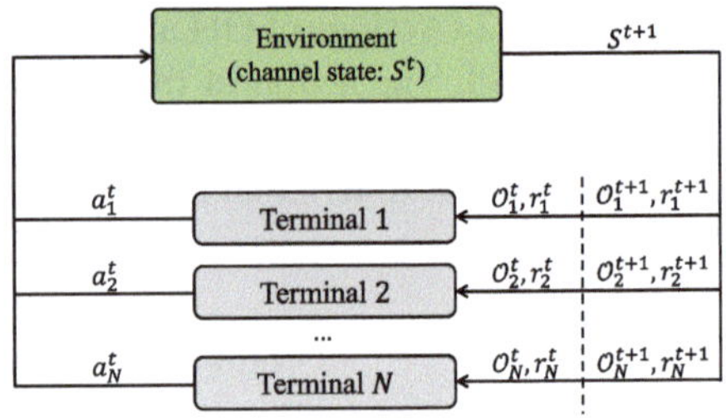

Fig. 4.3 Modelling the AutoCA problem with hidden terminals under the framework of MARL

denotes the immediate reward function of the n-th agent, which defines the reward that the environment assigns to an agent.

In the above context, the MARL problem associated with AutoCA is illustrated in Fig. 4.3. At any time step t, the action of the n-th terminal is defined as $a_n^t \in \mathcal{A} = \{'\text{transmit}', '\text{idle}'\}$; the state of the environment $S^t \in \mathcal{S}$ represents the channel usages and is the collection of actions of all the terminals. Each terminal has a local observation of S^t denoted by $\mathcal{O}_n^t$. In particular, $\mathcal{O}_n^t$ is determined by the carrier sensed channel state of the n-th terminal and the ACK feedback from the AP. Based on the observation $\mathcal{O}_n^t$ and the rewards received, each agent takes an action a_n^t. The set of actions $\{a_n^t\}_{n\in\mathcal{N}}$ steers the environment moving to a new state S^{t+1} and a reward r_n^t is assigned to each terminal to evaluate its last action. Over the course of MARL, the terminals aim to discover a set of transmission policies in a distributed manner – each terminal learns from its interactions with other terminals, its partially observed channel states, and the rewards it receives – to avoid collisions and maximize the BSS throughput, while guaranteeing fairness among each other.

Partial Observability

As stated above, the environment faced by a terminal in the IoT BSS is the other $N-1$ terminals. At any slot t, a decision (transmit or not) has to be made by each terminal based on the past behaviors of all terminals. In this context, we define the global state of the BSS at time slot t as the past actions of the terminals.

Definition 4.3 (Global State) The global state of an IoT BSS at slot t, denoted by S^t, is the transmission history of all the terminals in the last W slots:

$$S^t \triangleq \left\{a_n^\tau : n = 1, 2, \ldots, N, \tau = t-1, t-2, \ldots, t-W\right\}, \tag{4.4}$$

where W is the length of a look-back window, reflecting how many slots we look back into the past to make the transmission decision in slot t. The look-back window is necessary in that the terminals' behaviors can be non-Markovian in general, and hence, a decision has to be made based on a period of past behaviors such that the observed environment is approximately Markovian. Typically, W should be much larger than the packet length D.

The global state, however, is only partially observable to each terminal, as a terminal has only a local view/observation of other terminals' behaviors. Specifically, the observation of a terminal is subject to the following two constraints:[1]

- Constraint 1: A terminal cannot sense the channel while transmitting.
- Constraint 2: Hidden terminals are inaudible to each other.

Therefore, it is important to characterize the discrepancies between a terminal's local observations and the global state S^t.

Consider the n-th terminal. $a^t_{\mathrm{OH}(n)}$ is observable if and only if $a^t_n = 0$, due to Constraint 1. On the other hand, $a^t_{\mathrm{TH}(n)}$ is unobservable with carrier sensing. Nevertheless, notice that a successful transmission is always followed by an ACK, which is broadcasted by the AP to all the terminals in the BSS. This suggests that a terminal can further infer the behaviors of other terminals, including both its OH and TH neighbors, from the feedback it receives from the AP.

Definition 4.4 (Partial Observations of a Terminal) In an IoT BSS, the observation of the n-th terminal at the beginning of a slot t is defined as

$$O^t_n \triangleq \left\{a^\tau_n, o^\tau_{\mathrm{OH}(n)}, o^\tau_{\mathrm{TH}(n)} : \tau = t-1, t-2, \ldots, t-W\right\} \tag{4.5}$$

where $o^t_{\mathrm{OH}(n)}$ and $o^t_{\mathrm{TH}(n)}$ are estimates of $a^t_{\mathrm{OH}(n)}$ and $a^t_{\mathrm{TH}(n)}$, respectively, and are inferred by the n-th terminal from its carrier sensed channel states and the feedback from the AP.

In MADRL-HT, we put forth a look-back mechanism to determine $o^\tau_{\mathrm{OH}(n)}$ and $o^\tau_{\mathrm{TH}(n)}$ for the n-th terminal, $n = 1, 2, \ldots, N$. The look-back mechanism is summarized in Algorithm 6 and consists of two steps. At any time slot t, we feed the AP's feedback, i.e., ACK or NACK, into Algorithm 6 and estimate the transmission histories of the n-th terminal's OH and TH neighbors during the past D slots, i.e., $o^\tau_{\mathrm{OH}(n)}$ and $o^\tau_{\mathrm{TH}(n)}$, where $\tau \in \Omega \triangleq \{t-D+1, \ldots, t-1, t\}$. Note that D is the packet length measured in terms of the number of slots.

In the first step, the n-th terminal estimates the actions of its OH neighbors by carrier sensing: when $a^\tau_n = 0$, $\tau \in \Omega$, the carrier-sensed output is exactly $a^\tau_{\mathrm{OH}(n)}$, hence we set $o^\tau_{\mathrm{OH}(n)} = a^\tau_{\mathrm{OH}(n)}$; when $a^\tau_n = 1$, the terminal cannot carrier sense and we set $o^\tau_{\mathrm{OH}(n)} = \text{unk}$, where unk is a unique token standing for "unknown". On the other hand, the actions of the TH neighbors are unobservable from carrier sense, hence we simply set $o^\tau_{\mathrm{TH}(n)} = \text{unk}$ for the n-th terminal.

The second step is a look-back revision process. At the end of slot t, the terminal looks back and revises the carrier-sensed $o^\tau_{\mathrm{OH}(n)}$ and $o^\tau_{\mathrm{TH}(n)}$ if an ACK is received. Specifically,

1. if the n-th terminal was transmitting in the duration Ω, an ACK at slot t means the transmission is successful and no other terminals were transmitting. Thus,

[1] In practice, there can be more constraints, such that the power saving demands of terminals.

Algorithm 6 The look-back mechanism to infer the transmission histories of OH(n) and TH(n)

Input: AP's feedback at slot t, i.e., ACK or NACK.
Output: Estimated transmission history $o^{\tau}_{\mathrm{OH}(n)}$ and $o^{\tau}_{\mathrm{TH}(n)}$, where $\tau \in \Omega = \{t-D+1, \ldots, t-1, t\}$.
Carrier sensing
for $\tau \in \Omega$ **do**
    $o^{\tau}_{\mathrm{TH}(n)} = \mathrm{unk}$.
    if $a^{\tau}_{n} = 0$ **then** $o^{\tau}_{\mathrm{OH}(n)} = a^{\tau}_{\mathrm{OH}(n)}$;
    else $o^{\tau}_{\mathrm{OH}(n)} = \mathrm{unk}$.
Look-back revision
if ACK is received at slot t **then**
    if $a^{\tau}_{n} = 1, \forall \tau \in \Omega$ **then**
        $o^{\tau \in \Omega}_{\mathrm{OH}(n)} = 0,\ o^{\tau \in \Omega}_{\mathrm{TH}(n)} = 0$;
    else if $a^{\tau}_{n} = 0$ and $o^{\tau}_{\mathrm{OH}(n)} = 0, \forall \tau \in \Omega$ **then**
        $o^{\tau \in \Omega}_{\mathrm{TH}(n)} = 1$;
    else if $a^{\tau}_{n} = 0$ and $o^{\tau}_{\mathrm{OH}(n)} = 1, \forall \tau \in \Omega$ **then**
        $o^{\tau \in \Omega}_{\mathrm{TH}(n)} = 0$.

the actions of both OH and TH neighbors are revised to $o^{\tau}_{\mathrm{OH}(n)} = 0$, $o^{\tau}_{\mathrm{TH}(n)} = 0$, $\forall \tau \in \Omega$.

2. if the n-th terminal was idle in the duration Ω, the behaviors of its OH neighbors, $a^{\tau}_{\mathrm{OH}(n)}$, are fully known via carrier sensing. Therefore, if $o^{\tau}_{\mathrm{OH}(n)} = a^{\tau}_{\mathrm{OH}(n)} = 0$, i.e., none of its OH neighbors were transmitting during Ω, a received ACK suggests that one of its TH neighbors successfully transmits a packet. We thus revise $o^{\tau}_{\mathrm{TH}(n)} = 1$, $\forall \tau \in \Omega$.
3. Finally, if $a^{\tau}_{n} = 0$ and $o^{\tau}_{\mathrm{OH}(n)} = a^{\tau}_{\mathrm{OH}(n)} = 1$, i.e., at least one of the n-th terminal's OH neighbors was transmitting during Ω, the received ACK suggests that there was only one transmitting OH neighbor and all its TH neighbors were idle, because a collision would happen otherwise. Therefore, we revise $o^{\tau}_{\mathrm{TH}(n)} = 0$, $\forall \tau \in \Omega$.

To illustrate the look-back mechanism, an example is given in Fig. 4.4, where we consider a group of three IoT terminals A, B, and C in the BSS: A and B are OH neighbors; A and C are TH neighbors. Following Algorithm 6, terminal A determines the actions of its OH and TH neighbors in the following way. First, terminal A estimates $a^{t}_{\mathrm{OH}(A)}$ via carrier sensing. As can be seen from the carrier-sensed actions in Fig. 4.4, $\{a^{t}_{A}\}$ are fully known, $\{a^{t}_{\mathrm{OH}(A)}\}$ are known only when $a^{t}_{A} = 0$ (otherwise, $o^{t}_{\mathrm{OH}(A)}$ is set to unk); $\{a^{t}_{\mathrm{TH}(A)}\}$ are unknown. Then, terminal A revises its past observations based on the feedback of the AP. From the first ACK, terminal A knows that none of its OH and TH neighbors were transmitting in the last D slots, $\{o^{t}_{\mathrm{OH}(A)}\}$ and $\{o^{t}_{\mathrm{TH}(A)}\}$ are then modified to 0. From the second and third ACKs of the AP, terminal A knows that there is a successful transmission from its OH or TH neighbors. Combined with the carrier sensed channel states, the actions of the TH neighbors can be refined.

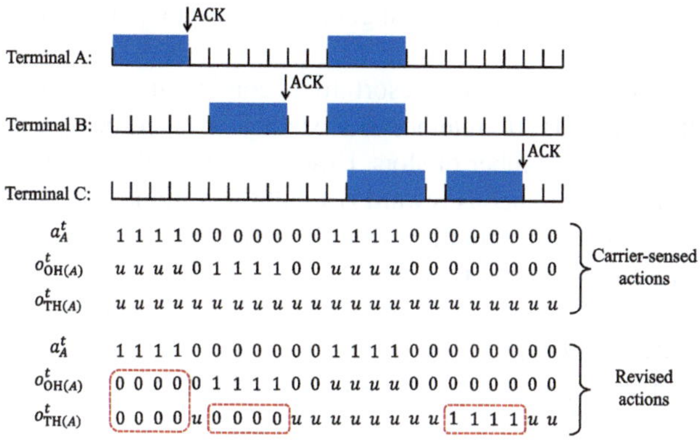

Fig. 4.4 An example illustrating how $o^t_{\text{OH}(A)}$ and $o^t_{\text{TH}(A)}$ are determined for a terminal A. We assume there are three terminals A, B, and C in the BSS, where A and B are OH neighbors; A and C are TH neighbors

An Alternative Definition for "Observation"

In prior works of AutoCA [3, 19] (without hidden terminals), the observation of a terminal is often designed as $\{a^t_n,\ \text{sensed channel states},\ \text{AP's feedback}\}$. When applied in our problem, the observation would have three rows: the first row consists of the actions of the n-th terminal itself, the second row consists of the observed actions of the OH neighbors $o^{\tau}_{\text{OH}(n)}$, and the third row consists of the feedback from the AP in the look-back window. Specifically, we would have

$$\widetilde{O}^t_n = \{a^{\tau}_n,\ o^{\tau}_{\text{OH}(n)},\ \text{ACK}^{\tau}, : \tau = t-1, t-2, \ldots, t-W\}. \tag{4.6}$$

The comparisons between (4.5) and (4.6) are given in Sect. 4.2.3.

Reward Design

In MADRL-HT, all agents learn simultaneously and their policies vary after each training. From a terminal's perspective, it faces a highly non-stationary environment, especially when there are hidden terminals whose dynamics are partially observable. To learn a good set of transmission policies, the reward design is of crucial importance.

Recall from (4.3) that our goal is to maximize the α-fairness function. This corresponds to an episodic MADRL task. That is, the terminals interact with each other in an episodic manner and an episode lasts for a fixed number of slots. A reward is generated at the end of each episode, which is the α-fairness measure computed from (4.2). The goal of MADRL is then to learn a set of policies for the terminals to maximize the reward generated in an episode. This formulation, however, is sample inefficient due to the very sparse reward—we empirically found that it is difficult for the agents to learn good transmission policies to maximize the α-fairness.

To mitigate this problem, we resort to a continuous MADRL formulation. Specifically, the terminals still interact with each other in an episodic manner and an episode lasts for a fixed number of slots. However, unlike the episodic formulation, we assign a global instantaneous reward r_t to all the terminals at each slot t during the episode, as opposed to generating a single reward at the end of the episode. The instantaneous reward r_t is specified below.

Let us define a flag $\delta_n^\tau \in \{0, 1\}$ to indicate whether a packet (of length D) is successfully transmitted in $[\tau, \tau+D-1]$ from the n-th terminal. At the beginning of a slot t, the number of packets successfully transmitted by the n-th terminal during the look-back window is given by

$$M_n^t = \sum_{\tau=t-W}^{t-1} \delta_n^\tau. \tag{4.7}$$

Notice that M_n^t, $n = 1, 2, 3, \ldots, N$, is known to the AP, hence, can be used to generate the reward. Let $\mathcal{M} \triangleq \{M_n^t : n = 1, 2, \ldots, N\}$, we further define

$$G^t \triangleq \max \mathcal{M} - \min \mathcal{M},$$

where G^t is the range of the set $\mathcal{M}$, i.e., the difference between the largest and smallest values of $\mathcal{M}$.

Then, the terminals make transmission decisions as long as they sense that the channel has been idle for at least a DIFS. The global reward incurred by their actions in slot t is generated at slot $t + D$ when δ_n^t is known. Let

$$F^t \triangleq \delta_1^t \vee \delta_2^t \vee \ldots \vee \delta_N^t,$$

where F^t indicates whether an ACK is received in slot $t + D$ ($F^t = 1$) or not ($F^t = 0$).

Definition 4.5 (Reward) The instantaneous global reward for slot t is computed by

$$r^t \triangleq \begin{cases} +1, & \text{if } F^t = 1 \text{ and } G^t \leq 1; \\ +1, & \text{if } F^t = 1, G^t > 1, \text{ and } \delta^t_{\arg\min \mathcal{M}} = 1; \\ -1, & \text{if } F^t = 1, G^t > 1, \text{ and } \delta^t_{\arg\min \mathcal{M}} = 0; \\ -1, & \text{if } F^t = 0; \\ 0, & \text{if no terminal transmits in slot } t. \end{cases} \tag{4.8}$$

In particular, $\delta^t_{\arg\min \mathcal{M}}$ indicates whether the $(\arg\min \mathcal{M})$-th terminal (i.e., the terminal with the least number of successful transmissions in the look-back window) successfully transmits a packet in slots $[t, t + D - 1]$.

The motivations behind (4.8) are as follows. First, our reward punishes actions $\{a^t_n : n = 1, 2, \ldots, N\}$ that lead to an unsuccessful transmission in slot t, i.e., the fourth line of (4.8) when $F^t = 0$. Second, for a successful transmission $F^t = 1$,

- If $G^t \leq 1$, we consider that the current policies have achieved a certain degree of fairness, and hence, a positive reward is assigned to all terminals.
- If $G^t > 1$, we consider that the current policies are unfair. In this case, if the transmitting terminal is the terminal with the least number of successful transmissions in the look-back window (i.e., $\delta^t_{\arg\min \mathcal{M}} = 1$), we reward the transmitting terminal as well as other terminals (for being idle)—a positive reward is assigned to all terminals. In contrast, if $\delta^t_{\arg\min \mathcal{M}} = 0$, we punish all terminals.

Finally, a zero reward is assigned to all terminals if no one transmits in slot t.

Overall, to attain a positive reward, the terminals must cooperate to (1) achieve an ACK instead of a NACK – they must learn to maximize the throughput while avoiding collisions; (2) balance the transmission opportunities to achieve an almost equal number of successful transmissions in the look-back window – intuitively, a terminal with less successful transmissions in the look-back window should learn to be more aggressive, while a terminal with more successful transmissions in the look-back window should learn to be less aggressive.

With the above design principles, our reward function is aligned with the optimization objective of (4.3), and hence, can potentially learn a good set of transmission policies.

The α-Reward

In addition to the reward given in Definition 4.5, we have considered many other forms of reward. In general, the reward can be global (where all terminals receive the same reward) or individual (where the terminals receive distinct rewards). We empirically found that global reward is a better fit for our design. For benchmark purposes, we introduce another global reward design, dubbed the α-reward, that utilizes the α-fairness of all terminals during the look-back window as the reward. Specifically, we define

$$\widetilde{r}_t = \sum_{n=1}^{N} f(\mathcal{T}^t_n), \tag{4.9}$$

where $\mathcal{T}^t_n$ is the throughput of the n-th terminal computed over the duration $[t - W, t - 1]$, i.e., the look-back window of slot t. As can be seen, the α-reward matches the objective (4.3) straightforwardly – a terminal is rewarded for the increase of throughput $\mathcal{T}_n^t$, but the reward gets smaller and smaller with the increase of $\mathcal{T}_n^t$ since f is a concave function. The performances of (4.8) and (4.9) will be compared in Sect. 4.2.3.

MADRL-HT

Given the definitions of partial observations and reward, this subsection presents the overall architecture of MADRL-HT for the AutoCA problem. Let us set out to define the actor and critic networks.

Definition 4.6 (Actor) The action of a terminal is determined by the transmission policy π_n. We parameterize this policy by a DNN, dubbed "actor", with a vector of parameters $\boldsymbol{\theta}_n$. The actor takes the observation o_n^t as input and outputs action a_n^t for slot t, that is,

$$a_n^t = \Psi_{\theta_n}(O_n^t). \tag{4.10}$$

Each terminal is equipped with an actor.

Definition 4.7 (Critic) We define a critic network to evaluate the value function of a global state $V(S^t)$. The critic network is parameterized by $\boldsymbol{\phi}$, giving

$$V(S^t) = \Psi_{\boldsymbol{\phi}}(S^t). \tag{4.11}$$

The critic is deployed at the AP.

Since the transmission history of each terminal is a temporal sequence, we design both actor and critic to be recurrent neural networks (RNNs) with bi-directional LSTM [34] as the recurrent units. As shown in Fig. 4.5, the actor and critic share the same architecture with the only difference being the input and output.

1. *Input.* For the actor at the n-th terminal, the input is a $3 \times W$ dimensional matrix consisting of the transmission history of the n-th terminal, OH(n), and TH(n) in the look-back window, as given in (4.6). For the critic, the input is a $N \times W$ dimensional matrix consisting of the transmission history of all the terminals in the look-back window, i.e., S^t in (4.4).

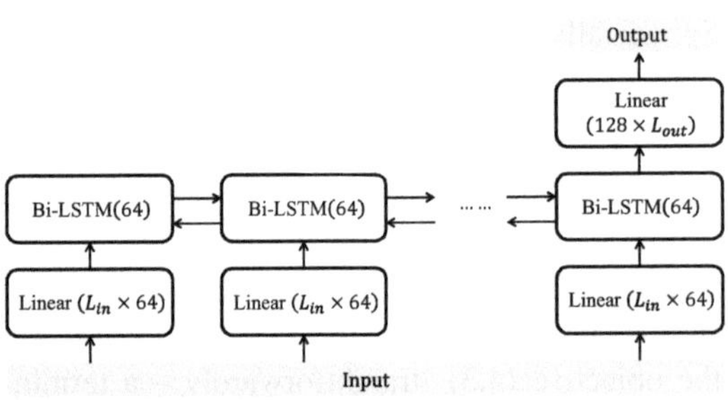

Fig. 4.5 The actor and critic networks are designed to be RNNs with Bi-LSTM units

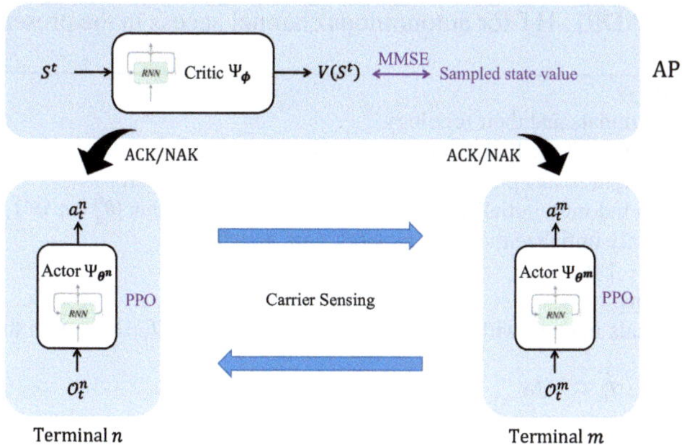

Fig. 4.6 The overall learning paradigm of MADRL-HT: terminals are equipped with actor networks and interact with each other in an episodic manner. The AP is equipped with a critic to estimate the state value function. At the end of an episode, the actors are trained by PPO and the critic is trained by regression on the MSE

2. *DNN*. The first layer is a linear layer of 64 neurons, where $L_{in} = 3$ for actors and $L_{in} = N$ for the critic. The second layer is a bi-directional LSTM layer with 64 hidden neurons. The third layer is a linear layer of 128 neurons, where $L_{out} = 2$ for actors and $L_{out} = 1$ for the critic. The ReLu activation function is applied to the output of each linear layer.
3. *Output*. For the actor at the n-th terminal, the output is a vector indicating the probability of transmission or being idle, i.e., output $= [\Pr(a_n^t = 0), \Pr(a_n^t = 1)]^\top$. For the critic, the output is the estimated value function $V(S^t)$.

The overall learning paradigm is illustrated in Fig. 4.6 and explained in detail in Algorithm 7. As shown, the terminals execute their actor/policy networks to interact with each other in an episodic manner. During the episode, the terminals continuously update their observations based on their carrier sensed information and the feedback broadcasted by the AP. The AP, on the other hand, computes the reward according to (4.8). Notice that a reward r_t is computable only D slots after the decision is made at the beginning of slot t.

At the end of an episode, the AP collects the actions of the terminals to form the global state S^t and estimates the state-value function of S^t from the current critic. Then, the AP computes the advantage function for each global state S^t, giving[2]

$$A_t = \sum_{t'>t} \gamma^{t'-t} r^{t'} - \Psi_{\theta_k}(S^t), \quad \forall t, \tag{4.12}$$

[2] In the implementation, we use generalized advantage estimation (GAE) [35] to compute the advantage functions to make the training more stable.

Algorithm 7 MADRL-HT for autonomous channel access in the presence of hidden terminals

```
Initialization:
Initialize N terminals and their topology;
Initialize the length of the look-back window W;
Initialize the length of an episode T_e and the number of episodes K;
The terminals randomly initialize the actor networks with weights {θ_n^0 : n = 1, 2, ..., N};
The AP randomly initialize the critic network with weights φ^0.
for k = 0 : K − 1 do
    Interactions:
    The terminals interact with each other for one episode of T_e slots, with the current actor networks Ψ_{θ_n^0}.
    for t = W : T_e − 1 do
        The terminals update their observations {o_n^t} from their carrier sensed information and the feedback from the AP, following Definition (4.5).
        The terminals perform the actions {a_n^t} and record the action chosen probabilities Ψ_{θ_n^k}(a_n^t | o_n^t), ∀n.
        The AP computes the reward.
    At the end of the episode, the AP computes the advantage function A_t, ∀t following (4.12) and broadcasts them to all terminals.
    DNN training:
    Upon receiving {A_t}, each terminal updates its actor following (4.13);
    The AP updates the critic by the regression on MSE, as shown in (4.14).
```

where γ is a discounting factor, and broadcasts them to all terminals.

The update of actor and critic networks follows the proximal policy optimization (PPO) algorithm [36]. Specifically, upon receiving $\{A_t\}$, each terminal updates its actor by maximizing a PPO-clip objective:

$$\boldsymbol{\theta}_n^{k+1} = \arg\max_{\boldsymbol{\theta}_n} \sum_{t=W}^{T_e-1} \min\left(\frac{\Psi_{\boldsymbol{\theta}_n}(a_n^t \mid o_n^t)}{\Psi_{\boldsymbol{\theta}_n^k}(a_n^t \mid o_n^t)} A_t, g(\epsilon, A_t) \right), \tag{4.13}$$

where

$$g(\epsilon, A) = \begin{cases} (1+\epsilon)A, & A \geq 0; \\ (1-\epsilon)A, & A < 0, \end{cases}$$

and ϵ is a clipping ratio.

On the other hand, the AP updates the critic by minimizing the MSE between the estimated state value $\Psi_{\boldsymbol{\phi}}(S^t)$ and the sampled state value from the interactions:

$$\boldsymbol{\phi}^{k+1} = \arg\min_{\phi} \sum_{t=W}^{T_e} \left(\sum_{t'>t} \gamma^{t'-t} r^{t'} - \Psi_{\phi}(S^t) \right). \tag{4.14}$$

It is worth noting that our MADRL-HT solution can also operate in a centralized training and distributed execution manner, in which case the AP trains both the actors and critic and periodically transmits the latest actor networks to the terminals.

In our MADRL-HT approach, a sequence of W slots of history is fed into the actor and decoder for decision making and value function estimation, respectively. In contrast, the widely-adopted learning architecture in the literature [28, 37, 38] assumes that the past history has been captured by the hidden state of the RNNs and feeds only the latest slot into the RNNs for decision making. We point out that this widely-used architecture cannot be used in our problem because of the partial observability of each terminal. In our solution, the observations and rewards of a terminal are updated based on the feedback from the AP (see Eq. (4.8) and Algorithm 6), which is delayed for D slots. As a result, we have to revisit and modify past observations and rewards upon receiving the ACK/NACK feedback.

4.2.3 *Numerical Evaluations*

This section performs extensive numerical experiments to verify the performance of MADRL-HT. We consider two benchmarks: one is the optimal set of transmission policies derived for the topologies considered in the experiments; the other is the CSMA/CA protocol, with which a terminal senses the common channel before accessing it and retransmits – in the case of a collision – after an exponential backoff.

4.2.3.1 Experimental Setup

The hyper-parameter settings for the experiments are shown in Table 4.1. We consider a slotted IoT BSS, where one slot lasts for 9 μs and a packet lasts for 5 slots. A terminal can be either an intelligent terminal operated with the actor of MADRL-HT or a standard terminal operated with CSMA/CA. The intelligent terminals make transmission decisions based on the transmission history of the last $W = 40$ slots. A standard terminal, on the other hand, follows the CSMA/CA protocol. In particular, we set the minimum and maximum contention windows of CSMA/CA to $\text{CW}_0 = 2$ slots and $\text{CW}_{\max} = 128$ slots, respectively. The DIFS is set to 1 slot.

The topologies of the BSS considered in the experiments are summarized in Table 4.2, in which uppercase letters stand for terminals in the BSS and we use vertical lines to separate TH neighbors. In Topo3', for example, $\{A, B \mid C\}$ means

Table 4.1 A list of hyper-parameters

	Hyperparameters	Values
System	Length of a slot	9 μs
	Packet length D	5 slots
	Look-back window W	40 slots
PPO	Number of epochs K	10,000
	Number of slots/episode T_e	100 slots
	Clipping ratio ε	0.2
	Learning rate for actor	0.001
	Learning rate for critic	0.0005
	Discount factor γ	0.99
	Lambda for GAE λ	0.95
CSMA/CA	Minimum CW: CW_0	2 slots
	Maximum CW: CW_{max}	128 slots
	DIFS	1 slot

Table 4.2 A summary of the topologies considered in the experiments

Abbreviations	Topological structures
Topo2	$\{A, B\}$
Topo2'	$\{A \mid B\}$
Topo3	$\{A, B, C\}$
Topo3'	$\{A, B \mid C\}$
Topo3"	$\{A, B \mid B, C\}$
Topo4	$\{A, B, C, D\}$
Topo4'	$\{A, B, C \mid D\}$
Topo4"	$\{A, B \mid B, C \mid D\}$

there are three terminals in the BSS: terminals A and B are OH neighbors to each other, while terminal C is a TH neighbor to both A and B.

For a given set of transmission policies $\boldsymbol{\pi}$, the key performance indicator is the α-fairness $F(\boldsymbol{\pi})$, where α is set to 1 (proportional fairness). Therefore, we have

$$F(\boldsymbol{\pi}) = \sum_{n=1}^{N} \log(\mathcal{T}_n(\boldsymbol{\pi}) + \varepsilon), \tag{4.15}$$

where $\varepsilon = 0.001$ is a constant to ensure that $F(\boldsymbol{\pi})$ does not go to $-\infty$. In addition to α-fairness, other metrics, such as throughput of individual terminals, packet collision rate, packet delay and jitter, are also considered to analyze the learned behaviors of the terminals.

Short-Term Fairness Versus Long-Term Fairness

It is worth noting that α-fairness is a function of the evaluation period T, as shown in (4.1). In general, the short-term fairness with a small T is much more difficult to achieve than the long-term fairness with a large T [39]. For a fair comparison in the experiments, the evaluation duration T is set to 0.01 s to compute the throughput $\mathcal{T}_n(\pi)$ and α-fairness for different schemes.

4.2.3.2 Reward Function and Observation Design

We first perform experiments to validate our designs of reward and observation, respectively.

In the first experiment, we evaluate our reward function r_t in (4.8) benchmarked against the α-reward $\widetilde{r}_t$ in (4.9). The experiments are performed on the topology Topo2, i.e., there are two terminals A and B in the IoT BSS and they are OH neighbors to each other.

Experimental results are presented in Fig. 4.8a, where the x-axis is time (i.e., the number of slots consumed in training) and the y-axis is the normalized α-fairness achieved by the learned actors at the corresponding time. In particular, the α-fairness is normalized to the region $[0, 1]$. The lower bound '0' corresponds to the performance of a set of never-transmit policies, under which the terminals will never transmit and $F_{\text{LB}}(\pi) = 2\log\varepsilon$. The upper bound '1', on the other hand, corresponds to the performance of the optimal policy. In Topo2, an optimal transmission pattern is 'A0B0A0B0A0B0…', where '0' stands for the DIFS. This is because terminals A and B are OH neighbors and are subject to the LBT constraint. Therefore, the optimal BSS throughput of Topo2 is $D/(D + \text{DIFS}) \approx 5/6$ and the optimal α-fairness measure is $F_{\text{UB}}(\pi) = 2\log(\frac{5}{12} + \varepsilon)$.

For both reward designs r_t and $\widetilde{r}_t$, we run the experiments 10 times (same for the experiments below) and present the mean (the solid lines in Fig. 4.8a) and standard deviation (the shaded areas in Fig. 4.8a) of the achieved α-fairness. As can be seen, MADRL-HT cannot learn a good set of transmission policies with the α-reward $\widetilde{r}_t$, despite the more intuitive interpretation of the reward function. In comparison, the proposed reward function r_t achieves the optimal α-fairness—the performance gain over the α-reward is up to 47% when the learning converges.

The throughput achieved by each terminal in Fig. 4.8a is more revealing to understand why the α reward does not yield good α-fairness. In Fig. 4.8b, c, we choose two typical experiments in Fig. 4.8a and study the throughput of each terminal over the course of training. In particular, Fig. 4.8b adopts the α-reward $\widetilde{r}_t$, and Fig. 4.8c adopts the proposed reward r^t. As shown, the BSS (sum) throughput can be maximized with both reward designs. Nevertheless, the α-reward leads to a set of policies that are quite unfair: terminal A monopolizes the channel all the time, while terminal B has no transmission opportunities. In comparison, our reward design yields fair transmission policies: the terminals have equal access to the common channel and the BSS throughput is maximized.

In the second experiment, we evaluate our observation design $\mathcal{O}_n^t$ in (4.5) benchmarked against the conventional design $\widetilde{\mathcal{O}}_n^t$ in (4.6). The experiments are

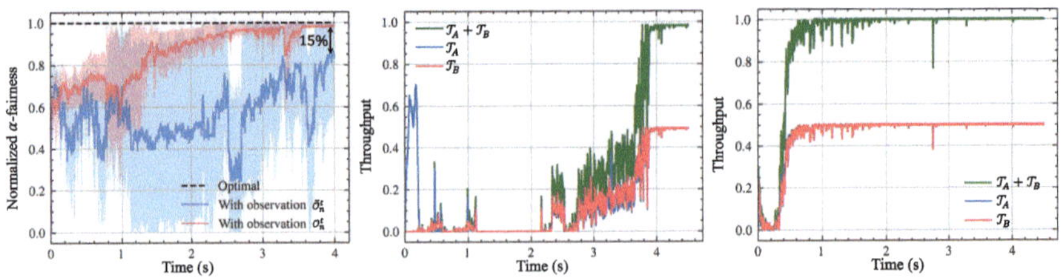

(a) Normalized α-fairness (b) Throughput with $\widetilde{O}_n^t$ (c) Throughput with O_n^t

Fig. 4.7 Performance of MADRL-HT in Topo2' with the proposed observation design O_n^t and the conventional observation design $\widetilde{O}_n^t$: (**a**) normalized α-fairness; (**b**) throughput with $\widetilde{O}_n^t$; (**c**) throughput with O_n^t

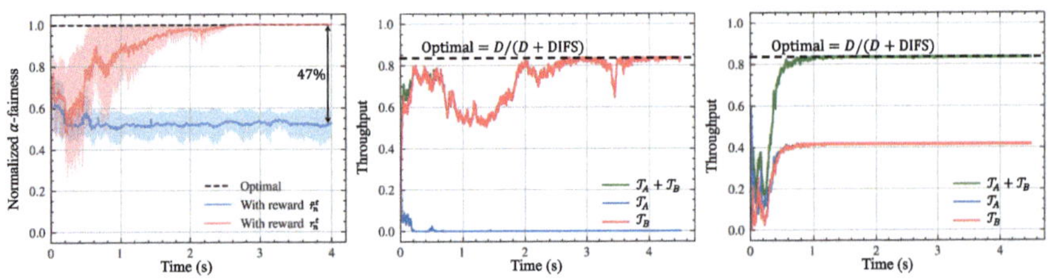

(a) Normalized α-fairness (b) Throughput with $\widetilde{r}_t$ (c) Throughput with r_t

Fig. 4.8 Performance of MADRL-HT in Topo2 with the proposed reward design r_t and the conventional reward design $\widetilde{r}_t$: (**a**) normalized α-fairness; (**b**) throughput with $\widetilde{r}_t$; (**c**) throughput with r_t

performed on the topology Topo2', i.e., there are two terminals A and B in the IoT BSS and they are TH neighbors to each other.

The normalized α-fairness versus time performance is shown in Fig. 4.7a, where the α-fairness is normalized to [0, 1] as Fig. 4.8a. Unlike Fig. 4.8a, the maximum achievable α-fairness in Topo2', i.e., what the normalized '1' stands for, is $F_{\text{UB}}(\boldsymbol{\pi}) = 2\log(\frac{1}{2} + \varepsilon)$. The reason is that an optimal transmission pattern in Topo2' is 'ABABAB…', since terminals A and B are TH neighbors – the sensed channel is always idle for each of them and it is possible for them to transmit alternately with no DIFS. Therefore, the optimal BSS throughput is 1 and the optimal α-fairness is $2\log(\frac{1}{2} + \varepsilon)$.

As can be seen from Fig. 4.7a, with the proposed observation design, MADRL-HT achieves much faster and more reliable learning. The average α-fairness performance of our design is 15% better than the conventional design when the learning converges. In addition to α-fairness, we choose two specific experiments and study the throughput of each terminal in Fig. 4.7b, c. In the chosen experiments, MADRL-HT can learn close-to-optimal policies with both observation designs, but our design yields much faster learning. We have further observed that, in more complex topologies that involve more terminals, the conventional observation design can lead to very poor performance.

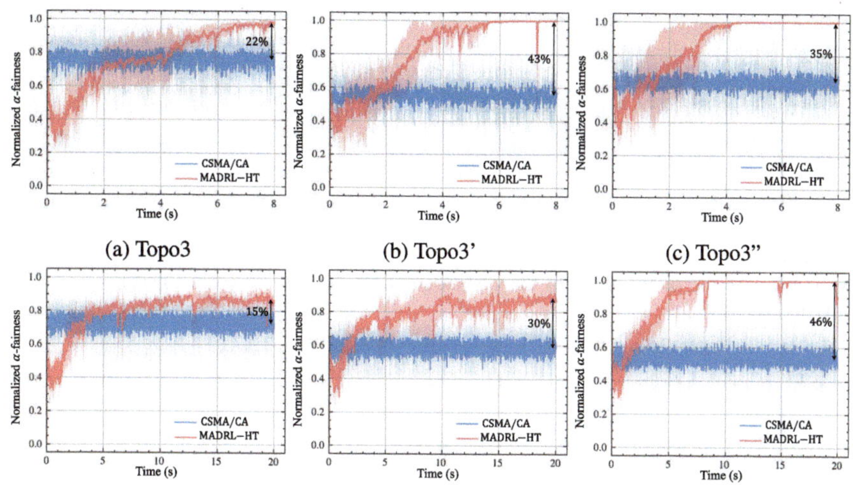

(a) Topo3 (b) Topo3' (c) Topo3"
(d) Topo4 (e) Topo4' (f) Topo4"

Fig. 4.9 The normalized α-fairness performance of a homogeneous MADRL-HT BSS and a standard CSMA/CA BSS in various topologies. (**a**) Topo3. (**b**) Topo3'. (**c**) Topo3". (**d**) Topo4. (**e**) Topo4'. (**f**) Topo4"

4.2.3.3 MADRL-HT Versus CSMA/CA

In this section, we perform extensive experiments under various topologies to evaluate the performance of our MADRL-HT solution to the AutoCA problem. Specifically, we consider a MADRL-HT BSS, where all terminals are intelligent, benchmarked against a standard CSMA/CA BSS, where all terminals are operated with the CSMA/CA protocol.

(1) Normalized α-Fairness Figure 4.9 presents the normalized α-fairness performance of both BSSs under various topologies listed in Table 4.2. As Fig. 4.8, we run each experiment multiple times and present the mean and standard deviation performances. With our MADRL-HT solution, the terminals in the BSS start from random channel-access policies and arrive at a set of policies that is remarkably better than the CSMA/CA protocol in various topologies. In terms of the average α-fairness performance, the gains of the MADRL-HT BSS over the CSMA/CA BSS are up to 46%.

To better understand the learned behaviors of the MADRL-HT terminals, next we focus on two specific topologies: Topo4 and Topo4', and analyze other quality-of-service (QoS) metrics, i.e., the achieved throughput of each terminal, packet collision rate, average packet delay and jitter, for both BSSs.

(2) Throughput of Individual Terminals We first study the achieved throughput of individual terminals. In Topo4, there is no hidden terminal. The achieved throughputs of individual terminals in both BSSs are shown in Fig. 4.10 by curves,

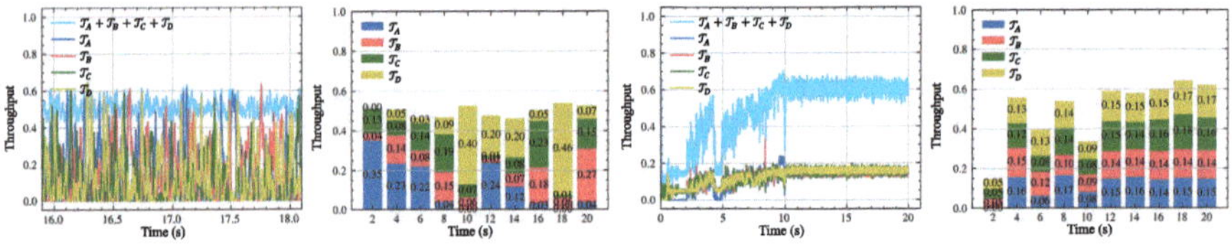

(a) CSMA/CA (b) CSMA/CA (c) MADRL-HT (d) MADRL-HT

Fig. 4.10 Topo4: the achieved throughput of each terminal in the BSS over the course of training. (**a**) and (**b**) are the performance of the CSMA/CA BSS; (**c**) and (**d**) are the performance of the MADRL-HT BSS

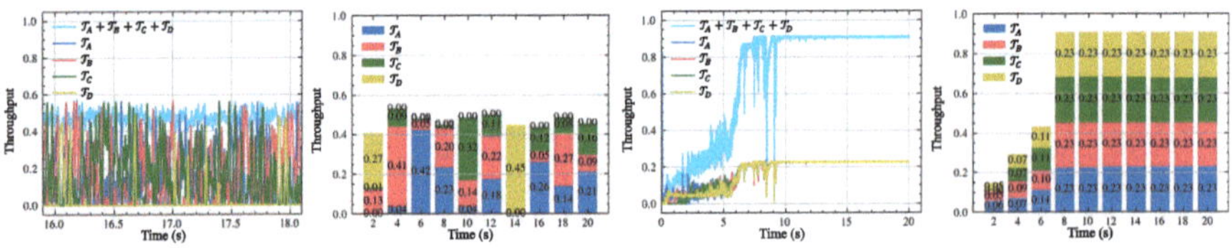

(a) CSMA/CA (b) CSMA/CA (c) MADRL-HT (d) MADRL-HT

Fig. 4.11 Topo4': the achieved throughput of each terminal in the BSS over the course of training. (**a**) and (**b**) are the performance of the CSMA/CA BSS; (**c**) and (**d**) are the performance of the MADRL-HT BSS

i.e., (a) and (c), and histograms i.e., (b) and (d). Likewise, the throughput of each terminal versus training time in Topo4' (with hidden terminal) is presented in Fig. 4.11.

We have two main observations from Figs. 4.10 and 4.11.

- For the CSMA/CA BSS, the channel-access policy of each terminal is fixed. The throughput of each terminal is characterized by fluctuations – the throughput of individual terminals exhibits an impulse characteristic, and is quite unfair in both Topo4 and Topo4'. Comparing Fig. 4.10b with Fig. 4.11b, hidden terminals degrade the BSS throughput and aggravate the unfairness among terminals.
- For the MADRL-HT BSS, the throughput achieved by individual terminals is almost equal, and the BSS throughput outperforms the CSMA/CA BSS in both Topo4 and Topo4'. An interesting observation is that MADRL-HT learns better in Topo4' than that in Topo4: the convergence is faster and the optimal BSS throughput is achieved.

(3) Packet Collision Rate (PCR) Figure 4.12a, b presents the PCR for both BSSs, where PCR is defined as the ratio between the number of collisions and the total transmission attempts. When hidden terminals are presented, the average PCR of the CSMA/CA BSS increases from 23% (in Fig. 4.12a) to 37% (in Fig. 4.12b). For the MADRL-HT BSS, on the other hand, the PCR reduces very quickly in the initial phase of training (within 0.2 s) in both Topo4 and Topo4'. However, as can be seen from Figs. 4.10c and 4.11c, the system throughput is quite low at that time. This

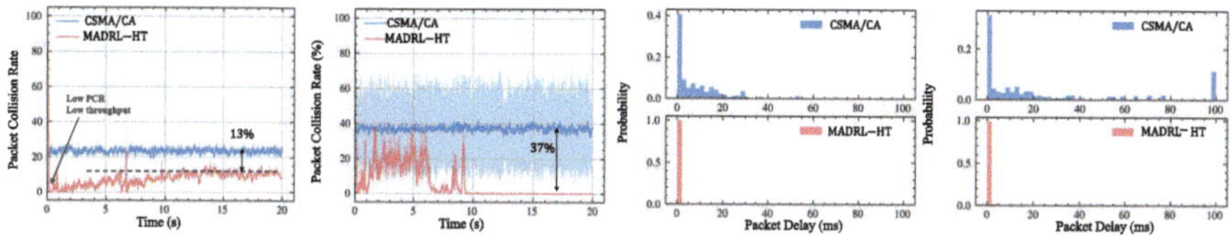

(a) Topo4, PCR (b) Topo4', PCR (c) Topo4, packet delay (d) Topo4', packet delay

Fig. 4.12 The performance of the CSMA/CA BSS and the MADRL-HT BSS: (**a** and **b**) the average packet collision rate (PCR), (**c** and **d**) the PDF of packet delay, where a packet is dropped if it cannot be transmitted in 100 ms. (**a**) Topo4, PCR. (**b**) Topo4', PCR. (**c**) Topo4, packet delay. (**d**) Topo4', packet delay

means that the terminals' transmission policies are conservative—they prefer not to transmit to avoid negative rewards caused by collisions. As learning proceeds, the terminals start to explore more active policies to reap positive rewards. Although the PCR increases, the system throughput steadily improves. When the learning converges, the average PCR of the MADRL-HT BSS is 13% lower than that of the CSMA/CA BSS in Topo4, and 37% lower in Topo4'.

(4) Delay and Jitter The final performance measure we shall evaluate is the packet delay and jitter [40] of the two BSSs. For both Topo4 and Topo4', we run the CSMA/CA BSS and the MADRL-HT BSS (using the trained model after 20 s) for 1 s, record the delay of all packets, and plot the probability density function (PDF) of the packet delay in Fig. 4.12c, d. In particular, a packet will be dropped if it is not transmitted within 100 ms. In Topo4, the average packet delay of the CSMA/CA BSS is 8.23 ms and the delay jitter is 10.17 ms. When it comes to Topo4' with a hidden terminal, the average packet delay increases to 25.06 ms, and the delay jitter increases to 32.93 ms. In contrast, the average packet delays of the MADRL-HT BSS under Topo4 and Topo4' are only 0.44 ms and 0.21 ms, respectively, while the jitters of the packet delay are 0.22 ms and 0.03 ms, respectively. To summarize, with our MADRL-HT solution, the average packet delay is reduced by 94.7% and 99.2%, and the delay jitter is reduced by 97.9% and 99.9%, in Topo4 and Topo4', respectively.

Overall, under the considered setup, our MADRL-HT solution attains excellent performance gains over the legacy CSMA/CA protocol in terms of various QoS metrics. Nevertheless, it is worth noting that our solution does not always converge to the optimal. Take Topo4' in Fig. 4.9e for example. There is still a 10% gap between the average α-fairness achieved by the MADRL-HT BSS and the global optimum. A more general observation is that obtaining a good set of transmission policies is becoming increasingly harder as the number of terminals increases. In actuality, this phenomenon is not surprising. More terminals lead to a more intricate transmission history. To exploit useful insights and patterns from history for decision-making, the learning task itself becomes much more complex. In our MADRL-HT design, the dimension of the input matrix to an actor is $3 \times W$. The

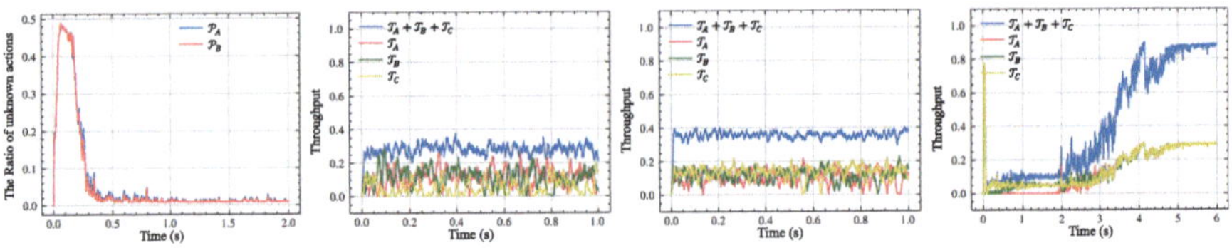

(a) Evaluating unknown actions (b) CSMA/CA (c) CSMA/CA with RT-S/CTS (d) MADRL-HT

Fig. 4.13 (**a**) The number of remaining unknown actions in the terminals' observations as a function of the training time in Topo2'. (**b–d**) A comparison among CSMA/CA, CSMA/CA with RTS/CTS, and MADRL-HT in Topo3'. (**a**) Evaluating unknown actions. (**b**) CSMA/CA. (**c**) CSMA/CA with RTS/CTS. (**d**) MADRL-HT

number of rows is fixed to 3 regardless of the number of terminals, thanks to the topological analysis in Sect. 4.2.2. The number of columns W (i.e., the look-back window length), however, depends on the number of terminals in the BSS—a BSS with more terminals requires a larger W to capture enough interactions among terminals to make the right decision. With the increase in W, the actors should be able to extract long-range correlations among the inputs, and this poses greater challenges to state-of-the-art DNNs. Empirically, with more terminals, the learning is more likely to converge to a local minimum. The above challenges call for more efficient schemes to compress the columns of the input matrix and extract useful features.

We point out that the learning performance of MADRL-HT can be reflected by the number of remaining unknown actions in the terminals' observations. One example is given in Fig. 4.13a, where we consider Topo2' and plot the portion of unknown actions in each terminal's observation. As can be seen, the portion of unknown actions is large at the beginning of training. In this phase, the system throughput is also very poor. As learning progresses, the number of unknowns in the observations starts to decrease, and finally, vanishes. Correspondingly, the system throughput reaches the global optimum.

Finally, we compare the performance of MADRL-HT with "CSMA/CA with RTS/CTS". Under our setup, the packet length is fixed to 5 slots. This corresponds to a packet of about 640 bytes, if we consider an OFDM system with 80 MHz bandwidth, 312.5 kHz subcarrier spacing, 16QAM modulation, and $1/2$ channel coding rate. For RTS/CTS, we assume that both RTS and CTS packets consume only one slot and there is no SIFS between them (i.e., the CTS feedback is instantaneous).[3] The simulation is performed in Topo3' and the results are shown in Fig. 4.13b–d. As can be seen,

[3] These are optimistic assumptions that would benefit the performance of CSMA/CA with RTS/CTS.

- Compared with CSMA/CA, the RTS/CTS mechanism resolves the hidden terminal problem, and hence, improves both the throughput performance and fairness among terminals. Nevertheless, the throughput gain is not significant, because the RTS and CTS packets themselves consume extra channel resources.
- Our MADRL-HT scheme achieves the best performance in terms of both throughput and fairness. When the training converges, the throughput and fairness are close to optimal.

4.2.4 Discussion

In this section, we developed a new learning-based multiple-access paradigm, termed MADRL-HT, for Autonomous Channel Access (AutoCA) in the presence of hidden terminals within IoT networks. By leveraging MADRL, our approach enables a group of IoT terminals to autonomously learn optimal transmission policies based on their local perceptions of the environment and feedback from the AP. This allows the terminals to adapt to each other's transmission behaviors, achieving a fair and high-throughput multiple-access protocol in a distributed and scalable manner.

The challenges addressed in this section are multifaceted. The hidden terminal problem introduces partial observability, as terminals cannot directly detect the presence or actions of hidden terminals. Effective transmission policies require terminals to infer the behaviors of these hidden nodes to avoid collisions and optimize network performance. Additionally, non-stationarity is inherent in multi-agent environments where agents (terminals) are part of each other's learning environment, and their policies evolve continuously during training. This results in a highly dynamic environment, especially in the initial phases of learning, further complicated by hidden non-stationarity due to the unseen actions of hidden terminals.

To overcome these challenges, this section made advances in several key areas:

- We formulated the AutoCA problem incorporating the practical challenge of hidden terminals and proposed an actor-critic MADRL paradigm to solve it. Through extensive numerical experiments, we validated the superior performance of our MADRL-HT solution, benchmarking it against the legacy CSMA/CA protocol across various Quality of Service (QoS) metrics.
- To develop a scalable learning paradigm, we leveraged topological insights to partition other terminals into two groups based on the number of communication hops from each terminal's perspective. By doing so, all other terminals can be effectively consolidated into two entities, regardless of their actual number, simplifying the learning process. This concept of terminal clustering based on communication hops is crucial for devising a scalable and compact learning paradigm and may offer new perspectives for designing efficient learning algorithms in large-scale networks.

- We introduced a look-back mechanism to mitigate partial observability and alleviate hidden non-stationarity. By revising the carrier-sensed channel state and inferring hidden terminals' behaviors from the ACK/NACK feedback of the AP, terminals gain a more accurate understanding of the network state. This mechanism can be readily extended to other wireless networking problems involving hidden terminals, enhancing its broader applicability.
- To maximize network throughput while ensuring fairness, we devised a novel reward function based on the relative number of successful transmissions over a look-back window. This reward function focuses on the terminals' relative performance; a positive global reward is achieved as long as the differences in their successful transmissions are not excessive. This promotes a more progressive and cooperative learning process. Empirically, we found that the window-based reward is particularly beneficial during the early stages of training when transmission policies are highly variable.

Overall, we demonstrate the significant potential of AI-driven multiple access protocols in enhancing the performance and scalability of IoT networks. By using AI to design new MAC protocols, we have shown that IoT terminals can make intelligent, autonomous decisions to optimize network performance. This not only addresses the current limitations of traditional protocols like CSMA/CA but also paves the way for more adaptive and efficient communication strategies.

4.3 DRL-based MAC Design for Feedback-Enhanced IoT

Chapter 2 presents a feedback-enhanced IoT (FE-IoT) communication framework and shows how we can leverage feedback for efficient coding to reduce the energy consumption of IoT devices. This section investigates the MAC layer design for FE-IoT, aiming to assess the potential enhancements in lifespan for an IoT cell comprising multiple devices. A key differentiator of the FE-IoT system at the MAC layer, when compared to traditional communication systems, is the incorporation of feedback as a new data type. This integration necessitates a reevaluation of resource allocation strategies, particularly the need for managing both uplink data and associated downlink feedback within a dual-channel framework.

4.3.1 A Half-Duplex Frequency Division Duplex MAC Design

We consider an IoT cell featuring an AP and L distributed IoT devices deployed for environmental sensing, as depicted in Fig. 4.14. The MAC operates in the half-duplex frequency division duplex (HD-FDD) mode. Time is segmented into cycles, denoted by $t = 0, 1, 2, \cdots$. In each cycle, IoT devices are required to transmit their sensing data, quantified in K bits, to the AP. Each cycle is structured into two

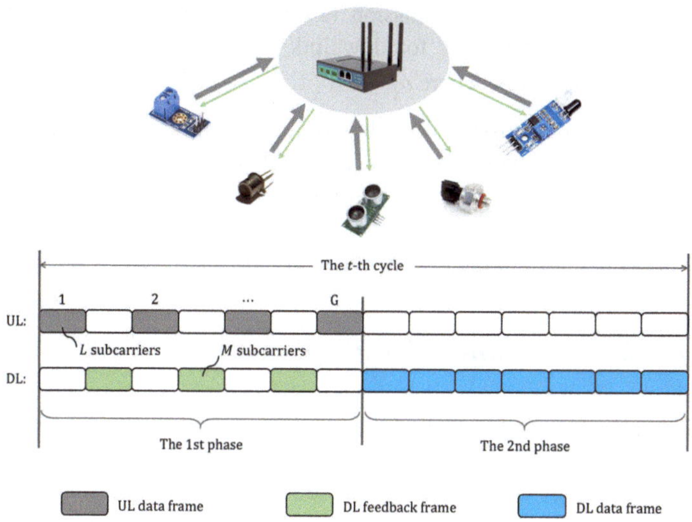

Fig. 4.14 The system model and duplexing mode of FE-IoT

phases: the 1st phase for uplink data transmission from the devices, and the 2nd phase for downlink data transmission from the AP. A key differentiator of FE-IoT from traditional communication systems is its approach to uplink data transmission. Hence, this section primarily concentrates on the 1st phase of each cycle.

Without loss of generality, we focus on the 1st phase of the t-th cycle, and details the process by which devices transmit their data to the AP. As illustrated in Fig. 4.14, FE-IoT divides the 1st phase into $2G - 1$ sequential data frames, with G frames allocated for uplink data transmission, termed uplink data frames, and the remaining $G - 1$ frames for downlink feedback, termed downlink feedback frames. As can be seen, this interspersing of downlink feedback frames amongst uplink data frames is a distinctive feature that sets FE-IoT apart from legacy communication systems. The purpose of the AP sending these feedback frames to the IoT devices is to enable them to ascertain the current decoding status of the AP. This knowledge allows the devices to modify their encoding schemes for subsequent transmissions. Such a mechanism significantly enhances the uplink channel coding efficiency. As a result, the IoT devices can achieve the same coverage area and communication performance as traditional systems but with reduced power consumption.

We assume that each uplink data frame and downlink feedback frame comprises Q OFDM symbols. Specifically,

- The uplink data transmission is operated with SC-FDMA. Unless stated otherwise, we assume in this section that each device occupies a single subcarrier, akin to the NB-IoT, to ensure sufficient coverage and signal penetration. The total count of subcarriers used for uplink data transmission is thus L. Adapting FE-IoT for cases where each device occupies multiple subcarriers is straightforward.

- The downlink feedback transmission is operated with OFDMA. We assume availability of M subcarriers for downlink feedback, and each IoT device is assigned $a_\ell^{(t)}$ subcarriers, where $a_\ell^{(t)} \in \{0, 1, 2, \cdots, M\}, \ell = 1, 2, \cdots, L$.

Overall, for the transmission of K bits in the 1st phase, each user sequentially transmits G uplink frames using the assigned subcarrier. The total count of real channel symbols each device transmits to the AP is $2QG$. In between two consecutive uplink frames, the AP feeds back a downlink frame. The total count of real symbols fed back to the ℓ-th device is $2Q(G-1)a_\ell^{(t)}$.

4.3.2 Feedback Channel Allocation

We consider a joint sensing task of IoT devices. Due to the distributed deployment of IoT devices, they experience diverse channel conditions with the AP. Over time, the communication energy consumption of these IoT devices varies, and the depletion of energy in any single device leads to the failure of the joint sensing task.

In legacy communication systems, devices under poorer channel conditions consume more energy for data transmission to achieve the desired decoding performance. Under such circumstances, the life expectancy of the sensing IoT network is capped by the device with the worst channel condition. FE-IoT offers an innovative approach to this dilemma: by strategically allocating more feedback to devices in poorer channel conditions and with lower energy reserves, we can boost their coding efficiency and reduce their energy consumption, ultimately extending the lifespan of the entire IoT cell. In the following, we shall formulate the feedback channel allocation problem and examine the potential gain of FE-IoT.

We start by introducing some definitions:

- The overall lifespan of the IoT cell is T cycles, and the initial energy reserve of the ℓ-th IoT device is ρ_ℓ.
- At the beginning of the t-th cycle, the remaining energy reserve of the ℓ-th IoT device is $\mathrm{E}_\ell^{(t)}$.
- In the t-th cycle, the energy consumed by the ℓ-th IoT device for transmitting, receiving, and sleep are denoted by $\mathrm{ET}_\ell^{(t)}$, $\mathrm{ER}_\ell^{(t)}$, and $\mathrm{ES}_\ell^{(t)}$, respectively.
- The receive power and sleep power of an IoT device are denoted by P_r and P_s, respectively.
- The duration of an OFDM symbol is T_{OFDM}.

Recall that the number of feedback subcarriers allocated to the ℓ-th IoT device in the t-th cycle is $a_\ell^{(t)} \in \mathbb{N}^+$, the feedback channel allocation problem can be formulated as follows:

$$\max_{\{a_\ell^{(t)}\}} \mathbb{E}\left[T : \mathrm{E}_\ell^{(T)} \geq 0, \forall \ell, \min_\ell \mathrm{E}_\ell^{(T+1)} < 0 \,\middle|\, \mathrm{E}_\ell^{(1)} = \rho_\ell\right] \tag{4.16a}$$

$$\text{s.t. } \mathrm{E}_\ell^{(t+1)} = \mathrm{E}_\ell^{(t)} - \mathrm{ET}_\ell^{(t)} - \mathrm{ER}_\ell^{(t)} - \mathrm{ES}_\ell^{(t)}, \tag{4.16b}$$

$$\mathrm{ET}_\ell^{(t)} = P_\ell^{(t)} T_{\mathrm{OFDM}} Q G, \tag{4.16c}$$

$$\mathrm{ER}_\ell^{(t)} = \left[1 - \delta(a_\ell^{(t)} = 0)\right] P_r T_{\mathrm{OFDM}} Q(G-1), \tag{4.16d}$$

$$\mathrm{ES}_\ell^{(t)} = \delta(a_\ell^{(t)} = 0) P_s T_{\mathrm{OFDM}} Q(G-1), \tag{4.16e}$$

$$P_\ell^{(t)} = \min\left\{ P_{\max}, \eta^* \left(\widetilde{h}_\ell^{(t)}, a_\ell^{(t)}\right) \frac{N_0}{\alpha_\ell |h_\ell^{(t)}|^2} \right\}, \tag{4.16f}$$

$$\sum_{\ell=1}^{L} a_\ell^{(t)} \leq M, \ \forall t. \tag{4.16g}$$

This formulation underscores our primary objective: to extend the operational lifespan of the IoT network, i.e., T in (4.16a), by judiciously distributing subcarriers among IoT devices. Equations (4.16b) to (4.16e) quantify the energy consumption of each IoT device within a given cycle, accounting for activities such as transmission, receiving, and periods of inactivity. In each cycle, the transmit power of an IoT device is configured based on the feedback channel quality and the number of subcarriers assigned to the device, as specified in (4.16f). Additionally, (4.16g) delineates the constraints on the total allocation of feedback subcarriers.

4.3.2.1 DP with Function Approximation

Given that the subcarriers are allocated across successive cycles, (4.16) is essentially a sequential decision-making problem. Consequently, it is feasible to model (4.16) as a Markov Decision Process (MDP) and approach its resolution through dynamic programming (DP). In the beginning of the t-th cycle, we characterize the state of the IoT cell as $\boldsymbol{s}^{(t)} \triangleq \left\{\left[\mathrm{E}_\ell^{(t)}, h_\ell^{(t)}, \widetilde{h}_\ell^{(t)}\right] : \forall \ell\right\}$. In particular, a state $\boldsymbol{s}^{(t)}$ is designated as terminal if and only if $\min_\ell \mathrm{E}_\ell^{(t)} < 0$. The collection of all terminal states is denoted by $\mathcal{S}_{\mathrm{ter}}$. The AP implements a subcarrier allocation policy π that dictates the allocation actions, i.e., the subcarrier distributions for the t-th cycle $\boldsymbol{a}^{(t)} \triangleq \left\{a_\ell^{(t)} : \forall \ell\right\}$, based on the current state. That is, $\boldsymbol{a}^{(t)} = \pi\left(\boldsymbol{s}^{(t)}\right)$. The cardinality of the action space $\mathcal{A}$ is $|\mathcal{A}| = \sum_{i=0}^{M} \binom{L+i-1}{L-1}$.

Upon executing action $a_\ell^{(t)}$, each IoT device tunes its transmit power $P_\ell^{(t)}$ based on the uplink and downlink channel gains, $h_\ell^{(t)}$ and $\widetilde{h}_\ell^{(t)}$, and communicates with the AP with the FE-IoT physical layer specified in Sect. 2.3.3.2. Transitioning to the start of the $(t+1)$-th cycle, the system evolves into a new state, $\boldsymbol{s}^{(t+1)}$, in accordance with the dynamics outlined in (4.16). Given that $\mathrm{E}_\ell^{(t)} < \mathrm{E}_\ell^{(t+1)}$, $\forall \ell$, it is inevitable for the system state to eventually reach a terminal state, positioning the MDP within the scope of a finite-horizon problem. A cycle yields an instantaneous reward of $r^{(t)} = 1$, provided that the system state $\boldsymbol{s}^{(t)}$ is not a terminal state.

Overall, the objective of the MDP is to maximize the cumulative reward, $\sum_{t=1}^{T} r^{(t)} = T$, across an episode. Here, T denotes the cycle at which the system first encounters a terminal state within an episode, aligning this optimization target with the goal of the original feedback channel allocation problem in (4.16). The optimal solution to the finite horizon MDP, and hence to (4.16), can be achieved through DP as follows:

$$q_*(\boldsymbol{s}, \boldsymbol{a}) = \delta(\boldsymbol{s} \notin \mathcal{S}_{\text{ter}}) + \mathbb{E}_{\boldsymbol{s}'}\left[\max_{\boldsymbol{a}'} q_*(\boldsymbol{s}', \boldsymbol{a}')\right], \tag{4.17}$$

where $q_*(\boldsymbol{s}, \boldsymbol{a})$ represents the optimal action-value function.

The continuous nature of the state space $\boldsymbol{s}$, however, renders the direct application of the value iteration algorithm (VIA) impractical due to the necessity of state space quantization, which is computationally prohibitive. To navigate this challenge, we adopt value function approximation, i.e., estimate the optimal action-value function through a function approximation, thereby facilitating efficient computation and circumventing the direct quantization of the continuous state space. Specifically, we approximate the optimal action-value function by:

$$q_*(\boldsymbol{s}, \boldsymbol{a}) \approx \Phi_\varphi(\boldsymbol{s}, \boldsymbol{a}), \tag{4.18}$$

where Φ_φ denotes a DNN parameterized by a vector of weights φ. The discovery of an appropriate vector of DNN weights φ that fulfills (4.18) paves the way to determining the optimal policy π_*. To this end, for a pair of state $\boldsymbol{s} = \left\{\left[\mathrm{E}_\ell, h_\ell, \widetilde{h}_\ell\right] : \forall \ell\right\}$ and action $\boldsymbol{a}$, we introduce a quadratic loss function $\mathcal{L}(\boldsymbol{s}, \boldsymbol{a}; \varphi)$ to facilitate the training of φ through gradient descent:

$$\boldsymbol{s}' = \left\{\left[\mathrm{E}_\ell - \mathrm{ET}_\ell - \mathrm{ER}_\ell - \mathrm{ES}_\ell, h'_\ell, \widetilde{h}'_\ell\right] : \forall \ell\right\}, \tag{4.19}$$

$$\mathcal{L}(\boldsymbol{s}, \boldsymbol{a}; \varphi) = \Bigg[\Phi_\varphi(\boldsymbol{s}, \boldsymbol{a}) - \delta(\boldsymbol{s} \notin \mathcal{S}_{\text{ter}}) - \int_{\{h'_\ell, \widetilde{h}'_\ell\}} \max_{\boldsymbol{a}'} \Phi_\varphi(\boldsymbol{s}', \boldsymbol{a}') \Pr\left(\{h'_\ell, \widetilde{h}'_\ell\}\right) d\{h'_\ell, \widetilde{h}'_\ell\}\Bigg]^2, \tag{4.20}$$

$$\varphi \leftarrow \varphi - \epsilon \nabla_\varphi \mathcal{L}(\boldsymbol{s}, \boldsymbol{a}; \varphi), \tag{4.21}$$

where ϵ is the step size. Following the practice of reinforcement learning (RL) [41], we approximate the Φ_φ within the max operator in (4.20) by another DNN. This dual network architecture helps to mitigate the correlation between successive state-action pairs' approximated q values, enhancing the stability and efficiency of the learning process.

4.3.3 *Index Table-Based Policy Gradient: A DRL Approach*

Employing function approximation to directly estimate the optimal action-value function in DP obviates the need for discretizing the state space, thereby avoiding the quantization errors. Furthermore, it capitalizes on the known dynamics of the system to forge a more robust solution.

Nonetheless, integrating DP with function approximation presents its own set of hurdles. A particularly daunting challenge is the requirement for exhaustive integral calculations across the uplink and downlink channel realizations for all IoT devices, as delineated in (4.20). These integrations call for significant computational effort, especially when the channel gain distributions are widespread and varied. This complexity stems from the inherent nature of full-width backup in DP—assessing the value of any state-action pair involves accounting for every conceivable subsequent state-action pairs. As the dimensionality of the problem escalates, the practicality of employing DP in conjunction with function approximation begins to wane due to these computational demands. To tackle this, we resort to sampling-based RL approaches.

For any subcarrier allocation policy π, we can sample a history of state transitions, known as a *trajectory* [42, 43]: $\tau = \{\boldsymbol{s}^{(1)}, \boldsymbol{a}^{(1)}, r^{(1)}, \boldsymbol{s}^{(2)}, \boldsymbol{a}^{(2)}, r^{(2)}, \ldots, \boldsymbol{s}^{(T)}, \boldsymbol{a}^{(T)}, r^{(T)}\}$. The objective is to refine the policy π to optimize the average sampled rewards. This approach is a cornerstone of policy gradient algorithms [42]. In this context, we can directly model the policy π by a DNN Ψ_{ψ}, characterized by a vector of parameters ψ. The policy is then iteratively updated to maximize the average total reward:

$$\psi \leftarrow \psi + \epsilon \nabla_{\psi} \mathbb{E}_{\tau \sim \pi_{\psi}}\big[T(\tau)\big], \tag{4.22}$$

where $T(\tau)$ denotes the lifespan of the IoT cell in a trajectory τ. It is easy to show that

$$\nabla_{\psi} \mathbb{E}_{\tau}\big[T(\tau)\big] = \mathbb{E}_{\tau}\left[\sum_{t=1}^{T(\tau)} \nabla_{\psi} \ln \pi_{\psi}\big(a^{(t)}(\tau)|s^{(t)}(\tau)\big) T(\tau)\right],$$

where

$$\Pr(\tau) = \sum_{t=1}^{T(\tau)} \pi_{\psi}\Big(a^{(t)}(\tau)|s^{(t)}(\tau)\Big) \Pr\Big(\{h_{\ell}^{(t+1)}(\tau), \widetilde{h}_{\ell}^{(t+1)}(\tau)\}\Big).$$

While this sampling-based Monte Carlo approach diverges from DP by not using the known system state transitions, it can implicitly learn about

(continued)

the system's dynamics through extensive and systematic sampling of state transitions, allowing for a more effective means of policy optimization to maximize rewards. Furthermore, the introduction of policy parameterization opens the door to new and low-complexity solutions, such as the index table-based scheme proposed below.

In DP with function approximation, the DNN Φ_ϕ is tasked with approximating the q values corresponding to specific state-action pairs: the input dimensionality is $3L + \sum_{i=0}^{M} \binom{L+i-1}{L-1}$, signifying the state and possible actions, while the output dimension is a singular value, representing the estimated q value. Conversely, within the framework of standard policy gradient methods, the DNN Φ_φ is employed to parameterize the policy itself. The input dimensionality of this configuration is $3L$, while the output dimensionality is $\sum_{i=0}^{M} \binom{L+i-1}{L-1}$, corresponding to the probability distribution over the action space. As can be seen, despite policy gradient methods circumventing the need to compute integrals numerically, the inherent training complexity remains directly proportional to the number of IoT devices L. This relationship poses substantial challenges in scenarios where L is large, complicating the training process due to the exponential increase in computational demands.

Addressing this challenge, this section puts forth a new, index table-based, low-complexity policy gradient method to solve (4.16) by leveraging insights from multi-armed bandit allocation indices [44]. The cornerstone of this approach is its emphasis on dimensionality reduction. By decoupling the L-dimensional MDP into M one-dimensional MDPs, we achieve a significant reduction in complexity. Specifically, this method ensures that the computational complexity grows linearly with L, and crucially, the training complexity remains invariant regardless of the number of IoT devices L.

Let us consider a single IoT device and examine how varying the number of feedback subcarriers influences its state transitions. Defining $\boldsymbol{s}_\ell^{(t)} \triangleq \left[\mathrm{E}_\ell^{(t)}, h_\ell^{(t)}, \widetilde{h}_\ell^{(t)}\right]$, the transition of $\boldsymbol{s}_\ell^{(t)}$ is a one-dimensional MDP. Therefore, the optimal action-value function can be written as

$$\begin{aligned} q_*(\boldsymbol{s}_\ell, a_\ell) &= \frac{1}{L}\delta(\mathrm{E}_\ell > 0) + \mathbb{E}_{\boldsymbol{s}'_\ell}\left[\max_{a'_\ell} q_*(\boldsymbol{s}'_\ell, a'_\ell)\right], \\ &\triangleq r_\ell + \mathbb{E}_{\boldsymbol{s}'_\ell}\left[v_*(\boldsymbol{s}'_\ell)\right], \end{aligned} \tag{4.23}$$

where the instantaneous reward of the IoT device is defined as $r_\ell = \frac{1}{L}\delta(\mathrm{E}_\ell > 0)$ because each device aims to maximize the operational longevity by ensuring its energy level remains above zero; $v_*(\boldsymbol{s}_\ell)$ is the value function of the state $\boldsymbol{s}'_\ell$, reflecting the long-term accumulated reward that can be reaped from being in state $\boldsymbol{s}'_\ell$. In the scenario of a single-device MDP, the state transitions are solely influenced by the actions taken for that particular device a_ℓ, but not $a_{\ell'}$, $\ell' \neq \ell$.

Lemma 4.1 (Monotonicity of the Value Function) *Consider any two states of the single-device MDP that share the same channel realizations:* $\boldsymbol{s}_\ell = \left[E_\ell, h_\ell, \widetilde{h}_\ell\right]$ *and* $\boldsymbol{s}'_\ell = \left[E'_\ell, h_\ell, \widetilde{h}_\ell\right]$. *If* $E_\ell \le E'_\ell$,

$$v_*(\boldsymbol{s}_\ell) \le v_*(\boldsymbol{s}'_\ell). \tag{4.24}$$

That is, $v_*(\boldsymbol{s}_\ell)$ *is monotonically non-decreasing in* E_ℓ *for a given pair of* h_ℓ *and* $\widetilde{h}_\ell$.

Proof For the state $\boldsymbol{s}_\ell$, we have

$$\begin{aligned} v_*(\boldsymbol{s}_\ell) &= \mathbb{E}_{\tau\sim\pi_*}\left[\sum_{t=1}^{T(\tau)} r_\ell^{(t)} \middle| \boldsymbol{s}_\ell^{(1)} = \boldsymbol{s}_\ell\right] \\ &= \int_{\left\{h_\ell^{(t)}(\tau), \widetilde{h}_\ell^{(t)}(\tau), a_\ell^{(t)}(\tau)\right\}_{t=1}^{T(\tau)}} T(\tau)\Pr(\tau)d\tau, \end{aligned} \tag{4.25}$$

where

$$\Pr(\tau) = \sum_{t=1}^{T(\tau)} \pi_*\left(a_\ell^{(t)}(\tau)|s_\ell^{(t)}(\tau)\right)\Pr\left(h_\ell^{(t+1)}(\tau), \widetilde{h}_\ell^{(t+1)}(\tau)\right).$$

To simply notation, we define $e_\tau(1, T(\tau)) \triangleq \left\{h_\ell^{(t)}(\tau), \widetilde{h}_\ell^{(t)}(\tau), a_\ell^{(t)}(\tau)\right\}_{t=1}^{T(\tau)}$.

For the state $\boldsymbol{s}'_\ell$, we construct a new policy π':

$$a_\ell^{(t)}(\tau) = \pi'\left(h_\ell^{(t)}(\tau), \widetilde{h}_\ell^{(t)}(\tau)\right),$$

which produces actions based only on the uplink and downlink channel realizations. In particular, the behavior of π' matches exactly with the trajectory τ in (4.25), and

$$v_{\pi'}(\boldsymbol{s}'_\ell) \le v_*(\boldsymbol{s}'_\ell). \tag{4.26}$$

Following (4.25), we have

$$\begin{aligned} &v_*(\boldsymbol{s}_\ell) - v_{\pi'}(\boldsymbol{s}'_\ell) \\ &= \int_{e_\tau(1,T(\tau))} T(\tau)\Pr(\tau)d\tau - \int_{e_{\tau'}(1,T(\tau'))} T(\tau')\Pr(\tau')d\tau'. \end{aligned} \tag{4.27}$$

Notice that the uplink and downlink channels are independent of the actions, and the policy π' is designed such that $a_\ell^{(t)}(\tau) = a_\ell^{(t)}(\tau')$ under the same channel conditions. There must exist a critical cycle T_c, for which (1) $e_\tau(1, T_c) = e_{\tau'}(1, T_c)$, and (2) either $\mathrm{E}_\ell^{(T_c+1)}(\tau) < 0$ or $\mathrm{E}_\ell^{(T_c+1)}(\tau') < 0$.

It can be verified from (4.16b) that the energy expenditures of τ and τ' in the first T_c cycles are exactly the same. Therefore,

$$\mathrm{E}_\ell^{(T_c+1)}(\tau') - \mathrm{E}_\ell^{(T_c+1)}(\tau) = \mathrm{E}_\ell^{(1)}(\tau') - \mathrm{E}_\ell^{(1)}(\tau) = \mathrm{E}'_\ell - \mathrm{E}_\ell \geq 0.$$

This suggests that τ enters the terminal state after T_c cycles. We can then partition the trajectory τ' into two phases: $\tau' = [\tau, \tau_c]$, where the number of cycles in τ is T_c.

The second term on the right hand side (RHS) of (4.27) can be refined as

$$\begin{aligned}
&\int_{e_{\tau'}(1,T(\tau'))} T(\tau')\Pr(\tau')d\tau' \qquad (4.28)\\
&= \int_{e_\tau(1,T(\tau))}\int_{e_{\tau'}(T(\tau)+1,T(\tau'))} \Big[T(\tau) + T(\tau_c)\Big]\Pr(\tau)\Pr(\tau_c)d\tau'\\
&= \int_{e_\tau(1,T(\tau))} T(\tau)\Pr(\tau)d\tau + \int_{e_{\tau'}(1,T(\tau'))} T(\tau_c)\Pr(\tau')d\tau'.
\end{aligned}$$

Substituting (4.28) into (4.27) yields

$$v_*(s_\ell) - v_{\pi'}(s'_\ell) = -\int_{e_{\tau'}(1,T(\tau'))} T(\tau_c)\Pr(\tau')d\tau' \leq 0. \qquad (4.29)$$

Combining (4.26) and (4.29), we have

$$v_*(s_\ell) \leq v_{\pi'}(s'_\ell) \leq v_*(s'_\ell). \qquad (4.30)$$

Lemma 4.1 is proved. □

To evaluate the effectiveness of an action a_ℓ when in a state s_ℓ, we set $a_\ell = 0$ as a reference point, and define an index for each action $a_\ell > 0$. The index will provide a quantitative measure to assess each action's relative benefit compared to the baseline.

Definition 4.8 (Index of an Action) In any state s_ℓ, we define an index for each action $a_\ell > 0$:

$$\mu(s_\ell, a_\ell) \triangleq q_*(s_\ell, a_\ell) - q_*(s_\ell, 0). \qquad (4.31)$$

Intuitively, the index of an action $\mu(a_\ell)$ can be understood as an allocating benefits, representing the benefits that can be derived from allocating a_ℓ subcarriers to the IoT device. Theorem 4.1 below summarizes the main properties of the indexes.

Theorem 4.1 (Properties of the Indexes) *Consider a single-device MDP and a state* $s_\ell = \big[E_\ell, h_\ell, \widetilde{h}_\ell\big]$.

- *Monotonicity. The indexes* $\mu(\mathbf{s}_\ell, a_\ell)$, $\forall a_\ell > 0$, *are monotonic: when the downlink channel SNR* $\widetilde{\eta}_{dB} \geq -\frac{u_1}{u_2}$, $\mu(\mathbf{s}_\ell, a_\ell)$ *is monotonically non-decreasing in* a_ℓ*; otherwise, when* $\widetilde{\eta}_{dB} < -\frac{u_1}{u_2}$, $\mu(\mathbf{s}_\ell, a_\ell)$ *is monotonically non-increasing in* a_ℓ.
- *Sign. The index* $\mu(\mathbf{s}_\ell, a_\ell) \leq 0$ *if and only if*

$$\eta^*\left(\widetilde{h}_\ell, a_\ell\right) \geq \min\left\{\eta_\ell^{max}, \eta_\ell^*(0)\right\} - \frac{G-1}{G}(P_r - P_s)\frac{\alpha_\ell |h_\ell|^2}{N_0}, \tag{4.32}$$

where $\eta_\ell^{max} \triangleq \frac{\alpha_\ell |h_\ell|^2}{N_0} P_{\max}$ *denotes the maximum uplink received SNR at the AP when the IoT device transmits at its maximum power* $P_{\max}$*;* $\eta_\ell^*(0)$ *denotes the minimum uplink SNR required for conventional forward channel codes to achieve a target PER.*[4]

Proof We first prove the monotonicity of the indexes. From the definition of the index in (4.31), we have

$$\begin{aligned}
\frac{\partial \mu(\mathbf{s}_\ell, a_\ell)}{\partial a_\ell} &= \frac{\partial q_*(\mathbf{s}_\ell, a_\ell)}{\partial a_\ell} \\
&= \int_{h'_\ell, \widetilde{h}'_\ell} \frac{\partial v_*\left(\mathbf{s}'_\ell = \left[\mathrm{E}'_\ell, h'_\ell, \widetilde{h}'_\ell\right]\right)}{\partial a_\ell} \Pr(h'_\ell, \widetilde{h}'_\ell) dh'_\ell d\widetilde{h}'_\ell \\
&= \mathbb{E}_{h'_\ell, \widetilde{h}'_\ell}\left[\frac{\partial v_*\left(\mathbf{s}'_\ell\right)}{\partial \mathrm{E}'_\ell}\frac{\partial \mathrm{E}'_\ell}{\partial a_\ell}\right].
\end{aligned} \tag{4.33}$$

From Lemma 4.1, we have $\frac{\partial v_*(\mathbf{s}'_\ell)}{\partial \mathrm{E}'_\ell} \geq 0$. Furthermore,

$$\begin{aligned}
\frac{\partial \mathrm{E}'_\ell}{\partial a_\ell} &= -\frac{\partial \mathrm{ET}'_\ell}{\partial a_\ell} \\
&= -\frac{\partial}{\partial a_\ell}\min\left\{P_{\max}, 10^{\frac{1}{10}\eta^*_{\mathrm{dB}}(\widetilde{h}_\ell, a_\ell)}\frac{N_0}{\alpha_\ell |h_\ell|^2}\right\} T_{\mathrm{OFDM}} Q G,
\end{aligned}$$

where

$$\frac{\partial \eta^*_{\mathrm{dB}}\left(\widetilde{h}_\ell, a_\ell\right)}{\partial a_\ell} = -\left[\frac{1}{\exp\left(u_0\widetilde{\eta}_{\mathrm{dB}} + u_1 a_\ell + u_2\widetilde{\eta}_{\mathrm{dB}} a_\ell + u_3\right) + u_4}\right]^2 \exp\left(u_0\widetilde{\eta}_{\mathrm{dB}} + u_1 a_\ell + u_2\widetilde{\eta}_{\mathrm{dB}} a_\ell + u_3\right)\left(u_1 + u_2\widetilde{\eta}_{\mathrm{dB}}\right).$$

[4] Given a specific uplink channel gain h_ℓ, both $\eta_\ell^{\max}$ and $\eta_\ell^*(0)$ are constants. The value of $\eta_\ell^*(0)$ is determined by the choice of the conventional forward channel code, e.g., Polar and Turbo codes.

Therefore, the sign of $\frac{\partial \mathrm{E}'_\ell}{\partial a_\ell}$ and $\frac{\partial \mu(s_\ell, a_\ell)}{\partial a_\ell}$ depends on $u_1 + u_2\widetilde{\eta}_{\mathrm{dB}}$. Since $u_2 > 0$, when $\widetilde{\eta}_{\mathrm{dB}} \geq -\frac{u_1}{u_2}$, $\frac{\partial \mu(s_\ell, a_\ell)}{\partial a_\ell} \geq 0$; when $\widetilde{\eta}_{\mathrm{dB}} < -\frac{u_1}{u_2}$, $\frac{\partial \mu(s_\ell, a_\ell)}{\partial a_\ell} \leq 0$. The monotonicity of the indexes is proved.

Next, we prove the sign of the indexes. The index of an action $a_\ell > 0$ can be further written as:

$$
\begin{aligned}
&\mu(s_\ell, a_\ell) = q_*(s_\ell, a_\ell) - q_*(s_\ell, 0) \\
&= \mathbb{E}_{h'_\ell, \widetilde{h}'_\ell}\left[v_*\left(\left[\mathrm{E}'_\ell(a_\ell), h'_\ell, \widetilde{h}'_\ell\right]\right) - v_*\left(\left[\mathrm{E}'_\ell(0), h'_\ell, \widetilde{h}'_\ell\right]\right)\right],
\end{aligned} \tag{4.34}
$$

where

$$
\begin{aligned}
\mathrm{E}'_\ell(a_\ell) &= \mathrm{E}_\ell - \mathrm{ET}_\ell(h_\ell, \widetilde{h}_\ell, a_\ell) - \mathrm{ER}_\ell, \\
\mathrm{E}'_\ell(0) &= \mathrm{E}_\ell - \mathrm{ET}_\ell(h_\ell, \widetilde{h}_\ell, 0) - \mathrm{ES}_\ell.
\end{aligned}
$$

Since $v_*(s_\ell)$ is monotonic in E_ℓ, the sign of $\mu(s_\ell, a_\ell)$ depends on the sign of $\mathrm{E}'_\ell(a_\ell) - \mathrm{E}'_\ell(0)$.

$$
\begin{aligned}
&\mathrm{E}'_\ell(a_\ell) - \mathrm{E}'_\ell(0) \\
&= \mathrm{ET}_\ell(h_\ell, \widetilde{h}_\ell, 0) - \mathrm{ET}_\ell(h_\ell, \widetilde{h}_\ell, a_\ell) + \mathrm{ES}_\ell - \mathrm{ER}_\ell \\
&= \min\left\{P_{\max}, \eta^*(0)\frac{N_0}{\alpha_\ell |h_\ell|^2}\right\} T_{\mathrm{OFDM}} Q G - \min\left\{P_{\max},\right. \\
&\left.\eta^*\left(\widetilde{h}_\ell, a_\ell\right)\frac{N_0}{\alpha_\ell |h_\ell|^2}\right\} T_{\mathrm{OFDM}} Q G - (P_r - P_s) T_{\mathrm{OFDM}} Q (G-1),
\end{aligned} \tag{4.35}
$$

where $\eta^*(0)$ denotes the minimum uplink SNR required for traditional forward channel codes to achieve the target PER.

We denote by $\eta_\ell^{\max} \triangleq \frac{\alpha_\ell |h_\ell|^2}{N_0} P_{\max}$ the maximum uplink received SNR at the AP when the IoT device transmits with the maximum power $P_{\max}$, and $\eta_{\ell,\mathrm{dB}}^{\max}$ the decibel form of $\eta_\ell^{\max}$. Multiplying the both sides of (4.35) by $\frac{\alpha_\ell |h_\ell|^2}{N_0 T_{\mathrm{OFDM}} Q G}$, we have

$$
\begin{aligned}
H &\triangleq \frac{\alpha_\ell |h_\ell|^2}{N_0 T_{\mathrm{OFDM}} Q G}\left[\mathrm{E}'_\ell(a_\ell) - \mathrm{E}'_\ell(0)\right] \\
&= \min\left\{\eta_\ell^{\max}, \eta_\ell^*(0)\right\} - \min\left\{\eta_\ell^{\max}, \eta^*\left(\widetilde{h}_\ell, a_\ell\right)\right\} \\
&\quad - \frac{G-1}{G}(P_r - P_s)\frac{\alpha_\ell |h_\ell|^2}{N_0}.
\end{aligned} \tag{4.36}
$$

To further simplify (4.36), we have to compare $\eta_\ell^{\max}$, $\eta_\ell^*(0)$, and $\eta^*\left(\widetilde{h}_\ell, a_\ell\right)$.

Case I: $\eta_\ell^*(0) \geq \eta_\ell^{\max}$ and $\eta^*\left(\widetilde{h}_\ell, a_\ell\right) \geq \eta_\ell^{\max}$. In this case,

$$H = \eta_\ell^{\max} - \eta_\ell^{\max} - \frac{G-1}{G}(P_r - P_s)\frac{\alpha_\ell |h_\ell|^2}{N_0} < 0.$$

Therefore, $\mathrm{E}'_\ell(a_\ell) < \mathrm{E}'_\ell(0)$ and $\mu(s_\ell, a_\ell) < 0$.
Case II: $\eta_\ell^*(0) < \eta_\ell^{\max}$ and $\eta^*\left(\widetilde{h}_\ell, a_\ell\right) \geq \eta_\ell^{\max}$. In this case,

$$H = \eta_\ell^*(0) - \eta_\ell^{\max} - \frac{G-1}{G}(P_r - P_s)\frac{\alpha_\ell |h_\ell|^2}{N_0} < 0.$$

Therefore, we also have $\mathrm{E}'_\ell(a_\ell) < \mathrm{E}'_\ell(0)$ and $\mu(s_\ell, a_\ell) < 0$.

The common thread in the first two scenarios is that $\eta^*\left(\widetilde{h}_\ell, a_\ell\right) \geq \eta_\ell^{\max}$, indicating that the uplink received SNR required by FE-IoT, $\eta^*\left(\widetilde{h}_\ell, a_\ell\right)$, exceeds what the uplink link can possibly provide, even at the maximum transmission power. Consequently, it is advantageous to opt for traditional forward error correction coding rather than feedback coding under these conditions. This rationale underpins the result that $\mu(s_\ell, a_\ell) < 0$.
Case III: $\eta_\ell^*(0) \geq \eta_\ell^{\max}$ and $\eta^*\left(\widetilde{h}_\ell, a_\ell\right) < \eta_\ell^{\max}$. In this case,

$$H = \eta_\ell^{\max} - \eta^*\left(\widetilde{h}_\ell, a_\ell\right) - \frac{G-1}{G}(P_r - P_s)\frac{\alpha_\ell |h_\ell|^2}{N_0}.$$

Thus, $H \leq 0$ if $\eta^*\left(\widetilde{h}_\ell, a_\ell\right) \geq \eta_\ell^{\max} - \frac{G-1}{G}(P_r - P_s)\frac{\alpha_\ell |h_\ell|^2}{N_0}$.
Case IV: $\eta_\ell^*(0) < \eta_\ell^{\max}$ and $\eta^*\left(\widetilde{h}_\ell, a_\ell\right) < \eta_\ell^{\max}$. In this case,

$$H = \eta_\ell^*(0) - \eta^*\left(\widetilde{h}_\ell, a_\ell\right) - \frac{G-1}{G}(P_r - P_s)\frac{\alpha_\ell |h_\ell|^2}{N_0},$$

Thus, $H \leq 0$ if $\eta^*\left(\widetilde{h}_\ell, a_\ell\right) \geq \eta_\ell^*(0) - \frac{G-1}{G}(P_r - P_s)\frac{\alpha_\ell |h_\ell|^2}{N_0}$.

Combing the four cases, $H \leq 0$ and $\mu(s_\ell, a_\ell) \leq 0$ if and only if

$$\eta^*\left(\widetilde{h}_\ell, a_\ell\right) \geq \min\left\{\eta_\ell^*(0), \eta_\ell^{\max}\right\} - \frac{G-1}{G}(P_r - P_s)\frac{\alpha_\ell |h_\ell|^2}{N_0}.$$

The second part of Theorem 4.1 is proved. □

The monotonicity of the indexes $\mu(s_\ell, a_\ell)$, $\forall a_\ell > 0$, reveals a preference within the single-device MDP framework for the AP to either allocate all M subcarriers to the device or limit the allocation to just one subcarrier. However, it is crucial to highlight that in scenarios where the downlink channel SNR $\widetilde{\eta}_{\mathrm{dB}} < -\frac{u_1}{u_2}$, the downlink channel SNR is so low that the

(continued)

performance of feedback codes is often inferior to that of forward error correction codes, such as Polar or Turbo codes. Under these conditions, we typically have $\mu(s_\ell, a_\ell) < 0$, making it more advantageous not to allocate any subcarriers to the IoT device. As such, the scenario where $\widetilde{\eta}_{\text{dB}} < -\frac{u_1}{u_2}$ becomes irrelevant for FE-IoT, focusing our attention on conditions where $\widetilde{\eta}_{\text{dB}} \geq -\frac{u_1}{u_2}$. This delineation ensures that the indexes, $\mu(s_\ell, a_\ell)$, would maintain a monotonically non-decreasing pattern if allocation decisions are made within the viable SNR region.

With the implementation of the action index, we have established a metric to quantify the AP's readiness to allocate resources to each IoT device. Leveraging this concept, we proceed to introduce our index table-based policy gradient (ITPG) algorithm.

At the start of cycle t, the IoT cell finds itself in state $\boldsymbol{s}^{(t)} = \left\{\boldsymbol{s}_\ell^{(t)} : \forall \ell\right\}$ and our goal is to determine $\boldsymbol{a}^{(t)} = \left\{a_\ell^{(t)} : \forall \ell\right\} \in \mathcal{A}$. We first parameterize the index function μ by a DNN Ω_ω with a vector of weights ω. When given the state of a device $\boldsymbol{s}_\ell^{(t)}$ as the input, Ω_ω generates the indexes for all possible actions $a_\ell^{(t)}$:

$$\mu\left(\boldsymbol{s}_\ell^{(t)}, a_\ell^{(t)}\right) = \Omega_\omega\left(\boldsymbol{s}_\ell^{(t)}\right), \ \forall a_\ell^{(t)}. \tag{4.37}$$

With Ω_ω, we compute indexes $\mu(\boldsymbol{s}_\ell^{(t)}, a_\ell^{(t)})$ for each user $\ell = 1, 2, \ldots, L$ and their possible actions $a_\ell^{(t)} = 0, 1, 2, \ldots, M$. Given a total budget of M subcarriers, we can construct an index table μ_* for $\boldsymbol{a}^{(t)} = \left\{a_\ell^{(t)} : \forall \ell\right\} \in \mathcal{A}$, and select the action with the largest index. Specifically, the index of $\boldsymbol{a}^{(t)}$ measures the total benefits that can be gained by executing $\boldsymbol{a}^{(t)}$. Thus, it is a sum of the indexes of each IoT device when assigning $\boldsymbol{a}_\ell^{(t)}$ subcarriers to them, giving

$$\mu_*(\boldsymbol{s}^{(t)}, \boldsymbol{a}^{(t)}) \triangleq \sum_{\ell=1}^{L} \mu(\boldsymbol{s}_\ell^{(t)}, a_\ell^{(t)}). \tag{4.38}$$

where $\boldsymbol{s}_\ell^{(t)} \subset \boldsymbol{s}^{(t)}$ and $a_\ell^{(t)} \subset \boldsymbol{a}^{(t)}$. The action $\left(\boldsymbol{a}^{(t)}\right)^*$ that has the maximum index is denoted by

$$\left(\boldsymbol{a}^{(t)}\right)^* = \arg\max_{\boldsymbol{a}^{(t)}} \mu_*(\boldsymbol{s}^{(t)}, \boldsymbol{a}^{(t)}). \tag{4.39}$$

The number of entries in the index table μ_* is the cardinality of the action space $\mathcal{A}$. However, according to Theorem 4.1, some entries can be excluded from comparison.

Corollary 4.1 *When the IoT cell is in state* $\mathbf{s}^{(t)} = \{s_\ell^{(t)} : \forall \ell\}$, *a subcarrier allocation* $\widehat{\mathbf{a}}^{(t)} = \{\widehat{a}_\ell^{(t)} : \forall \ell\}$ *does not need to be considered when constructing the index table* μ_*, *if a device* ℓ *satisfies either of the following three conditions:*

1. $\eta^*\left(\widetilde{h}_\ell, M\right) \geq \eta_\ell^{max}$ *and* $\widehat{a}_\ell^{(t)} \geq 1$.
2. $\eta_\ell^*(0) \leq \eta_\ell^{max}$ *and* $1 \leq \widehat{a}_\ell^{(t)} \leq \lfloor a_{th} \rfloor$, *where*

$$a_{th} = \frac{1}{u_1 + u_2 \widetilde{\eta}_{\ell,dB}} \ln\left(\frac{1}{\eta'_{\ell,dB} - u_5} - u_4\right) - \frac{u_3 + u_0 \widetilde{\eta}_{\ell,dB}}{u_1 + u_2 \widetilde{\eta}_{\ell,dB}},$$

and $\eta'_{\ell,dB} = 10\lg\left(\eta_\ell^*(0) - \frac{G-1}{G}(P_r - P_s)\frac{\alpha_\ell |h_\ell|^2}{N_0}\right)$.

3. $\eta_\ell^*(0) > \eta_\ell^{max}$ *and* $1 \leq \widehat{a}_\ell^{(t)} \leq \lfloor a'_{th} \rfloor$, *where*

$$a'_{th} = \frac{1}{u_1 + u_2 \widetilde{\eta}_{\ell,dB}} \ln\left(\frac{1}{\eta_{\ell,dB}^{max} - u_5} - u_4\right) - \frac{u_3 + u_0 \widetilde{\eta}_{\ell,dB}}{u_1 + u_2 \widetilde{\eta}_{\ell,dB}}.$$

Proof (sketch) We consider the four cases in the proof of Theorem 4.1. In the first two cases, we have $\eta^*\left(\widetilde{h}_\ell, a_\ell\right) \geq \eta_\ell^{\max}$. From the monotonicity of the index and comment 4.3.3, it is known that the index is monotonically non-decreasing. Therefore, if $\eta^*\left(\widetilde{h}_\ell, M\right) \geq \eta_\ell^{\max}$, then no action of the IoT device can achieve a positive index. This suggests that no subcarriers should be allocated to the IoT device, which constitutes the first condition in the corollary.

Then, for case IV, where $\eta_\ell^*(0) < \eta_\ell^{\max}$ and $\eta^*\left(\widetilde{h}_\ell, a_\ell\right) < \eta_\ell^{\max}$, both traditional forward channel coding and FE-IoT can meet the receiver's PER requirement. In this case, we have to find the critical action a_{th} from which the index μ changes from negative to positive. This can be obtained by setting $H = 0$. The final critical value is given in the third condition of the corollary.

Finally, for case III, where $\eta_\ell^*(0) \geq \eta_\ell^{\max}$ and $\eta^*\left(\widetilde{h}_\ell, a_\ell\right) < \eta_\ell^{\max}$, traditional forward channel coding can no longer meet the receiver's PER requirement, and we must use FE-IoT for transmission. Let $\eta_\ell^{\max} = \eta^*\left(\widetilde{h}_\ell, a_\ell\right)$, we can obtain the critical action a'_{th}, as indicated by the second condition in the corollary. □

The general principle of Corollary 4.1 is to exclude actions that yield negative indexes for any IoT device, because allocating feedback resources to them would be less beneficial than not allocating at all. That said, not all actions with indexes $\mu(s_\ell^{(t)}, a_\ell^{(t)}) \leq 0, \forall a_\ell^{(t)}$ should be excluded from consideration. This is because, under certain conditions, traditional forward channel coding may fail to achieve the target PER at the AP. In such scenarios, the only viable option is FE-IoT. Essentially, the conditions given in Corollary 4.1 serve as a

(continued)

Algorithm 8 Index table-based policy gradient (ITPG)

Continuously sample trajectories and the update the index DNN Ω_ω.
for each trajectory τ **do**
    for $t = 1 : T(\tau)$ **do**
        Compute the indexes for each device:

$$\mu\left(s_\ell^{(t)}, a_\ell^{(t)}\right) = \Omega_\omega\left(s_\ell^{(t)}\right), \ \forall \ell.$$

        Construct an index table μ_*:
        if a pair of $s^{(t)}$ and $a^{(t)}$ satisfies the three conditions in Corollary 4.1 **then**
            $\mu_*\left(s_\ell^{(t)}, a_\ell^{(t)}\right) = 0.$
        else
            $\mu_*(s^{(t)}, a^{(t)}) \triangleq \sum_{\ell=1}^{L} \mu(s_\ell^{(t)}, a_\ell^{(t)}).$
        $\pi_\omega\left(a^{(t)}(\tau)|s^{(t)}(\tau)\right) = \frac{\exp\{\mu_*(s^{(t)}(\tau), a^{(t)}(\tau))\}}{\sum_{s^{(t)}, a^{(t)}} \exp\{\mu_*(s^{(t)}, a^{(t)})\}}.$
    At the end of τ,

$$\omega \leftarrow \omega + \epsilon \nabla_\omega \mathbb{E}_{\tau \sim \pi_\omega}\left[T(\tau)\right].$$

preliminary filtering process. They preemptively eliminate devices that do not qualify for subcarrier allocation, thereby simplifying the complexity involved in the decision-making process of the index table.

Overall, the ITPG algorithm is summarized in Algorithm 8.

4.3.4 Performance

This section evaluates the efficacy of the ITPG algorithm within the FE-IoT MAC layer through simulation. Our simulation environment consists of an IoT cell with L devices, and the number of feedback subcarriers is matched to the number of devices, i.e., $M = L$. The FE-IoT physical layer configuration adheres to the parameter settings outlined in Table 4.3.

To address the feedback channel allocation challenge in (4.16), we employ both the DP with function approximation method and the ITPG approach. These methods are utilized to derive channel allocation policies for FE-IoT, with the aim of optimizing the average lifespan of the IoT cell, measured in the maximum number of cycles the IoT cell can operate before one device depletes its battery.

In implementing the ITPG algorithm, we introduce two enhancements to the optimization of the index network Ω_ω i.e., the last row of Algorithm 8:

Table 4.3 Parameter settings for FE-IoT

Parameter	Description	Value
K	#bits per IoT device	48
N	#channel uses per IoT device	144
R	Channel coding rate	1/3
κ	#bits per group	3
G	#uplink frames per cycle	9
Q	#OFDM symbols per frame	8
M	#downlink subcarriers	4
α	Path loss coefficient	10^{-5}
σ^2	Uplink channel fading	5
$\widetilde{\sigma}^2$	Downlink channel fading	5
$P_{\max}$	Uplink maximum transmit power	0.5 A
$\widetilde{P}_{\max}$	Downlink transmit power	4 A
N_0	AWGN power density	−90 dBm

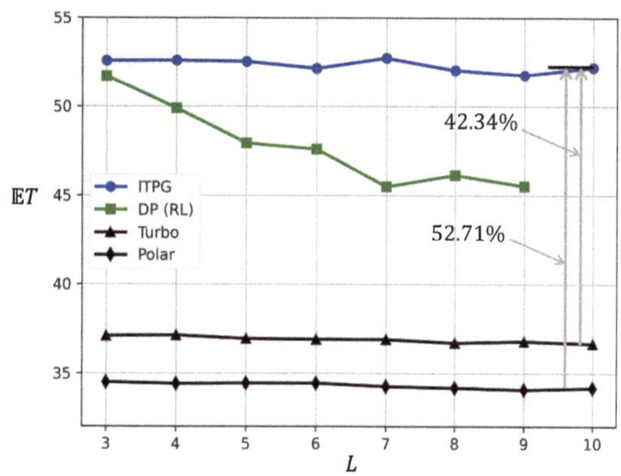

Fig. 4.15 The average lifespan T (# of cycles) of an IoT cell with L devices when operated with FE-IoT, benchmarked against traditional forward communication with Turbo and Polar codes

- We adopt a state-dependent instantaneous reward function $r^{(t)} = \frac{T}{\rho^T - 1}(\rho^t - \rho^{t-1})$, where $\rho = 1.07$ is a predefined constant, in place of the standard state-independent reward $r^{(t)} = 1$.
- We enhance the optimization process by substituting the gradient ascent method with the Proximal Policy Optimization (PPO) algorithm.

These advancements facilitate more stable and efficient policy updates, thereby improving the overall performance of the channel allocation process.

The simulation results, depicted in Fig. 4.15, benchmark the performance of FE-IoT (with DP and ITPG) against traditional forward communication methods employing Turbo and Polar codes.

First, we juxtapose the efficacy of DP with function approximation against the ITPG algorithm. The results illustrate that for a smaller number of IoT devices L, both ITPG and DP with function approximation deliver comparable average lifespans for the IoT cell. However, as L increases, the performance of DP with function approximation deteriorates markedly due to the rising complexity associated with identifying the optimal decision—a complexity that escalates exponentially with L. Conversely, ITPG's performance remains stable, unaffected by the increase in L, a testament to the efficiency of the index table-based approach in reducing the dimensionality of the MDP. Particularly, at $L = 10$, where the action space cardinality surges to 1.85×10^5, DP becomes impractical, whereas ITPG continues to excel.

When focusing on traditional forward communication utilizing Turbo and Polar codes, the average lifespan of the IoT cell diminishes as L grows. This is because, as the number of devices increases, the probability that at least one device experience deep fade—and thus deplete its battery prematurely—increases, adversely affecting the average lifespan of the IoT cell. Despite this, as depicted in Fig. 4.15, the observed performance reduction is relatively minor. Yet, the notable performance disparity between the ITPG method and traditional forward communication techniques underscores the considerable efficiency gains achievable with FE-IoT. Specifically, when employing ITPG, FE-IoT significantly enhances the IoT cell's performance, achieving a 42.34% improvement over traditional operations using Turbo codes, and a 52.71% enhancement over those with Polar codes.

Considering a scenario with $L = 4$ devices, Fig. 4.16 displays the energy consumption of each device over the lifespan of the IoT cell. The simulation results

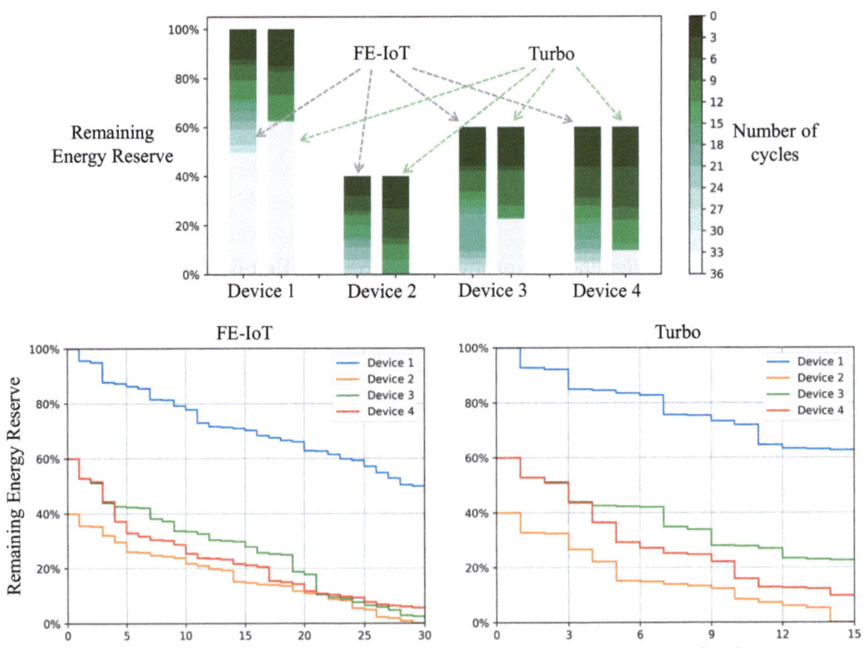

Fig. 4.16 The energy consumption of each device over the lifespan of the IoT cell with operated with FE-IoT and Turbo codes, where $L = 4$

clearly demonstrate how FE-IoT, through strategic allocation of feedback channel resources to devices with lower energy reserves, effectively balances the energy consumption across different devices. Compared to a traditional IoT cell utilizing Turbo codes, the FE-IoT system effectively doubles the lifespan of the cell.

4.4 Federated Learning-Driven Multiple Access

Having explored how AI can be utilized to design innovative multiple access protocols for IoT networks, we now shift our focus to how multiple access techniques can be specifically tailored to support AI applications within IoT environments. As AI becomes increasingly integrated into IoT systems, applications such as Federated Learning (FL) present unique challenges and opportunities for multiple access design. FL enables distributed IoT devices to collaboratively train machine learning models without sharing raw data, preserving privacy and reducing the need for centralized data storage. However, implementing FL in IoT networks introduces significant communication overhead and multiple-access bottlenecks due to the frequent exchange of model updates among numerous devices. In this part of the chapter, we delve into FL-driven multiple access, examining how specialized techniques can facilitate efficient and scalable FL in IoT edge computing.

4.4.1 IoT Federated Learning

The proliferation of IoT devices has led to an unprecedented generation of data at the network edge. This data holds immense potential for training machine learning models that can drive intelligent applications such as predictive maintenance, anomaly detection, and personalized services. However, transmitting vast amounts of raw data from distributed devices to a central server for model training poses significant challenges:

- **Bandwidth Limitations**: IoT networks often operate over constrained wireless channels, making it impractical to send large datasets to centralized locations.
- **Privacy Concerns**: Transmitting sensitive data can violate user privacy and regulatory requirements, particularly in domains like healthcare and finance.
- **Scalability Issues**: Centralized training becomes infeasible as the number of devices scales into the millions or billions.

Federated Learning emerges as a promising solution to these challenges. Introduced by Google in 2016, FL is a decentralized machine learning paradigm where devices collaboratively train a shared global model while keeping their local data on-device. The key characteristics of FL include privacy preservation, as raw data remains on the devices, reducing the risk of privacy breaches. Communication efficiency is also enhanced since only model updates—such as gradients or weights—

are transmitted, which are typically smaller than the raw data. Additionally, FL leverages edge computation, with devices performing local computations using their processing capabilities, thereby reducing the burden on central servers.

In the context of IoT, FL enables the harnessing of distributed data for machine learning without compromising privacy or overloading the network infrastructure. Applications of FL in IoT are diverse. In smart homes, devices like thermostats and security cameras collaboratively learn patterns to optimize energy usage or detect anomalies. In industrial IoT (IIoT), sensors and machinery in manufacturing plants contribute to models for predictive maintenance and operational efficiency. In healthcare monitoring, wearable devices train models for health monitoring and early detection of medical conditions while preserving patient data privacy.

The FL process typically involves the following steps:

1. **Initialization**: A central server initializes a global model and distributes it to participating IoT devices (clients).
2. **Local Training**: Each IoT device updates the global model using its local data by performing one or more iterations of training (e.g., stochastic gradient descent).
3. **Model Update Transmission**: IoT devices send their updated model parameters (e.g., weight updates) back to the central server.
4. **Aggregation**: The central server aggregates the updates from all IoT devices to create a new global model. Common aggregation methods include weighted averaging, where each device's contribution is weighted based on factors like the number of local data samples.
5. **Iteration**: Steps 2–4 are repeated for multiple rounds until the model converges or achieves satisfactory performance.

While FL offers significant advantages, implementing it in IoT networks introduces several challenges, particularly related to communication and multiple access. One major challenge is the communication overhead. Federated learning requires devices to frequently send model updates to the central server, leading to high communication demands. IoT networks often have limited bandwidth, and transmitting model updates from numerous devices can saturate the network. Moreover, communication is energy-intensive, which is a critical concern for battery-powered IoT devices.

Another challenge is the multiple-access bottleneck. Many devices attempt to transmit their model updates simultaneously, causing collisions and increased latency. Traditional multiple access protocols struggle to efficiently handle the massive number of devices participating in FL, and contention-based access methods can lead to underutilization of the available spectrum due to collisions and retransmissions.

4.4.2 *Computation-Oriented Multiple Access for IoT Federated Edge Learning*

To address the communication bottleneck in FL for IoT networks, analog over-the-air computation (AirComp) offers an innovative solution by exploiting the superposition property of wireless channels [45–56]. The basic idea of AirComp is to create and leverage inter-user interferences over the multiple-access channel rather than trying to avoid it. When operated with AirComp, devices send their model updates in an uncoded fashion by directly mapping each model parameter to a channel symbol: each device first precodes the transmitted symbols by the inverse of the UL channel gain (assumed to be known to the transmitter in advance) and then transmits the precoded symbols to the receiver in an analog fashion. All the participating devices transmit simultaneously in the same communication link such that their signals overlap at the PS. Provided that the channel-gain precoding and transmission timing are accurate, the fading multiple-access channel reduces to a Gaussian multiple-access channel and the signal overlapped from the devices to the PS over-the-air naturally produces the arithmetic sum of the local model-updates [52].

Compared with the traditional digital multiple-access schemes, wherein the communication and computation constitute separate processes, AirComp is a joint computation-and-communication scheme exploiting the fact that the multiple-access channel inherently yields an additive superposed signal.

4.4.2.1 System Model

We consider FL with the help of a wireless parameter server (PS) where IoT devices with distinct local datasets collaborate over the shared wireless medium to train a common model, as shown in Fig. 4.17. The learning process goes through many iterations. Without loss of generality, let us focus on one of the iterations, wherein M active devices participate in the training. The iteration proceeds as follows [57]:

1. DL broadcast: at the beginning of the iteration, the PS broadcasts the global model $\boldsymbol{\theta} \in \mathbb{R}^d$ to the M IoT devices;
2. Local training: each of the M devices trains the global model $\boldsymbol{\theta}$ on its local dataset $\mathcal{B}_m$ of size B_m and obtains a new model $\tilde{\boldsymbol{\theta}}_m \in \mathbb{R}^d$;
3. UL aggregation: each device scales the local model update $\tilde{\boldsymbol{\theta}}_m - \boldsymbol{\theta}$ by B_m and transmits the scaled model update $\boldsymbol{\theta}'_m = B_m(\tilde{\boldsymbol{\theta}}_m - \boldsymbol{\theta}) \in \mathbb{R}^d$ back to the PS;

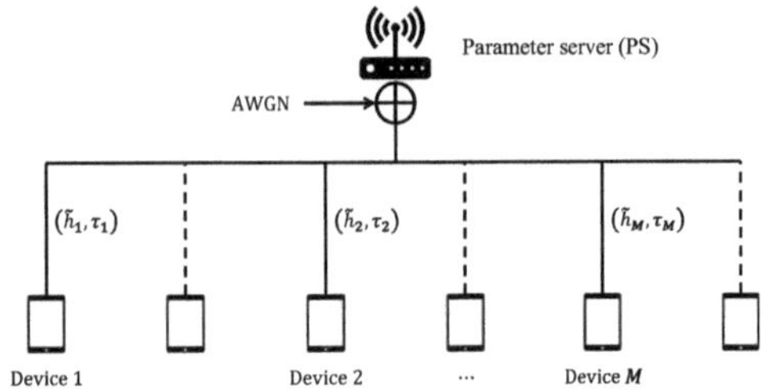

Fig. 4.17 In FL, IoT devices collaboratively train a shared model with the help of a wireless access point acting as a PS

4. Arithmetic-sum estimation: the PS estimates the arithmetic sum of the transmitted model-updates $\boldsymbol{\theta}'_m$ from the IoT devices:

$$\boldsymbol{\theta}_+ = \sum_{m=1}^{M} \boldsymbol{\theta}'_m; \tag{4.40}$$

5. Model update: the PS updates the global model by

$$\boldsymbol{\theta}_{\text{new}} = \boldsymbol{\theta} + \frac{1}{\sum_m B_m} \boldsymbol{\theta}_+. \tag{4.41}$$

The updated global model $\boldsymbol{\theta}_{\text{new}}$ is then broadcasted in the next iteration and the cycle continues.

To compute (4.41), the sum of the dataset sizes $\sum_m B_m$ has to be known to the PS a priori. Thus, we let each device transmit the local dataset size B_m to the PS reliably in advance of data transmission (in a digital manner over a control channel, with channel coding and automatic repeat request, for example).

Among the above five steps, the uplink model aggregation poses the greatest challenge to the multiple-access edge network. In this step, each device has to transmit d real numbers to the PS, where d can run into hundreds of millions or more. We consider analog AirComp to realize the uplink model aggregation.

When operated with AirComp, IoT devices transmit their raw model updates $\boldsymbol{\theta}'_m$ simultaneously to the PS in an analog manner (without digital modulation and channel coding). The PS, on the other hand, estimates the sum of the model updates $\boldsymbol{\theta}_+$ directly from the received overlapped signal.

The signal flow is detailed as follows. Each device partitions its sequence of scaled model update $\boldsymbol{\theta}'_m \in \mathbb{R}^d$ to two subsequences $\boldsymbol{\theta}'_m = [(\boldsymbol{s}^{\mathtt{r}}_m)^\top, (\boldsymbol{s}^{\mathtt{i}}_m)^\top]^\top$, where $\boldsymbol{s}^{\mathtt{r}}_m, \boldsymbol{s}^{\mathtt{i}}_m \in \mathbb{R}^{d/2}$, and constructs a complex sequence $\boldsymbol{s}_m \in \mathbb{C}^{d/2}$: $\boldsymbol{s}_m = \boldsymbol{s}^{\mathtt{r}}_m + j\boldsymbol{s}^{\mathtt{i}}_m$, that is, the raw model update information $\boldsymbol{\theta}'_m$ is carried on both the real and imaginary parts of $\boldsymbol{s}_m$.

Time is divided into slots, and each device transmits a packet of L symbols in each slot where $\boldsymbol{s}_m = [s_m[1], s_m[2], \ldots, s_m[L]]^\top$. Therefore, to transmit the total number of $d/2$ complex symbols, $\lceil d/2L \rceil$ slots are needed. Without loss of generality, we focus on the signal processing in one slot.

The time-domain signal transmitted by the m-th device in one slot is given by

$$x_m(t) = \alpha_m \sum_{\ell=1}^{L} s_m[\ell] p(t - \ell T), \tag{4.42}$$

where (1) $p(t) = 1/2 \left[\text{sgn}(t + T) - \text{sgn}(t) \right]$ is a rectangular pulse of duration T; (2) α_m is the channel precoding factor. Given an estimated channel coefficient $\bar{h}_m$ at the m-th device, α_m is designed to be $\alpha_m = 1/\bar{h}_m$.[5] Each of the M IoT devices then calibrates the transmission timing, based on its distance from the PS and its moving speed, so that the signals from different devices arrive at the PS simultaneously.

In practice, however, both the channel-gain precoding and transmission-timing calibration can be imperfect due to the non-ideal hardware and inaccurate estimation of the channel gains and transmission delays. The received signal $r(t)$ at the PS can be written as

$$r(t) = \sum_{m=1}^{M} \tilde{h}_m x_m(t - \tau_m) + z(t), \tag{4.43}$$

where

- $\tilde{h}_m$ is the time-domain complex channel response. We consider flat fading (frequency nonselective) and slow fading (time nonselective) channels [58]. That is, for the channel between each device and the PS, the maximum delay spread is less than the symbol period T so that there is only one resolvable path with the channel response $\tilde{h}_m$ at the receiver, and $\tilde{h}_m$ remains constant over one packet.
- Without loss of generality, we sort the M devices so that the symbols from the devices with smaller indexes arrive at the receiver earlier. The delay of the first device is set to $\tau_1 = 0$, and the relative delay of the m-th device with respect to the first device is denoted by τ_m. We assume the time offsets τ_m, $\forall\, m$, are less than the symbol duration T, as shown in Fig. 4.18. In the ideal case where the timing calibrations are perfect, the relative delays among packets are $\tau_m = 0$, $\forall m$.

---

[5] In practice, the channel-gain precoding is limited by the maximum transmission power of the IoT devices. In the case of deep fading α_m would be very large, and we would have to clip α_m to satisfy the peak or average transmission-power constraint. In our formulation, this may be one cause of the residual channel gains at the receiver.

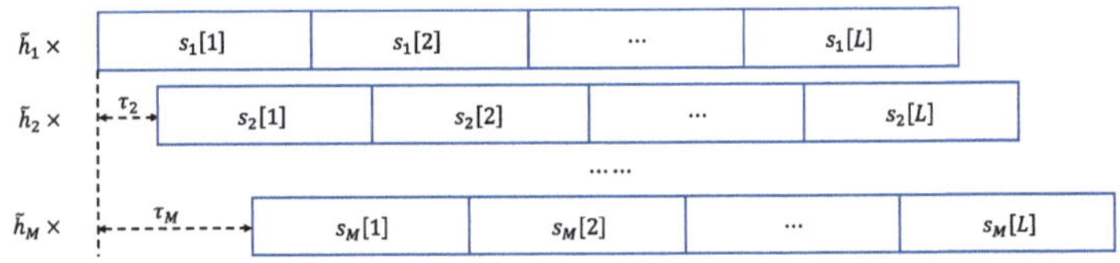

Fig. 4.18 In each slot, the transmitted packets from different devices overlap at the PS with channel misalignments and relative time offsets

- $z(t)$ is the zero-mean baseband complex additive white Gaussian noise (AWGN), the double-sided power spectral densities of each of its real and imaginary parts is $N_0/2$ for an aggregate of N_0.

A Generalized System Model

To ease exposition and conserve space, here we consider only the channel-gain misalignment, time misalignment, and slow fading channel. However, the system model can be easily generalized to AirComp with residual carrier frequency offset (CFO) and fast fading channel. Interested readers may refer to Appendix C.1 for more details.

Substituting (4.42) into (4.43) gives us

$$r(t) = \sum_{m=1}^{M} \tilde{h}_m \alpha_m \sum_{\ell=1}^{L} s_m[\ell] p(t - \tau_m - \ell T) + z(t)$$
$$= \sum_{\ell=1}^{L} \sum_{m=1}^{M} h'_m s_m[\ell] p(t - \tau_m - \ell T) + z(t), \tag{4.44}$$

where $h'_m = \tilde{h}_m / \bar{h}_m$ is the residual channel-fading coefficient between the m-th device and PS.

Succinctly speaking, there can be two kinds of misalignments among the signals transmitted from different devices: (1) channel-gain misalignment h'_m caused by inaccurate channel-gain precoding; and (2) time misalignment τ_m caused by inaccurate calibration of the transmission timing.

The objective of the PS is to estimate $\boldsymbol{\theta}_+ = \sum_{m=1}^{M} \boldsymbol{\theta}'_m$, i.e., the arithmetic sum of the local model updates. This is equivalent to estimating the arithmetic sum of the transmitted complex symbols $\boldsymbol{s}_+$, where $s_+[i] \triangleq \sum_{m=1}^{M} s_m[i]$, because $\boldsymbol{\theta}'_m$ are carried on the real and imaginary parts of the $\boldsymbol{s}_m$ sequence. In other words, given the estimated sequence $\hat{\boldsymbol{s}}_+$, the estimated arithmetic sum of the local model updates is

$\hat{\boldsymbol{\theta}}_+ = \left[(\hat{\boldsymbol{s}}_+^{\mathfrak{r}})^\top, (\hat{\boldsymbol{s}}_+^{\mathfrak{i}})^\top \right]^\top$. Therefore, we shall focus on the estimation of the complex symbols $\boldsymbol{s}_+$ in the following sections.

Advantages of Analog AirComp

Compared with digital communications, analog AirComp has two main advantages. First, the communication link is greatly simplified. In the perfectly-aligned case, the receiver directly receives the arithmetic-sum of the transmitted symbols—the complexity is minimal in this case since no other processing is needed. When there is time or channel-gain misalignments, the complexity is inevitably higher since an additional estimator is required at the receiver to tackle the misalignment. However, as will be shown later, the complexity of the estimators devised in this section for misaligned AirComp is at most $\mathcal{O}(L)$. In contrast, in digital communications, the complexity of the channel encoding module alone is often $\mathcal{O}(L^2)$.

The second advantage of analog AirComp is its low latency. AirComp allows simultaneous transmissions from all IoT devices, while digital communications require orthogonal transmissions. As a result, for the FL among M devices, the uplink model aggregation in each iteration is M times faster with AirComp compared to that with digital communications.

We first consider the case where there is neither channel-gain misalignment nor time misalignment, which we refer to as the *aligned AirComp*. In this case, we have $\alpha_m = 1/\tilde{h}_m$ and $\tau_m = 0, \forall m$, and the received signal is given by

$$r(t) = \sum_{\ell=1}^{L} \sum_{m=1}^{M} s_m[\ell] p(t - \ell T) + z(t). \tag{4.45}$$

Matched filtering $r(t)$ by the same rectangular pulse $p(t)$ and sampling at $t = iT$, $i = 1, 2, \ldots, L$, gives us

$$r[i] = \frac{1}{T} \int_{(i-1)T}^{iT} r(t) dt = \sum_{m=1}^{M} s_m[i] + z[i] = s_+[i] + z[i], \tag{4.46}$$

where the noise sequence $z[i]$ in the samples is independent and identically distributed (i.i.d.), $z[i] \sim \mathcal{CN}(0, \frac{N_0}{T})$.

As can be seen, the target signal $s_+[i]$ appears explicitly on the right hand side (RHS) of (4.46). In this context, the fading multiple-access channel degenerates to a Gaussian multiple-access channel and the M devices can be abstracted as a single device transmitting the arithmetic sum of the local model updates directly to the PS. In practice, however, the channel-gain precoding and the calibration of transmission timing can be inaccurate. With either channel-gain or time misalignment, clean samples as in (4.46) with s_+ explicitly present are no longer available.

4.4.3 Misaligned Multiple Access: A Practical Concern

The successful operation of AirComp hinges on accurate channel-gain precoding and strict synchronization among the participating devices [46, 47]. In practice, however, both requirements may not be perfectly fulfilled. On the one hand, the channel-gain precoding at the edge devices can be imperfect due to the inaccurate channel estimation and non-ideal hardware. The consequence is that there can be residual channel-gain mismatch in the overlapped signals. On the other hand, to meet the synchronization requirement, each device has to carefully calibrate the transmission timing—based on its distance from the PS and its moving speed—so that their signals overlap exactly with each other at the PS. This strict synchronization across different devices is very expensive to realize in practice, and there can be residual asynchronies among the signals from different devices [59].

We refer to the AirComp with residual channel-gain variation and symbol-timing asynchrony among devices as the *misaligned AirComp*, wherein an open problem is how to estimate the arithmetic sum of the transmitted symbols from different devices. This section fills this gap and addresses the key problem in the misaligned AirComp on how to devise the maximum likelihood (ML) estimator in the face of the channel-gain and time misalignments among signals.

With channel-gain and time misalignments, the received signal $r(t)$ is given in (4.44) and illustrated in Fig. 4.18. Let us first follow the standard signal processing flow in digital communications to process the received signal. Specifically, we first matched filter $r(t)$ by the rectangular pulse $p(t)$ and then oversample the matched filtered signal at $\{iT + \tau_k : i = 1, 2, \ldots, L; k = 1, 2, \ldots, M\}$ to collect sufficient statistics [60]. In so doing, the samples we get, denoted by $\{r_k[i] : k = 1, 2, \ldots, M; i = 1, 2, \ldots, L\}$, can be written as

$$
\begin{aligned}
r_k[i] &= \frac{1}{T}\int_{(i-1)T+\tau_k}^{iT+\tau_k} r(t) * p(t)\, dt \qquad (4.47)\\
&= \frac{1}{T}\sum_{m=1}^{M}\int_{(i-1)T+\tau_k}^{(i-\mathbb{1}_{m>k})T+\tau_m} h'_m s_m[i-\mathbb{1}_{m>k}]\, d\zeta \, + \\
&\qquad \frac{1}{T}\sum_{m=1}^{M}\int_{(i-\mathbb{1}_{m>k})T+\tau_m}^{iT+\tau_k} h'_m s_m[i+\mathbb{1}_{m<k}]\, d\zeta + \frac{1}{T}\int_{(i-1)T+\tau_k}^{iT+\tau_k} z(\zeta)\, d\zeta \\
&\triangleq \sum_{m=1}^{M} c_{m,k}[i] s_m[i-\mathbb{1}_{m>k}] + \sum_{m=1}^{M} c'_{m,k}[i] s_m[i+\mathbb{1}_{m<k}] + z_k[i],
\end{aligned}
$$

where $\mathbb{1}$ is the indicator function and $c_{m,k}[i] = \frac{h'_m}{T}[(1-\mathbb{1}_{m>k})T + \tau_m - \tau_k]$, $c'_{m,k}[i] = \frac{h'_m}{T}[\mathbb{1}_{m>k}T + \tau_k - \tau_m]$. When there is residual CFO and the channel is fast fading, the discrete samples can be written in the same form as (4.47). The details are presented in Appendix C.1.

The noise sequence $\{z_k[i]\}$ in (4.47) is colored since

$$
\begin{aligned}
\mathbb{E}[z_k[i]z_{k'}[i']] &= \frac{1}{T^2}\int_{(i-1)T+\tau_k}^{iT+\tau_k}\int_{(i'-1)T+\tau_{k'}}^{i'T+\tau_{k'}} z(\zeta)z(\zeta')\,d\zeta\,d\zeta' \\
&= \begin{cases} \frac{N_0}{T^2}\lambda(i,i',k,k') & \text{if } \lambda(i,i',k,k') \in [0,T), \\ \frac{N_0}{T^2}[2T-\lambda(i,i',k,k')] & \text{if } \lambda(i,i',k,k') \in [T,2T), \\ 0 & \text{otherwise,} \end{cases}
\end{aligned} \tag{4.48}
$$

where $\lambda(i,i',k,k') = (i'-i-1)T + \tau_{k'} - \tau_k$.

Given the samples $\{r_k[i]\}$ in (4.47), we now set out to estimate the desired arithmetic sum s_+. First, the sequence of samples $y_k[i]$ can be written in a more compact form as

$$
\boldsymbol{r} = \boldsymbol{A}\boldsymbol{s} + \boldsymbol{z}, \tag{4.49}
$$

where the dimensionalities of $\boldsymbol{r}$, $\boldsymbol{s}$, and $\boldsymbol{z}$ are $ML \times 1$, and the dimensionality of matrix $\boldsymbol{A}$ is $ML \times ML$ (the detailed form of (4.49) is given below). Denoting by Σ_z the covariance matrix of the noise sequence $\boldsymbol{z}$, each element of Σ_z can then be computed from (4.48).

$$
\begin{bmatrix} r_1[1] \\ \vdots \\ r_M[1] \\ r_1[2] \\ \vdots \\ r_M[2] \\ \vdots \\ r_1[L] \\ \vdots \\ r_M[L] \end{bmatrix}
=
\left[\begin{array}{cccccccccccccc}
c_{1,1}[1] & c'_{2,1}[1] & c'_{3,1}[1] & \cdots & c'_{M,1}[1] & & & & & & & & & \\
c_{1,2}[1] & c_{2,2}[1] & c'_{3,2}[1] & \cdots & c'_{M,2}[1] & c'_{1,2}[2] & & & & & & & & \\
\vdots & \vdots & \vdots & \vdots & \vdots & \vdots & \vdots & & & & & & & \\
c_{1,M}[1] & c_{2,M}[1] & c_{3,M}[1] & \cdots & c_{M,M}[1] & c'_{1,M}[2] & c'_{2,M}[2] & \cdots & c'_{M-1,M}[2] & & & & & \\
 & c_{2,1}[1] & c_{3,1}[1] & \cdots & c_{M,1}[1] & c_{1,1}[2] & c'_{2,1}[2] & c'_{3,1}[2] & \cdots & c'_{M,1}[2] & & & & \\
 & & c_{3,2}[1] & \cdots & c_{M,2}[1] & c_{1,2}[2] & c_{2,2}[2] & c'_{3,2}[2] & \cdots & c'_{M,2}[2] & c'_{1,2}[3] & & & \\
 & & \vdots & \vdots & \vdots & \vdots & \vdots & \vdots & \vdots & \vdots & \vdots & & & \\
 & & & & c_{1,M}[2] & c_{2,M}[2] & c_{3,M}[2] & \cdots & c_{M,M}[2] & c'_{1,M}[3] & c'_{2,M}[3] & \cdots & c'_{M-1,M}[3] & \\
 & & & & & \vdots & \vdots & \vdots & \vdots & \vdots & \vdots & \vdots & \vdots & \cdots
\end{array}\right]
\begin{bmatrix} s_1[1] \\ \vdots \\ s_M[1] \\ s_1[2] \\ \vdots \\ s_M[2] \\ \vdots \\ s_1[L] \\ \vdots \\ s_M[L] \end{bmatrix}
+
\begin{bmatrix} z_1[1] \\ \vdots \\ z_M[1] \\ z_1[2] \\ \vdots \\ z_M[2] \\ \vdots \\ z_1[L] \\ \vdots \\ z_M[L] \end{bmatrix}
$$

The desired sequence s_+, on the other hand, can be written as a linear transformation of the complex vector $\boldsymbol{s}$:

$$
\boldsymbol{s}_+ = \boldsymbol{V}\boldsymbol{s}, \tag{4.50}
$$

where the $L \times ML$ matrix $\boldsymbol{V}$ is given by

$$
\boldsymbol{V} = \begin{bmatrix} \mathbf{1}_{1\times M} & & & \\ & \mathbf{1}_{1\times M} & & \\ & & \cdots & \\ & & & \mathbf{1}_{1\times M} \end{bmatrix},
$$

in which $\mathbf{1}_{1\times M}$ represents an $1 \times M$ all-ones matrix.

Multiplying both sides of (4.49) by $\boldsymbol{V}\boldsymbol{A}^{-1}$ gives us

$$\boldsymbol{V}\boldsymbol{A}^{-1}\boldsymbol{r} = \boldsymbol{s}_+ + \boldsymbol{V}\boldsymbol{A}^{-1}\boldsymbol{z}, \tag{4.51}$$

based on which an ML estimator can be devised, as in Definition 4.9.

Definition 4.9 (ML Estimation for Misaligned AirComp) Given a sequence of samples $\boldsymbol{r} \in C^{ML}$ in (4.49), the ML estimate of sequence $\boldsymbol{s}_+ \in C^L$ is

$$\hat{\boldsymbol{s}}_+^{\mathrm{ml}} = \boldsymbol{V}\boldsymbol{A}^{-1}\boldsymbol{r}. \tag{4.52}$$

Equation (4.52) follows directly from (4.51) since the likelihood function of $\boldsymbol{s}_+$ is an L-dimensional Gaussian distribution. Specifically, given $\boldsymbol{r}$, the likelihood function of $\boldsymbol{s}_+$ is

$$f(\boldsymbol{V}\boldsymbol{A}^{-1}\boldsymbol{r}|\boldsymbol{s}_+) \propto \mathcal{CN}(\boldsymbol{V}\boldsymbol{A}^{-1}\boldsymbol{r}, \boldsymbol{V}\boldsymbol{A}^{-1}\Sigma_z\boldsymbol{A}^{-H}\boldsymbol{V}^H).$$

Differentiating $f(\boldsymbol{V}\boldsymbol{A}^{-1}\boldsymbol{r}|\boldsymbol{s}_+)$ with respect to $\boldsymbol{s}_+$ gives us the ML estimate $\hat{\boldsymbol{s}}_+^{\mathrm{ml}}$ in (4.52).

An important implication of (4.52) is that, the maximum-likelihood $\boldsymbol{s}_+$ can be obtained by first finding the maximum-likelihood transmitted vector $\hat{\boldsymbol{s}}^{\mathrm{ml}} = \boldsymbol{A}^{-1}\boldsymbol{r}$, and then performing the arithmetic sum $\hat{\boldsymbol{s}}_+^{\mathrm{ml}} = \boldsymbol{V}\hat{\boldsymbol{s}}^{\mathrm{ml}}$. Said in another way, the ML estimation of $\boldsymbol{s}_+$ boils down to multi-user estimation/detection (MUE) when we have misaligned channel gains at the receiver.

> The transmitted vector $\boldsymbol{s}$ carries the local weight-updates of a neural network. Therefore, $\boldsymbol{s}$ is by no means i.i.d. considering the strong correlations among the weights of the neural network. This also suggests that the prior information of $\boldsymbol{s}$ is hard to obtain and is unlikely to be known to the PS in advance. As a result, ML estimation is the only choice at the receiver.

More About the ML Estimation

The result that the ML estimation of $\boldsymbol{s}_+$ boils down to MUE is exclusive to the misaligned AirComp system, wherein the transmitted symbols are continuous valued and the channel gains are misaligned. In digital communications, we do not have such result for ML estimation. The reason for this divergence is as follows. AirComp is an analog communication system wherein the transmitted symbols $\boldsymbol{s}$ are continuous complex values. In contrast, the transmitted symbols in digital communications are discrete constellations. Whether the prior probability distribution of the transmitted symbols is available to the receiver or not, the constellation itself serves as a kind of prior information as the detection space is naturally narrowed down to the possible constellation points. As a result, when we perform ML estimation in digital communications, we inherently assume all the constellations are equiprobable. In that case, the likelihood function $f(\boldsymbol{r}|\boldsymbol{s})$ is a Gaussian mixture instead of Gaussian and the MUE-and-sum estimation is no longer optimal if we were to estimate the arithmetic sum $\boldsymbol{s}_{+}$.

On the other hand, when we perform ML estimation in AirComp, all the complex space is assumed to be equiprobable. The only information we have is the noise-contaminated sample and the likelihood function is a Gaussian centered around the noisy sample. Two implications about the ML estimation in AirComp are thus (1) it faces an infinitely large estimation space; (2) it can be very susceptible to noise.

The ML estimator in (4.52) is not a practical estimator due to the prohibitive computational complexity of matrix inversion. To invert an n by n matrix, the best proven lower bound of the computational complexity is $\Omega(n^2 \log n)$ [61]. Notice that the dimensionality of $\boldsymbol{A}$ is ML by ML. Thus, the computational complexity of (4.52) is $\Omega(L^2M^2 \log(LM))$. In practical AirComp systems, M cannot be too large due to the saturation of the receiver (that is, the received signal power can exceed the dynamic range of the receiver if M is too large), but the packet length L can be extremely large. Let us fix M as a constant, the computational complexity of (4.52) is then $\Omega(L^2 \log L)$.

To address this problem and devise an ML estimator with acceptable computational complexity, we put forth in Sect. 4.4.3.1 a factor-graph based ML estimator by exploiting the sparsity of the coefficient matrix. Compared with the ML estimator in (4.52), the computational complexity of the factor-graph based ML estimator is only $\Omega(L)$.

4.4.3.1 A Sum-Product ML Estimator and The Aligned-Sample Estimator

Before we dive deeper to dissect the inner structure of the coefficient matrix, let us first introduce a new matched filtering and sampling scheme that gives us oversampled but independent samples, which we refer to as the *whitened matched filtering and sampling* (WMFS). Two benefits of the WMFS scheme are (1) the independent samples obtained from the scheme allows us to construct a factor graph with a simple structure, based on which a low-complexity SP-ML estimator can be devised; (2) the whitened scheme yields a subsequence of samples in which the indexes of symbols from different devices are consistent—in these samples, symbols from different devices are aligned in time, as shown in (4.64). This admits an aligned-sample estimator for the misaligned AirComp.

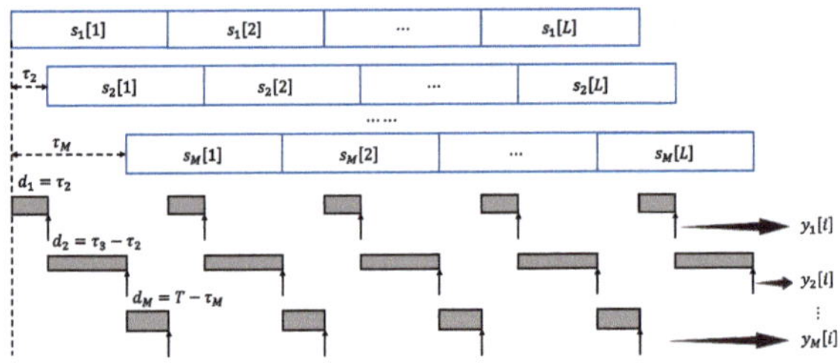

Fig. 4.19 Matched filtering the received signal by a bank of M filters of lengths $d_k = \tau_{k+1} - \tau_k$

WMFS

The key idea of the WMFS scheme is to use a bank of matched filters of different lengths to collect power judiciously from $r(t)$. Specifically, instead of using the rectangular pulse $p(t)$ as the matched filter as in (4.47), we define M matched filters $\{p'_k(t) : k = 1, 2, \ldots, M\}$ as follows:

$$p'_k(t) = \frac{1}{2}\big[\text{sgn}(t+T) - \text{sgn}(t+T-d_k)\big], \tag{4.53}$$

where the length of the k-th matched filter is $d_k = \tau_{k+1} - \tau_k$, $k = 1, 2, \ldots, M$. For completeness, we define $\tau_{M+1} = T$.

The matched filtering and sampling processes are illustrated in Fig. 4.19. As shown, the signal filtered by the k-th matched filter is given by $y_k(t) = \frac{1}{d_k}\int_{-\infty}^{\infty} r(\zeta)p'_k(t-\zeta)\,d\zeta$, and we sample $y_k(t)$ at $(i-1)T + \tau_{k+1} : i = 1, 2, \ldots, L+1$, giving

$$\begin{aligned} y_k[i] &= y_k(t = (i-1)T + \tau_{k+1}) \\ &= \frac{1}{d_k}\int_{(i-1)T+\tau_k}^{(i-1)T+\tau_{k+1}} \sum_{m=1}^{M} h'_m s_m[i - \mathbb{1}_{m>k}]\,d\zeta + \frac{1}{d_k}\int_{(i-1)T+\tau_k}^{(i-1)T+\tau_{k+1}} z(\zeta)\,d\zeta \\ &\triangleq \sum_{m=1}^{M} h'_m s_m[i - \mathbb{1}_{m>k}] + \tilde{z}_k[i], \end{aligned} \tag{4.54}$$

where we have defined $s_m[0] = s_m[L+1] = 0$, $\forall m$, for completeness.

An important observation from (4.54) is that the noise term $\tilde{z}_k[i] \sim \mathcal{CN}(0, N_0/d_k)$ is independent for different k and i. This is because

$$\begin{aligned} \mathbb{E}\left[\tilde{z}_k[i]\tilde{z}_{k'}[i']\right] &= \mathbb{E}\left[\frac{1}{d_k}\int_{(i-1)T+\tau_k}^{(i-1)T+\tau_{k+1}} \tilde{z}(\zeta)\,d\zeta \frac{1}{d_{k'}}\int_{(i'-1)T+\tau_{k'}}^{(i'-1)T+\tau_{k'+1}} \tilde{z}(\zeta')\,d\zeta'\right] \\ &= \frac{1}{d_k d_{k'}}\int_{(i-1)T+\tau_k}^{(i-1)T+\tau_{k+1}} \int_{(i'-1)T+\tau_{k'}}^{(i'-1)T+\tau_{k'+1}} \mathbb{E}\left[\tilde{z}(\zeta)\tilde{z}(\zeta')\right] d\zeta\, d\zeta' \end{aligned}$$

$$= \frac{1}{d_k d_{k'}} \int_{(i-1)T+\tau_m}^{(i-1)T+\tau_{m+1}} N_0 \delta\left((i-i')(k-k')\right) d\varsigma$$

$$= \frac{N_0}{d_k} \delta\left((i-i')(k-k')\right). \tag{4.55}$$

Similarly to (4.49), we can rewrite (4.54) in a matrix form as

$$\boldsymbol{y} = \boldsymbol{D}\boldsymbol{s} + \tilde{\boldsymbol{z}}, \tag{4.56}$$

where the sequence of transmitted symbols $\boldsymbol{s}$ is the same as that in (4.49). Unlike (4.49), the vectors $\boldsymbol{y}$ and $\tilde{\boldsymbol{z}}$ in (4.56) are $M(L+1)-1$ by 1 dimensional, giving

$$\boldsymbol{y} = \Big[y_1[1], y_2[1], \ldots, y_M[1], y_1[2], y_2[2], \ldots, y_M[2], \ldots,$$

$$y_1[L], y_2[L], \ldots, y_M[L], y_1[L+1], y_2[L+1], \ldots, y_{M-1}[L+1] \Big]^{\top},$$

$$\tilde{\boldsymbol{z}} = \Big[\tilde{z}_1[1], \tilde{z}_2[1], \ldots, \tilde{z}_M[1], \tilde{z}_1[2], \tilde{z}_2[2], \ldots, \tilde{z}_M[2], \ldots, \tilde{z}_1[L], \tilde{z}_2[L], \ldots,$$

$$\tilde{z}_M[L], \tilde{z}_1[L+1], \tilde{z}_2[L+1], \ldots, \tilde{z}_{M-1}[L+1] \Big]^{\top},$$

and the coefficient matrix $\boldsymbol{D}$ is $M(L+1)-1$ by ML, giving[6]

$$\boldsymbol{D} = \begin{bmatrix}
h'_1 & & & & & & & & \\
h'_1 & h'_2 & & & & & & & \\
\cdots & h'_2 & \cdots & & & & & & \\
h'_1 & \cdots & \cdots & h'_M & & & & & \\
 & h'_2 & \cdots & h'_M & h'_1 & & & & \\
 & & \cdots & \cdots & h'_1 & h'_2 & & & \\
 & & & h'_M & \cdots & h'_2 & \cdots & & \\
 & & & & h'_1 & \cdots & \cdots & h'_M & \\
 & & & & & h'_2 & \cdots & h'_M & \cdots \\
 & & & & & & \cdots & \cdots & \cdots \\
 & & & & & & & h'_M & \cdots \\
 & & & & & & & & \cdots
\end{bmatrix}. \tag{4.57}$$

The desired sequence at the PS is $\boldsymbol{s}_+ = \boldsymbol{V}\boldsymbol{s}$, as in (4.50).

Equation (4.55) validates that the noise sequence $\tilde{\boldsymbol{z}}$ is white, that is, (4.56) can be viewed as a whitened model of (4.49), and hence the name WMFS.

We emphasize that the signal model in (4.56) is equivalent to that in (4.49) since they are sampled from the same received signal $r(t)$ and no information is lost [62]. More specifically, (4.56) can be transformed back to (4.49) after some

---

[6] When there is residual CFO and the channel is fast fading, matrix $\boldsymbol{D}$ can be written in the same form as (4.57). See Appendix C.1 for more details.

elementary row operations and row deletions. Nevertheless, the model in (4.56) is more favorable than (4.49) thanks to the following factors:

1. The whitened noise. As will be shown later, white noise admits a sample-by-sample factorization of the likelihood function and a much simpler structure of the factor graph.
2. The alleviated inter-symbol and inter-user interferences. A sample $y_k[i]$ is related to only M complex symbols, each of which comes from a different device. In contrast, a sample $r_k[i]$ in (4.49) is related to $2M - 1$ symbols.

Since (4.49) and (4.56) are equivalent, we can also design the ML estimator from (4.56). Denoting the covariance matrix of $\tilde{z}$ by $\boldsymbol{\Sigma}_{\tilde{z}}$ (it is a diagonal matrix since $\tilde{z}$ is white), the ML estimator is given by

$$\hat{s}_+^{\mathrm{ml}} = \boldsymbol{V}(\boldsymbol{D}^H \boldsymbol{\Sigma}_{\tilde{z}}^{-1} \boldsymbol{D})^{-1} \boldsymbol{D}^H \boldsymbol{\Sigma}_{\tilde{z}}^{-1} \boldsymbol{y}. \tag{4.58}$$

Again, the inversion of $\boldsymbol{D}^H \boldsymbol{\Sigma}_{\tilde{z}}^{-1} \boldsymbol{D}$ is computationally demanding, as in (4.52).

A Factor Graph Approach

A possible way to reduce the complexity of the ML estimation is to exploit the sparsity of the coefficient matrix $\boldsymbol{D}$. To this end, let us focus on the ML estimate of a single entry in the desired sequence $\boldsymbol{s}_+$, i.e., $\hat{s}_+^{\mathrm{ml}}[i]$.

Let $\boldsymbol{s}[\boldsymbol{i}] = \left[s_1[i], s_2[i], \ldots, s_M[i]\right]^\top$. Given an observed sample sequence $\boldsymbol{y}$, we have

$$\hat{s}_+^{\mathrm{ml}}[i] = \arg\max_{s_+[i]} f(\boldsymbol{y}|s_+[i]) = \arg\max_{s_+[i]} \int_{\boldsymbol{1}^\top \boldsymbol{s}[\boldsymbol{i}]=s_+[i]} f(\boldsymbol{y}|\boldsymbol{s}[\boldsymbol{i}])\, d\boldsymbol{s}[\boldsymbol{i}].$$

In particular, $f(\boldsymbol{y}|\boldsymbol{s}[\boldsymbol{i}])$ is a marginal function of $f(\boldsymbol{y}|\boldsymbol{s})$. Thus, to find the ML estimate $\hat{s}_+^{\mathrm{ml}}[i]$, a first step is to analyze $f(\boldsymbol{y}|\boldsymbol{s})$. In the following analysis, we will call $f(\boldsymbol{y}|\boldsymbol{s})$ the global likelihood function and $f(\boldsymbol{y}|\boldsymbol{s}[\boldsymbol{i}])$ the marginal likelihood function.

In ML estimation, the transmitted symbols $\boldsymbol{s}$ are treated as constants. Randomness is only introduced by the noise sequence $\tilde{z}$. Therefore, in the whitened model (4.56), the elements of $\boldsymbol{y}$ are independent of each other. We can then factorize the likelihood function $f(\boldsymbol{y}|\boldsymbol{s})$ as

$$f(\boldsymbol{y}|\boldsymbol{s}) \propto \prod_{k=1}^{M} \prod_{i=1}^{L+1} f(y_k[i]|\boldsymbol{s}) \overset{(a)}{=} \prod_{k=1}^{M} \prod_{i=1}^{L+1} f(y_k[i]|\mathcal{V}(y_k[i])), \tag{4.59}$$

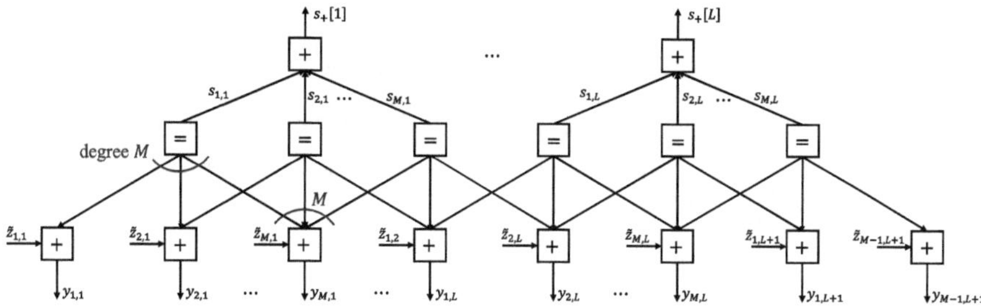

Fig. 4.20 A graphical interpretation of the factorization in (4.59). To simplify notations, we denote $y_k[i]$, $z_k[i]$ and $s_m[i]$ by $y_{k,i}$, $z_{k,i}$ and $s_{m,i}$ in the figure, respectively

where (a) follows because a sample $y_k[i]$ is related to only M complex symbols in $\boldsymbol{s}$. As per (4.54), we denote these symbols by $\mathcal{V}(y_k[i]) = \{s_1[i], s_2[i], \ldots, s_k[i], s_{k+1}[i-1], s_{k+2}[i-1], \ldots, s_M[i-1]\}$ and call them the *neighbor symbols* of $y_k[i]$. A sample $y_k[i]$ is then fully determined by the values of its neighbor symbols. Note that the number of non-zero symbols in $\mathcal{V}(y_k[i])$ is the number of non-zero elements in the corresponding row of $\boldsymbol{D}$, giving

$$|\mathcal{V}(y_k[i])| = \begin{cases} k, & \text{when } i = 1; \\ M, & \text{when } 1 \leq i \leq L; \\ M-k, & \text{when } i = L+1. \end{cases} \tag{4.60}$$

Based on the factorization in (4.59), $f(\boldsymbol{y}|\boldsymbol{s})$ can be depicted by a graphical model [63–65]. As shown in Fig. 4.20, we use a Forney-style factor graph [64] to represent the factorization. In particular, each edge in the graph corresponds to a variable in (4.59), i.e., a complex symbol $s_m[\ell]$, an observation $y_k[i]$, or a noise term $\tilde{z}_k[i]$. To simplify notations, we denote them by $s_{m,i}$, $y_{k,i}$, and $\tilde{z}_{k,i}$ in Fig. 4.20, respectively.

As can be seen, each sample $y_k[i]$ is related to a set of symbols $\mathcal{V}(y_k[i])$; each complex symbol $s_k[i]$, on the other hand, is related to M samples. Thus, we duplicate each symbol $s_{m,i}$ for M times and connect them to M consecutive samples – the equality constraint function "=" means that the values of the variables connecting to this function must be equal. The output degree of each equality constraint function is M, so is the input degree of each plus function "+" (except for the $M-1$ samples at both ends of the packet). The target symbols $\boldsymbol{s}_+$ are shown at the top of Fig. 4.20.

The marginal likelihood function $f(\boldsymbol{y}|\boldsymbol{s[i]})$ can be obtained from the global likelihood function $f(\boldsymbol{y}|\boldsymbol{s})$ by a marginalization process operated on the factor graph, which can be implemented efficiently via the sum-product (SP) algorithm. However, note in Fig. 4.20 that we have a loopy graph. Moreover, the girth of the graph (the length of the shortest loop) is 4. Such short loops prevent the sum-product algorithm from converging [66]. Even if they converge, the performance of the sum-product algorithm often degrades greatly, and the equilibrium posterior distribution

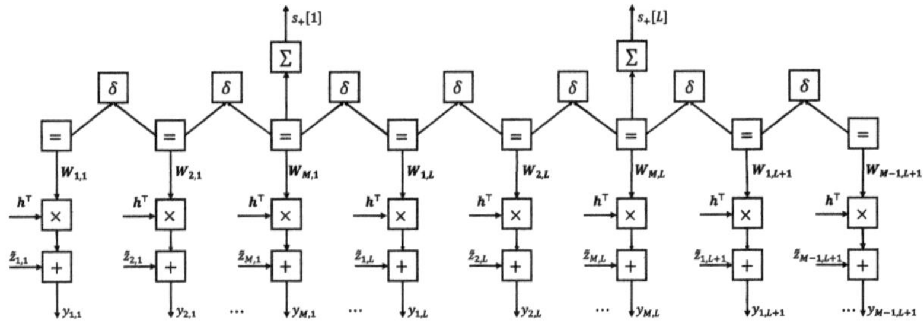

Fig. 4.21 An equivalent tree structure to the loopy graph in Fig. 4.20. The variables connected to the same sample are clustered as a new high-dimensional variable. Each observation node is connected to a single variable node after clustering

is only an approximation of the true posterior distribution of the variables. Said in another way, the independence assumption of the extrinsic information passed along the edges no longer holds because the messages can circulate indefinitely around the loops.

To circumvent this problem, below we transform the loopy graph in Fig. 4.20 to a loop-free graph, at the expense of increasing the dimension of the variables [63]. For each sample $y_k[i]$, we cluster all its neighbor symbols (i.e., the edges/variables that connect to $y_k[i]$ in Fig. 4.20) and construct a new higher-dimensional variable. Let us denote the new high-dimensional variable by $\boldsymbol{W}_{k,i} = \mathcal{V}(y_k[i])$. As illustrated in Fig. 4.21, each sample $y_k[i]$ is now connected to a single high-dimensional variable $\boldsymbol{W}_{k,i}$ after clustering and the loops are removed.

After clustering, the new high-dimensional variables $\boldsymbol{W}_{k,i}$ are correlated with each other because they contain common symbols. For example, $\boldsymbol{W}_{1,1} = \{s_1[1]\}$, $\boldsymbol{W}_{2,1} = \{s_1[1], s_2[1]\}$, and $\boldsymbol{W}_{3,1} = \{s_1[1], s_2[1], s_3[1]\}$, the common symbol between $\boldsymbol{W}_{1,1}$ and $\boldsymbol{W}_{2,1}$ is $s_1[1]$, that between $\boldsymbol{W}_{1,1}$ and $\boldsymbol{W}_{3,1}$ is $s_1[1]$, and that between $\boldsymbol{W}_{2,1}$ and $\boldsymbol{W}_{3,1}$ are $s_1[1]$, $s_2[1]$. Therefore, we have to add constraints among the new high-dimensional variables to ensure that the values of the common symbols are consistent across different variables.

In Fig. 4.21, the compatibility function δ is added on the adjacent variables to represent the above constraints. Specifically, for the two adjacent variables $\boldsymbol{W}$ and $\boldsymbol{W}'$ connected to the same delta function, the compatibility function $\delta(\boldsymbol{W}, \boldsymbol{W}')$ is defined as

$$\delta(\boldsymbol{W}, \boldsymbol{W}') = \begin{cases} 1, \text{if the values of all common symbols between } \boldsymbol{W} \text{ and } \boldsymbol{W}' \text{ are equal;} \\ 0, \text{otherwise.} \end{cases}$$

That is, function δ is an on-off function that ensures that the messages passed from $\boldsymbol{W}$ to $\boldsymbol{W}'$ and from $\boldsymbol{W}'$ to $\boldsymbol{W}$ satisfy the constraint that the values of the common symbols between $\boldsymbol{W}$ and $\boldsymbol{W}'$ are equal.

Compatibility Functions

It is worth noting that adding compatibility functions between two adjacent variables is enough to depict all the constraints. For example, let us consider $\boldsymbol{W}_{1,1}$, $\boldsymbol{W}_{2,1}$ and $\boldsymbol{W}_{3,1}$. In Fig. 4.21, we only add compatibility functions $\delta(\boldsymbol{W}_{1,1}, \boldsymbol{W}_{2,1})$ and $\delta(\boldsymbol{W}_{2,1}, \boldsymbol{W}_{3,1})$. There is no need to add an extra compatibility function $\delta(\boldsymbol{W}_{1,1}, \boldsymbol{W}_{3,1})$ between $\boldsymbol{W}_{1,1}$, $\boldsymbol{W}_{3,1}$ although they do have a common symbol $s_1[1]$. This is because $\boldsymbol{W}_{1,1}$ and $\boldsymbol{W}_{3,1}$ are independent conditioned on the compatibility functions $\delta(\boldsymbol{W}_{1,1}, \boldsymbol{W}_{2,1})$ and $\delta(\boldsymbol{W}_{2,1}, \boldsymbol{W}_{3,1})$. More specifically, $\delta(\boldsymbol{W}_{1,1}, \boldsymbol{W}_{2,1})$ has ensured that the value of $s_1[1]$ in $\boldsymbol{W}_{1,1}$ equals that in $\boldsymbol{W}_{2,1}$ and $\delta(\boldsymbol{W}_{2,1}, \boldsymbol{W}_{3,1})$ has ensured that the values of $s_1[1]$ in $\boldsymbol{W}_{2,1}$ equals that in $\boldsymbol{W}_{3,1}$. Thus, the values of $s_1[1]$ in $\boldsymbol{W}_{1,1}$ and $\boldsymbol{W}_{3,1}$ must be equal.

Overall, the factor graph in Fig. 4.21 presents a tree structure. Compared with Fig. 4.20, although the dimensions of the variables are M times larger, the marginal likelihood function $f(\boldsymbol{y}|\boldsymbol{s}[i])$ can now be computed exactly via the sum-product algorithm—the tree structure guarantees that a root node collects all the messages from its adjoining nodes before further passing the message on [67].

Analog Message Passing and the SP-ML Estimator

In standard sum-product algorithms, the messages passed on the edges are the probability mass function (PMF) of the variables associated with the edges, i.e., a probability vector of finite length [63, 65]. This is because the transmitted symbols in digital communications are chosen from finite-size constellations. However, in the case of AirComp, the transmitted symbols $\boldsymbol{s}_m$ are continuous complex numbers. Hence, the messages passed on the edges should be the probability density functions (PDFs) of the associated variables, which is a continuous function rather than a finite-length vector.

To enable message passing, a straightforward idea is to quantize the PDFs so that we can employ digital message passing. The output of the SP algorithm is then the marginal PDF in a quantized form. However, quantization suffers from the "curse of dimensionality": when the dimension of the variables increases, the volume of the space increases exponentially fast. In order to get a sound result, a significantly larger number of quantization levels is required in each dimension compared with the low-dimensional case [68]. In our problem, we have deliberately increased the dimensionality of the variables to remove the loops in Fig. 4.20. Thus, the standard sum-product algorithm with quantization cannot be used considering its prohibitive complexity.

Now that quantization is not an option, the problem is how to pass continuous PDFs along the edges of Fig. 4.21. A natural idea is then to parameterize the PDFs and pass their parameters [64, 69]. In actuality, the main effort of this section is to show the following: given an observed sample sequence, all the messages passed on the tree are multivariate Gaussian distributions. A multivariate Gaussian PDF can be parameterized by its mean vector and covariance matrix—passing these parameters is equivalent to passing the continuous PDF.

Theorem 4.2 (Analog Message Passing) *The following results holds:*

1. *The marginal likelihood function $f(\mathbf{y}|\mathbf{s}[i])$ is a multivariate complex Gaussian distribution of dimension M. The likelihood function $f(\mathbf{y}|s_+[i])$ is a single-variate complex Gaussian distribution.*
2. *A complex random variable can be viewed as a pair of real random variables (i.e., the real part and the imaginary part of the complex random variable). Thus, we can denote the M-dimensional complex Gaussian $f(\mathbf{y}|\mathbf{s}[i])$ by a $2M$-dimensional real Gaussian*

$$f(\mathbf{y}|\mathbf{s}[i]) \sim \mathcal{N}\left(\mathbf{s}[i], \boldsymbol{\mu}_{\mathbf{s}[i]} = \begin{bmatrix} \boldsymbol{\mu}^{\mathfrak{r}}_{\mathbf{s}[i]} \\ \boldsymbol{\mu}^{\mathfrak{i}}_{\mathbf{s}[i]} \end{bmatrix}, \boldsymbol{\Sigma}_{\mathbf{s}[i]} = \begin{bmatrix} \boldsymbol{\Sigma}^{\mathfrak{rr}}_{\mathbf{s}[i]} & \boldsymbol{\Sigma}^{\mathfrak{ri}}_{\mathbf{s}[i]} \\ \boldsymbol{\Sigma}^{\mathfrak{ir}}_{\mathbf{s}[i]} & \boldsymbol{\Sigma}^{\mathfrak{ii}}_{\mathbf{s}[i]} \end{bmatrix}\right), \tag{4.61}$$

where $\boldsymbol{\mu}_{\mathbf{s}[i]}$ is a $2M$ by 1 real vector consisting of the real and imaginary parts of the mean of $\mathbf{s}[i]$. That is, $\boldsymbol{\mu}^{\mathfrak{r}}_{\mathbf{s}[i]}$ and $\boldsymbol{\mu}^{\mathfrak{i}}_{\mathbf{s}[i]}$ are the real and imaginary parts of sequence $\mathbb{E}[\mathbf{s}[i]]^\top$. The covariance matrix $\boldsymbol{\Sigma}_{\mathbf{s}[i]}$ is a $2M$ by $2M$ covariance matrix. The moment parameters $(\boldsymbol{\mu}_{\mathbf{s}[i]}, \boldsymbol{\Sigma}_{\mathbf{s}[i]})$ can be computed by a sum-product process described in Algorithm 9.
3. *Given the marginal likelihood function $f(\mathbf{y}|\mathbf{s}[i])$ in (4.61), $f(\mathbf{y}|s_+[i])$ can be constructed by*

$$f(\mathbf{y}|s_+[i]) \sim \mathcal{N}\left(s_+[i], \boldsymbol{\mu}_{s_+[i]} = \begin{bmatrix} \mu^{\mathfrak{r}}_{s_+[i]} \\ \mu^{\mathfrak{i}}_{s_+[i]} \end{bmatrix}, \boldsymbol{\Sigma}_{s_+[i]} = \begin{bmatrix} \Sigma^{\mathfrak{rr}}_{s_+[i]} & \Sigma^{\mathfrak{ri}}_{s_+[i]} \\ \Sigma^{\mathfrak{ir}}_{s_+[i]} & \Sigma^{\mathfrak{ii}}_{s_+[i]} \end{bmatrix}\right), \tag{4.62}$$

where $\mu^{\mathfrak{r}}_{s_+[i]} = \mathbf{1}^\top \boldsymbol{\mu}^{\mathfrak{r}}_{\mathbf{s}[i]}$, $\mu^{\mathfrak{i}}_{s_+[i]} = \mathbf{1}^\top \boldsymbol{\mu}^{\mathfrak{i}}_{\mathbf{s}[i]}$, $\Sigma^{\mathfrak{rr}}_{s_+[i]} = \mathbf{1}^\top \boldsymbol{\Sigma}^{\mathfrak{rr}}_{\mathbf{s}[i]}\mathbf{1}$, $\Sigma^{\mathfrak{ri}}_{s_+[i]} = \mathbf{1}^\top \boldsymbol{\Sigma}^{\mathfrak{ri}}_{\mathbf{s}[i]}\mathbf{1}$, $\Sigma^{\mathfrak{ir}}_{s_+[i]} = \mathbf{1}^\top \boldsymbol{\Sigma}^{\mathfrak{ir}}_{\mathbf{s}[i]}\mathbf{1}$, $\Sigma^{\mathfrak{ii}}_{s_+[i]} = \mathbf{1}^\top \boldsymbol{\Sigma}^{\mathfrak{ii}}_{\mathbf{s}[i]}\mathbf{1}$.

Proof See Appendix C.2. □

Following (4.62), an SP-ML estimator can be designed, as given in Definition 4.10.

Definition 4.10 (SP-ML Estimation) The WMFS scheme gives us the whitened samples $\boldsymbol{y}$ in (4.56). To estimate the designed sequence $\boldsymbol{s}_+$, an SP-ML estimator first computes the moment parameters of the multivariate Gaussian $f(\mathbf{y}|\mathbf{s}[i])$, $\forall i$, by an analog sum-product process, and then estimates each element of $\boldsymbol{s}_+$ by

$$\hat{s}^{\mathrm{ml}}_+[i] = \mathbf{1}^\top \boldsymbol{\mu}^{\mathfrak{r}}_{\mathbf{s}[i]} + j\mathbf{1}^\top \boldsymbol{\mu}^{\mathfrak{i}}_{\mathbf{s}[i]}, \tag{4.63}$$

where $\boldsymbol{\mu}_{\mathbf{s}[i]} = [\boldsymbol{\mu}^{\mathfrak{r}}_{\mathbf{s}[i]}, \boldsymbol{\mu}^{\mathfrak{i}}_{\mathbf{s}[i]}]^\top$ is the mean of $f(\mathbf{y}|\mathbf{s}[i])$.

Algorithm 9 Analog message passing for ML estimation

Input: Samples $\boldsymbol{y}$ and coefficient matrix $\boldsymbol{D}$.
Output: The marginal likelihood function $f(\boldsymbol{y}|\boldsymbol{s}[\boldsymbol{i}])$.
Initialization:
for $k = 1, 2, \ldots, M$ and $i = 1, 2, \ldots, L + 1$ **do**
    $\boldsymbol{w}_{k,i} = \mathcal{V}(y_k[i])$;
    Compute the information about $\boldsymbol{w}_{k,i}$ contained in each sample $y_k[i]$, i.e., $f_b(\boldsymbol{w}_{k,i})$, following (C.9).
Forward message passing:
for $i = 1, 2, \ldots, L + 1$ **do**
    for $k = 1, 2, \ldots, M$ **do**
        Compute the information about $\boldsymbol{w}_{k,i}$ contained in all the samples $\{y_{k'}[i'] : k' < k, i' < i\}$, i.e., $f_\ell(\boldsymbol{w}_{k,i})$, following (C.15) and (C.16).
Backward message passing:
for $i = L + 1, L, \ldots, 2, 1$ **do**
    for $k = M, M - 1, \ldots, 2, 1$ **do**
        Compute the information about $\boldsymbol{w}_{k,i}$ contained in all the samples $\{y_{k'}[i'] : k' > k, i' > i\}$, i.e., $f_r'(\boldsymbol{w}_{k,i})$, following (C.17) and (C.18).
Marginalization:
for $i = 1, 2, \ldots, L$ **do**
    Compute the marginal likelihood function $f(\boldsymbol{y}|\boldsymbol{s}[\boldsymbol{i}])$, as per (C.20).

The reason behind (4.63) is as follows. It has been shown that $f(\boldsymbol{y}|s_+[i])$ is conditionally Gaussian. As per the ML rule, we should choose the mean of $f(\boldsymbol{y}|s_+[i])$ as the estimate of $s_+[i]$ as it maximizes the likelihood function. This gives us (4.63).

Maximum Likelihood Sequence Estimation (MLSE) Versus Bahl-Cocke-Jelinek-Raviv (BCJR) Estimation

The ML estimators in (4.52) and (4.58) aim to find the ML sequence $\boldsymbol{s}_+$ in the space $\mathbb{C}^L$. In the language of digital communications, they are ML optimal in the sense of MLSE [70]. On the other hand, the ML estimator in (4.63) aims to maximize the likelihood function of each element of $\boldsymbol{s}_+$. Thus, it is ML optimal in the BCJR [71] sense.

When we perform ML estimation in digital communications, MLSE-optimal and BCJR-optimal are different criteria because the former minimizes the block error rate (BLER) while the latter minimizes the bit error rate (BER). For the ML estimation in AirComp, however, the two criteria are equivalent. The reason for this discrepancy is again that the discrete constellations used in digital communications serve as a kind of prior information to the receiver while an AirComp receiver has no prior information at all.

More specifically, let us consider the message passing in Fig. 4.21. For ML estimation in AirComp, we have shown that all the messages passed on the graph, including $f(\boldsymbol{y}|\boldsymbol{s}_+)$ and $f(\boldsymbol{y}|s_+[i])$, are Gaussian. Thus, the ML sequence $\boldsymbol{s}_+$ also gives us the ML $s_+[i]$, $\forall i$, after marginalization. As a result, MLSE-optimal and BCJR-optimal are equivalent, and the ML estimators in (4.52), (4.58), and (4.63) are identical. In contrast, when we perform ML estimation in digital communications, the prior information imposes each transmitted symbol to belong to a finite constellation. As a consequence, the messages passed on the graph are Gaussian mixtures, and neither $f(\boldsymbol{y}|\boldsymbol{s}_+)$ nor $f(\boldsymbol{y}|s_+[i])$ is Gaussian distributed. As a result, MLSE-optimal and BCJR-optimal are different criteria.

Computational Complexity
Finally, we evaluate the computational complexity of the SP-ML estimator. With analog/parametric message passing, messages passed on the graph are simply the parameters of the Gaussian distributions instead of continuous Gaussian PDFs. Computations involved in analog message passing are simply (1) the sum of 2M-dimensional vectors/matrices, and (2) 2M-dimensional matrix inversion. Therefore, the computational complexity of the SP-ML estimator is $\Omega(LM^2 \log M)$. If we fix M as a constant, the SP-ML estimator significantly reduces the computational complexity of ML estimation from $\Omega(L^2 \log L)$ to $\Omega(L)$.

Aligned-Sample Estimator
As stated in the beginning of this section, another benefit of the WMFS scheme is that it yields a sequence of samples wherein the indexes of symbols from different devices are consistent. Specifically, let us consider the outputs of the M-th matched filter.

Let $k = M$ in (4.54), we have $\mathcal{V}(y_M[i]) = \{s_1[i], s_2[i], \ldots, s_M[i]\}$, and

$$y_M[i] = \sum_{m=1}^{M} h'_M[i]s_m[i] + z_M[i], \tag{4.64}$$

where $z_M[i] \sim \mathcal{CN}(0, N_0/d_M)$ and d_M is the duration of the M-th matched filter. As can be seen, unlike the outputs of other matched filters, the neighbor symbols of $y_M[i]$ have the same index (that is, the symbol indexes are aligned within the integral interval of the M-th matched filter). Therefore, we can utilize the outputs of the M-th matched filter to devise an aligned-sample estimator.

Definition 4.11 (Aligned-Sample Estimator for Misaligned AirComp) Given the outputs of the WMFS $\{y_k[i]\}$, aligned-sample estimator estimates the desired sequence $\boldsymbol{s}_+ \in \mathcal{C}^L$ symbol-by-symbol by

$$\hat{s}_+[i] = y_M[i]. \tag{4.65}$$

Equation (4.64) is an underdetermined equation since we have one equation for M unknowns, and the estimator in (4.65) is our best prediction about $\hat{s}_+[i]$. When there is no or mild channel-gain misalignment (i.e., $h'_M \to 1$), the estimator (4.65) is supposed to perform very well.

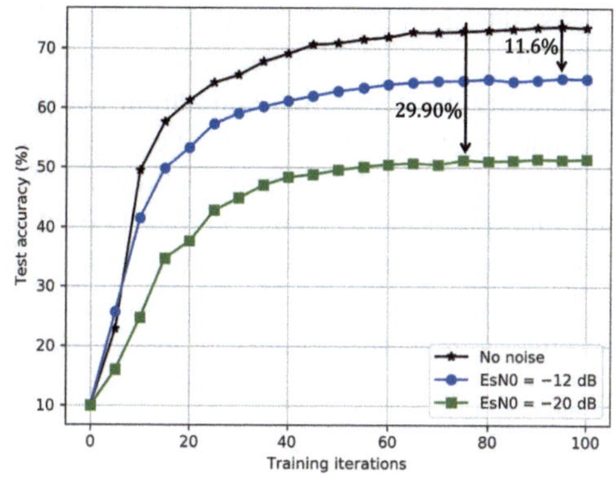

Fig. 4.22 Test accuracy of the learned model over the course of training, with and without noise. There is no time or channel-gain misalignment. We use the ML estimator at the PS

4.4.3.2 Simulation Results

This section evaluates the system performance of FL with misaligned AirComp considering two estimators at the receiver: the ML estimator and the aligned-sample estimator. In particular, for ML estimation, we use our SP-ML estimator since the ML estimators in (4.52) and (4.58) are computationally prohibitive.[7] We implement a FL system wherein 40 devices collaboratively train a convolution neural network (CNN) to solve the CIFAR-10 classification task. The CIFAR-10 dataset has a training set of 50,000 examples and a test set of 10,000 examples in 10 classes. Each example is a 32×32 colour image. The non-i.i.d. training examples are assigned to the 40 devices in the following manner: (1) we first let each device randomly sample 1000 samples from the dataset; (2) for the remaining 10,000 examples, we sort them by their labels and group into 40 shards of size 250 [57]. Each device is then assigned one shard.

The implemented CNN is a ShuffleNet V2 network [72] with $d = 1.26 \times 10^6$ parameters (this corresponds to 6.32×10^5 complex values). In each iteration, we assume $M = 4$ devices are active and participate in the training. Each device will train the global model locally for 5 epochs and then transmit the model-update to the PS in 44 packets (the packet length is $L = 1.44 \times 10^4$) in each iteration. The packets from different transmitters overlap at the PS with time and channel-gain misalignments, and the PS employs the ML and aligned-sample estimators to estimate the arithmetic sum of transmitted symbols, i.e., $\boldsymbol{s}_+$ (and hence, $\boldsymbol{\theta}_+$). The estimated arithmetic sum $\boldsymbol{\theta}_+$ is then used to update the global model, as per (4.41).

The metric we use to assess the performance of an estimator is the *test accuracy*. Specifically, when operated with a given estimator, we will train the global model for 100 iterations and take the prediction accuracy of the learned model on the test set as the performance indicator of the estimator. An example is given in Fig. 4.22.

[7] We have performed additional simulations to validate that the three estimators in (4.52), (4.58), and (4.63) are equivalent (using a much shorter packet length L). The simulation results are omitted here to conserve space.

As shown in Fig. 4.22, we run the FL system for 100 iterations with and without noise, and plot the test accuracy over the course of training. There are no misalignments in this simulation and ML estimator is used at the PS. First, the dark curve corresponds to the noiseless case and the test accuracy after 100 iterations is 73.43%. We point out that this is the global optimal test accuracy since the multiple-access channel is ideal, i.e., there is no time misalignment, channel-gain misalignment, or noise in the multiple-access channel. The other two curves in Fig. 4.22 correspond to the learning performance when noise is presented in the received signal. In particular, noise is added according to a given EsN0, i.e., the average received energy per symbol to noise power spectral density ratio, defined as

$$\text{EsN0} = \frac{\mathbb{E}_i\left[\left|\sum_{m=1}^{M} h'_m s_m[i]\right|^2\right]}{N_0}. \tag{4.66}$$

As shown, noise is detrimental to the test accuracy after convergence. Compared with the noiseless case, the test accuracy drops by 11.6% with an EsN0 of −12 dB, and by 29.9% with an EsN0 of −20 dB.

In addition to noise, we next introduce time misalignment into the received signal. The received signal is given in (4.54), where the noise term is $z[i] \sim \mathcal{CN}(0, N_0/d_k)$, and we set the residual channel gain to $h'_m = 1, \forall\, m$. Without loss of generality, symbol duration is set to $T = 1$. As can be seen from (4.54), time offsets $\{\tau_m : m = 1, 2, \ldots, M\}$ determine the noise power of samples from different matched filters (since $d_k = \tau_{k+1} - \tau_k$). In the simulation, the time offsets $\tau_m, \forall\, m$, are set in the following manner: first, we fix a maximum time offset τ_M (and hence d_M); then, we generate the other time offsets uniformly in $(0, \tau_M)$.

The simulation results are presented in Fig. 4.23. When $\tau_M = 0$, there is neither time nor channel-gain misalignment in the received signal. The ML estimator and the aligned-sample estimator are equivalent in this case, and yield the same test accuracy. As we increase τ_M, the performance of both estimators deteriorate.

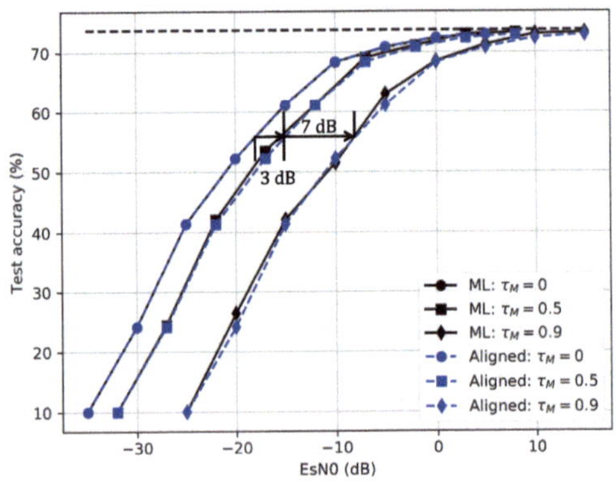

Fig. 4.23 Test accuracy (after 100 iterations) of the asynchronous AirComp under various EsN0. There is no channel-gain misalignment in the simulation and we use both the ML estimator and the aligned-sample estimator

1. For the aligned-sample estimator, the performance deterioration is easy to understand since the inputs to the estimator are the outputs of the M-th matched filter $y_M[i]$. As a result, the performance of the aligned-sample estimator is governed by the maximum time offset τ_M—the larger the τ_M, the worse the performance. As shown in Fig. 4.23, the introduction of time misalignment results in an EsN0 penalty for the aligned-sample estimator. When $\tau_M = 0.5$, the EsN0 penalty is 3 dB because d_M is reduced by a factor of 2 (from 1 to 0.5). Likewise, the EsN0 penalty is 10 dB when $\tau_M = 0.9$ since d_M is reduced by 10 times (from 1 to 0.1).
2. For different τ_M, the performance gain of the ML estimator over the aligned-sample estimator is negligible. The aligned-sample estimator utilizes only the outputs of the M-th matched filter. The ML estimator, on the other hand, utilizes the outputs of all matched filters and attempts to estimate the mostly likely s_+. It turns out that both estimators yield nearly the same performance when there is only time misalignment and noise.

It should be noted that the above result does not imply that the samples of the matched filters other than the M-th one are useless, because the ML estimator cannot achieve the same performance as the aligned-sample estimator using only the outputs of the M-th matched filter: recall from (4.65) that the outputs of the M-th matched filter results in a set of underdetermined equations. If we perform ML estimation based on the samples in (4.65), the estimation error can be arbitrarily large as $s_+[i]$ can be any value.

Error propagation: In the misaligned AirComp, ML estimation boils down to MUE and faces an infinitely large estimation space. It is then very susceptible to noise and suffers from error propagation. Take the SP-ML estimator for instance. In the forward message passing, the successful estimation of a likelihood function (of a multivariate variable) hinges on the accurate estimations of the likelihood functions on the left. When a sample is contaminated by noise, the mean of the likelihood function deviates from the true value of the noiseless sample. This estimation error will be propagated along the tree all the way to the rightmost leaf, because there are no known messages (i.e., prior information) in between to alleviate/correct the error. This can be one cause of the results in Fig. 4.23.

In the third simulation, let us further introduce channel-gain misalignment into the received signal (4.54). For each device, the residual channel gain is $h_m = |h_m|e^{j\phi_m}$. We set $|h_m| = 1, \forall m$, and focus on the impact of the phase offsets ϕ_m only. In particular, we assume $\{\phi_m : m = 1, 2, \ldots, M\}$ are uniformly distributed in $(0, \phi)$, where ϕ is the maximum phase offset (i.e., $\phi_m \sim U(0, \phi)$). It is worth noting that ϕ_m can be any distribution in general.

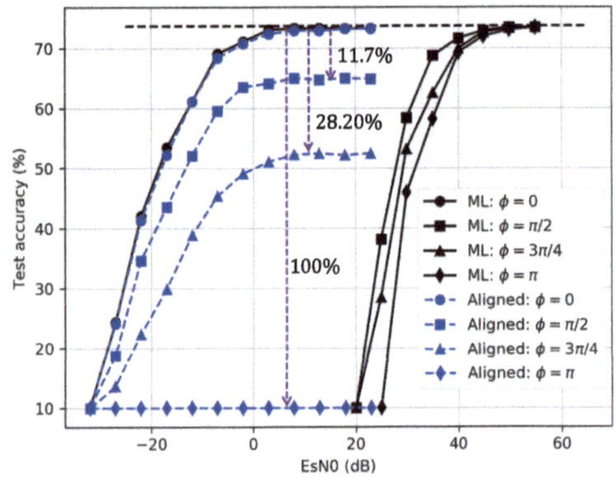

Fig. 4.24 Test accuracies of the ML and aligned-sample estimators under different degrees of phase misalignments. The maximum time offset $\tau_M = 0.5$ and the maximum phase misalignment $\phi = 0$ (no), $\pi/2$ (mild), $3\pi/4$ (moderate), and π (severe), respectively

Figure 4.24 presents the test accuracy of the ML and aligned-sample estimators versus EsN0 (in dB), wherein the maximum time offset τ_M is fixed to 0.5 and the maximum phase offset $\phi = 0$ (no phase misalignment), $\pi/2$ (mild), $3\pi/4$ (moderate), and π (severe), respectively.

A caveat here is that ϕ is the maximum phase offset—the phase offsets of all devices are uniformly distributed in $[0, \phi]$. If we look at the phase misalignment between any two devices, however, the average pairwise-phase-misalignment is only $\phi/3$. That is why we classify $\pi/2$ as mild phase misalignment because the average pairwise-phase-misalignment is only $\pi/6$.

We have the following observations from Fig. 4.24:

1. When there is no phase misalignment ($\phi = 0$), the two curves coincide just as in Fig. 4.23.
2. When there is mild phase misalignment ($\phi = \pi/2$), the aligned-sample estimator suffers from two penalties: (i) a small EsN0 penalty for about 5 dB, i.e., we need a 5 dB higher EsN0 to achieve the same test accuracy; (ii) a 11.7% test-accuracy loss, i.e., the test accuracy after convergence is 11.7% less than the phase-aligned case.

   The ML estimator, on the other hand, suffers from a large EsN0 penalty. The reason behind is that the phase misalignment enhances the error/noise propagation in ML estimation, which we refer to as *noise enhancement*. As a result, ML estimation does not work in the low-EsN0 regime when there is phase misalignment. On the bright side, ML estimator performs very well in the high-EsN0 regime: it suffers from no test-accuracy loss—the test accuracy after convergence is the same as the phase-aligned case.
3. When we further increase the maximum phase misalignment ϕ, the aligned-sample estimator suffers from larger EsN0 and test-accuracy penalties. In the

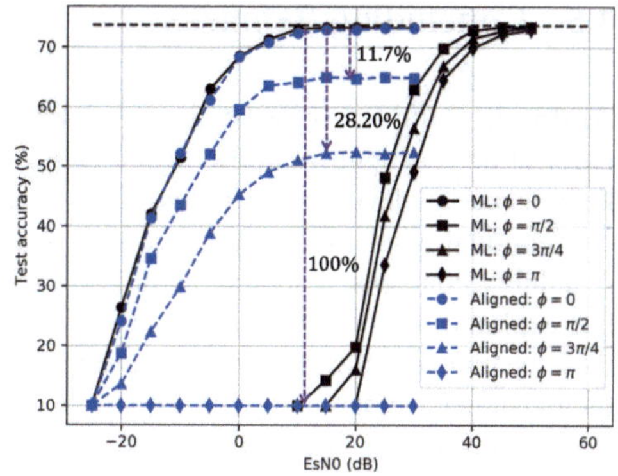

Fig. 4.25 Test accuracies of the ML and aligned-sample estimators under different degrees of phase misalignments. The maximum time offset $\tau_M = 0.9$ and the maximum phase misalignment $\phi = 0$ (no), $\pi/2$ (mild), $3\pi/4$ (moderate), and π (severe), respectively

case of moderate phase misalignment ($\phi = 3\pi/4$), the test-accuracy penalty is up to 28.2%. In the case of severe phase misalignment ($\phi = \pi$), the learning diverges with the aligned-sample estimator.

In contrast, the ML estimator is more robust to moderate and severe phase misalignments in the high-EsN0 regime – there is no test-accuracy loss and only a small EsN0 penalty.

Figure 4.24 studies the impact of phase misalignment under a fixed maximum time offset $\tau_M = 0.5$. Next, we consider a larger time offset $\tau_M = 0.9$ and repeat the simulations in Fig. 4.24. The simulation results are presented in Fig. 4.25.

For the aligned-sample estimator, time offset only incurs an EsN0 penalty. Thus, each performance curve of the aligned-sample estimator in Fig. 4.25 is simply a right-shift of the corresponding curve in Fig. 4.24 by 7 dB.

On the other hand, we observe that ML estimation benefits from larger time misalignment when there is phase misalignment. For example, with a mild phase misalignment ($\phi = \pi/2$), the EsN0 gain is about 2 dB when the maximum time offset τ_M is increased from 0.5 to 0.9, as shown in Figs. 4.24 and 4.25. In contrast, when there is no phase misalignment, ML estimation suffers from larger time misalignment, as shown in Fig. 4.23.

To conclude this section, we summarize the main simulation results as follows.

1. When there is no phase misalignment, the ML and aligned-sample estimators are on equal footing as far as the learning performance is concerned.
2. When there is mild or moderate phase misalignment, the aligned-sample estimator outperforms the ML estimator in the low-EsN0 regime, but is worse than the ML estimator in the high-EsN0 regime.
3. When there is severe phase misalignment, the aligned estimator leads to divergence of learning, but ML estimation still works in the high-EsN0 regime.

4.5 Concluding Remarks

AI-driven multiple access represents a fundamental transformation in IoT communication, enabling networks to efficiently handle massive-scale, dynamic, and heterogeneous device ecosystems. Traditional MA techniques, while effective in conventional wireless systems, face serious limitations when applied to IoT due to scalability concerns, resource inefficiency, and high energy demands. This chapter has explored how machine learning and deep reinforcement learning can address these challenges, allowing IoT devices to intelligently and autonomously optimize their channel access.

By introducing Multi-Agent Deep Reinforcement Learning (MADRL) for Autonomous Channel Access, this chapter demonstrated how IoT devices can collaboratively learn optimal transmission policies, reducing collisions, latency, and energy consumption while enhancing network throughput and fairness. Additionally, the discussion on Over-the-Air Computation (AirComp) for Federated Learning (FL) illustrated how AI-driven MA schemes can be tailored to support distributed AI applications, optimizing communication efficiency for large-scale learning tasks.

Beyond these contributions, the integration of AI into multiple access protocols signifies a shift toward self-learning, adaptive, and intelligent network architectures. Future research should focus on real-world deployments of AI-driven MA, the development of lightweight AI models for resource-constrained IoT devices, and hybrid MA schemes that combine AI-based optimization with traditional techniques. Moreover, edge intelligence and federated learning-based MA architectures will play a key role in enabling secure, scalable, and low-latency IoT networks.

Ultimately, AI-driven multiple access not only optimizes communication efficiency but also lays the foundation for autonomous, self-evolving IoT ecosystems, paving the way for the next generation of smart, interconnected environments.

References

1. Y. Cai, Z. Qin, F. Cui, G.Y. Li, J.A. McCann, Modulation and multiple access for 5G networks. IEEE Commun. Surv. Tutor. **20**(1), 629–646 (2017)
2. M.B. Shahab, R. Abbas, M. Shirvanimoghaddam, S.J. Johnson, Grant-free non-orthogonal multiple access for IoT: a survey. IEEE Commun. Surv. Tutor. **22**(3), 1805–1838 (2020)
3. Y. Yu, T. Wang, S.C. Liew, Deep-reinforcement learning multiple access for heterogeneous wireless networks. IEEE J. Sel. Areas Commun. **37**(6), 1277–1290 (2019)
4. Z. Guo, Z. Chen, P. Liu, J. Luo, X. Yang, X. Sun, Multi-agent reinforcement learning based distributed channel access for next generation wireless networks. IEEE J. Sel. Areas Commun. **40**(5), 1587–1599 (2022)
5. Y. Shao, Y. Cai, T. Wang, Z. Guo, P. Liu, J. Luo, D. Gündüz, Learning-based autonomous channel access in the presence of hidden terminals. IEEE Trans. Mob. Comput. **23**(5), 3680–3695 (2023)
6. S. Gronauer, K. Diepold, Multi-agent deep reinforcement learning: a survey. Artif. Intell. Rev. **55**(2), 895–943 (2022)

7. Y. Shao, D. Gündüz, S.C. Liew, Federated edge learning with misaligned over-the-air computation. IEEE Trans. Wirel. Commun. **21**(6), 3951–3964 (2021)
8. E. Ziouva, T. Antonakopoulos, CSMA/CA performance under high traffic conditions: throughput and delay analysis. Comput. Commun. **25**(3), 313–321 (2002)
9. J. Tan, L. Zhang, Y.-C. Liang, D. Niyato, Intelligent sharing for LTE and Wi-Fi systems in unlicensed bands: a deep reinforcement learning approach. IEEE Trans. Commun. **68**(5), 2793–2808 (2020)
10. H. Li, Multiagent-learning for ALOHA-like spectrum access in cognitive radio systems. EURASIP J. Wirel. Commun. Netw. **2010**, 1–15 (2010)
11. W. Wydmański, S. Szott, Contention window optimization in IEEE 802.11 ax networks with deep reinforcement learning, in *IEEE Wireless Commun. Netw. Conf. (WCNC)* (IEEE, 2021), pp. 1–6
12. Y. Shao, Q. Cao, S.C. Liew, H. Chen, Partially observable minimum-age scheduling: the greedy policy. IEEE Trans. Commun. **70**(1), 404–418 (2021)
13. X. Yang, N. Vaidya, On physical carrier sensing in wireless ad-hoc networks, in *Proc. IEEE 24th Annual Joint Conference of the IEEE Computer and Communications Societies.*, vol. 4 (IEEE, 2005), pp. 2525–2535
14. S. Safaric, K. Malaric, ZigBee wireless standard, in *Proc. ELMAR* (IEEE, 2006), pp. 259–262
15. Y. Shao, S.C. Liew, J. Liang, Sporadic ultra-time-critical crowd messaging in V2X. IEEE Trans. Commun. **69**(2), 817–830 (2020)
16. A. Kumar, G. Verma, C. Rao, A. Swami, S. Segarra, Adaptive contention window design using deep Q-learning, in *IEEE Int. Conf. Acoustics, Speech and Signal Proc. (ICASSP)* (IEEE, 2021), pp. 4950–4954
17. Y. Yu, S.C. Liew, T. Wang, Non-uniform time-step deep Q-network for carrier-sense multiple access in heterogeneous wireless networks. IEEE Trans. Mob. Comput. **20**(9), 2848–2861 (2020)
18. Y. Yu, S.C. Liew, T. Wang, Multi-agent deep reinforcement learning multiple access for heterogeneous wireless networks with imperfect channels. IEEE Trans. Mob. Comput. **21**(10), 3718–3730 (2021)
19. O. Naparstek, K. Cohen, Deep multi-user reinforcement learning for distributed dynamic spectrum access. IEEE Trans. Wirel. Commun. **18**(1), 310–323 (2018)
20. K.-L.A. Yau, P. Komisarczuk, D.T. Paul, Enhancing network performance in distributed cognitive radio networks using single-agent and multi-agent reinforcement learning, in *IEEE Local Computer Netw. Conf.* (IEEE, 2010), pp. 152–159
21. C. Zhong, Z. Lu, M.C. Gursoy, S. Velipasalar, A deep actor-critic reinforcement learning framework for dynamic multichannel access. IEEE Trans. Cogn. Commun. Netw. **5**(4), 1125–1139 (2019)
22. S.Q. Jalil, S. Chalup, M.H. Rehmani, Cognitive radio spectrum sensing and prediction using deep reinforcement learning, in *Int. Joint Conf. Neural Networks (IJCNN)* (IEEE, 2021), pp. 1–8
23. Y. Shao, A. Rezaee, S.C. Liew, V.W.S. Chan, Significant sampling for shortest path routing: a deep reinforcement learning solution. IEEE J. Sel. Areas Commun. **38**(10), 2234–2248 (2020)
24. A. Tsertou, D.I. Laurenson, Revisiting the hidden terminal problem in a CSMA/CA wireless network. IEEE Trans. Mob. Comput. **7**(7), 817–831 (2008)
25. S. Khurana, A. Kahol, A.P. Jayasumana, Effect of hidden terminals on the performance of IEEE 802.11 MAC protocol, in *Proc. Annual Conference on Local Computer Networks* (IEEE, 1998), pp. 12–20
26. S. Ray, J.B. Carruthers, D. Starobinski, Evaluation of the masked node problem in ad-hoc wireless LANs. IEEE Trans. Mob. Comput. **4**(5), 430–442 (2005)
27. J.L. Sobrinho, R. De Haan, J.M. Brazio, Why RTS-CTS is not your ideal wireless LAN multiple access protocol, in *IEEE Wireless Commun. Netw. Conf.*, vol. 1 (IEEE, 2005), pp. 81–87

28. T. Rashid, M. Samvelyan, C. Schroeder, G. Farquhar, J. Foerster, S. Whiteson, Qmix: monotonic value function factorisation for deep multi-agent reinforcement learning, in *Int. Conf. Machine Lear.*, PMLR (2018), pp. 4295–4304
29. A. Valcarce, J. Hoydis, Toward joint learning of optimal MAC signaling and wireless channel access. IEEE Trans. Cogn. Commun. Netw. **7**(4), 1233–1243 (2021)
30. M.P. Mota, A. Valcarce, J.-M. Gorce, J. Hoydis, The emergence of wireless MAC protocols with multi-agent reinforcement learning, in *IEEE Globecom Workshops* (2021)
31. T. Lan, D. Kao, M. Chiang, A. Sabharwal, An axiomatic theory of fairness in network resource allocation (2009). arXiv:0906.0557
32. J. Majkowski, F.C. Palacio, Dynamic TXOP configuration for Qos enhancement in IEEE 802.11 e wireless LAN, in *Int. Conf. Software Telecom. Computer Netw.* (IEEE, 2006), pp. 66–70
33. K. Zhang, Z. Yang, T. Başar, Multi-agent reinforcement learning: a selective overview of theories and algorithms, in *Handbook of Reinforcement Learning and Control* (2021), pp. 321–384
34. S. Hochreiter, J. Schmidhuber, Long short-term memory. Neural Comput. **9**(8), 1735–1780 (1997)
35. J. Schulman, P. Moritz, S. Levine, M. Jordan, P. Abbeel, High-dimensional continuous control using generalized advantage estimation (2015). arXiv:1506.02438
36. J. Schulman, F. Wolski, P. Dhariwal, A. Radford, O. Klimov, Proximal policy optimization algorithms (2017). arXiv:1707.06347
37. M. Hausknecht, P. Stone, Deep recurrent Q-learning for partially observable MDPs, in *AAAI fall Symposium Series* (2015)
38. J. Foerster, G. Farquhar, T. Afouras, N. Nardelli, S. Whiteson, Counterfactual multi-agent policy gradients, in *Proc. AAAI Conf. Artificial Intelligence*, vol. 32 (2018)
39. P. Shen, Y. Shao, H. Pan, L. Lu, Y.C. Eldar, Channel cycle time: a new measure of short-term fairness. IEEE Trans. Mob. Comput. **24**(3), 1386–1401 (2024)
40. I.I.M. Society, IEEE standard jitter and phase noise. IEEE Std (2020), pp. 1–42
41. V. Mnih, K. Kavukcuoglu, D. Silver, A.A. Rusu, J. Veness, M.G. Bellemare, A. Graves, M. Riedmiller, A.K. Fidjeland, G. Ostrovski, et al., Human-level control through deep reinforcement learning. Nature **518**(7540), 529 (2015)
42. D. Silver, G. Lever, N. Heess, T. Degris, D. Wierstra, M. Riedmiller, Deterministic policy gradient algorithms, in *Int. Conf. Mach. Learning (ICML)* (2014)
43. Y. Shao, S.C. Liew, T. Wang, Alphaseq: sequence discovery with deep reinforcement learning. IEEE Trans. Neural Netw. Learn. Syst. **31**(9), 3319–3333 (2019)
44. J. Gittins, K. Glazebrook, R. Weber, *Multi-Armed Bandit Allocation Indices* (John Wiley & Sons, New York, 2011)
45. M.M. Amiri, D. Gündüz, Machine learning at the wireless edge: distributed stochastic gradient descent over-the-air. IEEE Trans. Signal Process. **68**, 2155–2169 (2020)
46. M.M. Amiri, D. Gündüz, Federated learning over wireless fading channels. IEEE Trans. Wirel. Commun. **19**(5), 3546–3557 (2020)
47. G. Zhu, K. Huang, Broadband analog aggregation for low-latency federated edge learning. IEEE Trans. Wirel. Commun. **19**, 491–506 (2020)
48. T. Sery, N. Shlezinger, K. Cohen, Y. Eldar, Over-the-air federated learning from heterogeneous data (2020). arXiv preprint arXiv:2009.12787
49. K. Yang, T. Jiang, Y. Shi, Z. Ding, Federated learning via over-the-air computation. IEEE Trans. Wirel. Commun. **19**(3), 2022–2035 (2020)
50. G. Zhu, Y. Du, D. Gunduz, K. Huang, One-bit over-the-air aggregation for communication-efficient federated edge learning: design and convergence analysis. IEEE Trans. Wirel. Commun. **20**(3), 2120–2135 (2020)
51. L. You, X. Zhao, R. Cao, Y. Shao, L. Fu, Broadband digital over-the-air computation for wireless federated edge learning. IEEE Trans. Mob. Comput. **23**(5), 5212–5228 (2023)
52. M.M. Amiri, T.M. Duman, D. Gündüz, S.R. Kulkarni, H.V. Poor, Blind federated edge learning. IEEE Trans. Wirel. Commun. **20**(8), 5129–5143 (2021)

53. Y. Sun, S. Zhou, D. Gündüz, Energy-aware analog aggregation for federated learning with redundant data, in *IEEE International Conference on Communications (ICC)* (2020), pp. 1–7
54. J. Dong, Y. Shi, Z. Ding, Blind over-the-air computation and data fusion via provable Wirtinger flow. IEEE Trans. Signal Process. **68**, 1136–1151 (2020)
55. Y. Shao, D. Gündüz, S.C. Liew, Bayesian over-the-air computation. IEEE J. Sel. Areas Commun. **41**(3), 589–606 (2022)
56. Y. Shao, S.C. Liew, D. Gündüz, Denoising noisy neural networks: a bayesian approach with compensation. IEEE Trans. Signal Process. **71**, 2460–2474 (2023)
57. B. McMahan, E. Moore, D. Ramage, S. Hampson, B.A. y Arcas, Communication-efficient learning of deep networks from decentralized data, in *AI Statistics*, PMLR (2017), pp. 1273–1282
58. D. Tse, P. Viswanath, *Fundamentals of Wireless Communication* (Cambridge University Press, Cambridge, 2005)
59. Q. Yang, S.C. Liew, L. Lu, Y. Shao, Symbol misalignment estimation in asynchronous physical-layer network coding. IEEE Trans. Veh. Technol. **66**(3), 2844–2852 (2016)
60. S. Verdu, *Multiuser Detection* (Cambridge University Press, Cambridge, 1998)
61. A. Tveit, On the complexity of matrix inversion. Mathematical Note (2003)
62. R. Zhang, Y. Shao, Y.C. Eldar, Polarization aware movable antenna (2024). arXiv preprint arXiv:2411.06690
63. F.R. Kschischang, B.J. Frey, H.-A. Loeliger, Factor graphs and the sum-product algorithm. IEEE Trans. Inf. Theory **47**(2), 498–519 (2001)
64. H.-A. Loeliger, J. Dauwels, J. Hu, S. Korl, L. Ping, F.R. Kschischang, The factor graph approach to model-based signal processing. Proc. IEEE **95**(6), 1295–1322 (2007)
65. Y. Shao, S.C. Liew, L. Lu, Asynchronous physical-layer network coding: symbol misalignment estimation and its effect on decoding. IEEE Trans. Wirel. Commun. **16**(10), 6881–6894 (2017)
66. K. Murphy, Y. Weiss, M.I. Jordan, Loopy belief propagation for approximate inference: an empirical study (2013). arXiv preprint arXiv:1301.6725
67. J. Pearl, Reverend Bayes on inference engines: a distributed hierarchical approach. Cognitive Systems Laboratory (1982)
68. N. Noorshams, M.J. Wainwright, Belief propagation for continuous state spaces: stochastic message-passing with quantitative guarantees. J. Mach. Learn. Res. **14**(1), 2799–2835 (2013)
69. T. Wang, L. Shi, S. Zhang, H. Wang, Gaussian mixture message passing for blind known interference cancellation. IEEE Trans. Wirel. Commun. **18**(9), 4268–4282 (2019)
70. G. Forney, Maximum-likelihood sequence estimation of digital sequences in the presence of inter-symbol interference. IEEE Trans. Inf. Theory **18**(3), 363–378 (1972)
71. L. Bahl, J. Cocke, F. Jelinek, J. Raviv, Optimal decoding of linear codes for minimizing symbol error rate. IEEE Trans. Inf. Theory **20**(2), 284–287 (1974)
72. N. Ma, X. Zhang, H.-T. Zheng, J. Sun, Shufflenet v2: practical guidelines for efficient cnn architecture design, in *ECCV* (2018), pp. 116–131

Chapter 5
AI-Enhanced Network Control for IoT

Machines learn, networks adapt, and intelligence flows seamlessly.

5.1 Overview

The IoT represents a transformative shift in how devices communicate, interact, and operate, profoundly impacting industries, economies, and daily life. As IoT networks expand, connecting billions of devices ranging from sensors and actuators to complex machinery, the need for intelligent network control mechanisms becomes paramount [1–4]. Traditional network management approaches are increasingly insufficient to handle the scale, diversity, and dynamic nature of IoT environments. This has led to the exploration of AI as a means to empower IoT communication and network control.

This chapter explores the application of advanced AI techniques in IoT network control, specifically focusing on deep reinforcement learning (DRL) [5], graph neural networks (GNNs) [6], and large language models (LLMs) [7]. By examining these technologies, we aim to illustrate how AI can address the inherent challenges of IoT networks, enhance their performance, and pave the way for intelligent and autonomous networking.

5.1.1 AI in IoT Network Control

The integration of AI into IoT network control represents a significant evolution in network management strategies. AI brings the capability to process vast amounts of data, learn from interactions, and make intelligent decisions autonomously. This capability is particularly crucial in IoT networks, where the heterogeneity of devices and the complexity of interactions demand sophisticated control mechanisms.

One of the primary benefits of incorporating AI into IoT network control is adaptive decision-making. AI models can adjust to changing network conditions,

Y. Shao, *AI-Empowered IoT Communications*, Wireless Networks,
https://doi.org/10.1007/978-3-031-98816-5_5

optimizing performance in real time. This adaptability ensures that the network maintains optimal operation despite fluctuations in device behavior, network traffic, or environmental conditions.

Scalability is another advantage offered by AI techniques. As IoT networks grow, managing large-scale networks efficiently becomes increasingly challenging. AI algorithms can handle the complexity by accommodating growth in device numbers and data traffic without a proportional increase in management overhead. This efficiency is essential for the sustainability of expanding IoT ecosystems.

Enhanced user interaction is also facilitated through AI, particularly via natural language processing. AI allows for more intuitive user interfaces, enabling users to interact with the network using everyday language. This development makes network management more accessible and user-friendly, reducing the need for specialized technical knowledge or complex command inputs.

In network control, AI can automate tasks such as resource allocation, interference management, and policy optimization. By leveraging learning algorithms, networks can anticipate and respond to events proactively, improving reliability and efficiency. This proactive approach reduces downtime and enhances the overall performance of the network. For example,

- DRL enables the learning of optimal policies through interactions with the environment, even when system models are unknown. This feature is particularly useful in dynamic and complex IoT environments where traditional modeling is challenging.
- GNNs are designed to handle graph-structured data, making them ideal for networks where relationships between nodes (devices) are significant. They capture the underlying structure of the network, enabling more efficient processing and decision-making based on the topology and connectivity of devices.
- LLMs facilitate natural language understanding and generation, enabling more seamless human-machine interactions. By interpreting and generating human-like language, LLMs can translate user intentions into actionable network commands, bridging the gap between user requirements and technical implementations.

5.1.2 Challenges in IoT Network Management

Managing IoT networks involves overcoming several inherent challenges that stem from their dynamic and complex nature. Understanding these challenges is crucial for developing effective AI-driven solutions that can enhance network performance, reliability, and scalability.

5.1.2.1 Dynamic Environments and Adaptability

IoT networks can operate in highly dynamic environments characterized by constantly changing conditions. One of the primary challenges is the variable network topologies resulting from devices frequently joining or leaving the network. This fluctuation is due to factors such as device mobility, power constraints, or intermittent connectivity, which can disrupt established communication paths and necessitate continuous reconfiguration of the network.

Another challenge is the fluctuating traffic patterns inherent in IoT applications. Data transmission rates can vary widely based on sensor readings, user interactions, or time-dependent activities. For instance, environmental sensors might generate more data during specific conditions, or user devices may have peak usage times, leading to uneven network load and potential congestion.

Additionally, environmental influences such as physical interference, obstacles, and weather conditions can significantly impact network performance. Signal degradation due to obstacles or atmospheric conditions can lead to increased error rates or connectivity loss, requiring the network to adapt promptly to maintain communication quality.

These dynamic factors necessitate network control mechanisms that are:

- **Responsive**: Capable of adjusting operational parameters in real-time to maintain optimal performance amid changing conditions.
- **Predictive**: Able to anticipate future changes based on historical data and trends, enabling proactive adjustments before issues arise.
- **Resilient**: Robust to unexpected events or anomalies, ensuring continuous operation despite unforeseen challenges.

Traditional static or rule-based control systems often fall short in such environments because they lack the flexibility to adapt quickly to changes. Their inability to respond promptly can lead to inefficiencies, including increased latency, reduced throughput, and unnecessary energy consumption. Consequently, there is a pressing need for intelligent control systems that can dynamically adjust to the evolving state of the network.

5.1.2.2 Centralized vs. Decentralized Control

A fundamental consideration in IoT network management is the architecture of control mechanisms, specifically the choice between centralized and decentralized control approaches.

In a centralized control architecture, a central controller maintains a global view of the entire network, facilitating coordinated decision-making and simplifying policy enforcement and updates. This centralized entity can optimize resource allocation and network performance by considering the network as a whole. However, centralized control presents the following disadvantages:

- **Scalability Issues**: As the network grows, the central controller can become a bottleneck, struggling to process the increased volume of data and control requests efficiently.

- **Latency Concerns**: Communicating with a central point can introduce delays, especially in large or widely distributed networks, adversely affecting time-sensitive applications.
- **Single Point of Failure:** The central controller represents a critical vulnerability; if it fails or is compromised, the entire network's operation can be disrupted.

Alternatively, decentralized control distributes control tasks across multiple nodes, allowing each node to make decisions based on local information. This approach offers several advantages. For example, distributing control reduces the burden on any single node, allowing the network to grow without overwhelming central resources. Also, local decision-making minimizes the need for long-distance communication, enabling faster responses to local events. Furthermore, eliminating reliance on a central controller mitigates the risk of a single point of failure, improving the network's resilience to faults or attacks.

However, decentralized control also faces challenges such as:

- **Limited Global Awareness**: Individual nodes may lack comprehensive information about the entire network, potentially leading to decisions that are locally optimal but globally suboptimal.
- **Coordination Challenges**: Ensuring consistent policies and avoiding conflicts between nodes require sophisticated coordination mechanisms, which can be complex to design and implement.

Finding the optimal balance between centralized and decentralized control is critical. Hybrid approaches aim to leverage the benefits of both architectures, combining global oversight with local autonomy. For example, a central controller might set high-level policies while allowing nodes to make local adjustments. However, hybrid models introduce additional complexity in coordination and implementation, necessitating advanced control algorithms and communication protocols to ensure seamless integration.

5.1.3 Chapter Organization

This chapter is structured to provide a comprehensive exploration of how advanced AI techniques can be harnessed to address the challenges inherent in IoT network control and management. We begin by delving into the application of DRL in adaptive network control. In this section, we examine how DRL enables IoT networks to learn optimal policies for decision-making in complex, dynamic networks without relying on precise system models.

Next, we shift our focus to GNNs and their role in facilitating decentralized network control. Recognizing the scalability and robustness challenges of centralized architectures, we explore how GNNs can be utilized to distribute control tasks across the network. By representing the network as a graph where nodes correspond to devices and edges represent interactions, GNNs enable each node to make informed decisions based on local information and neighborhood relationships. We delve into the specific application of channel allocation in IoT networks, demonstrating how GNNs can effectively minimize interference among APs by allowing for local decision-making that collectively leads to globally efficient outcomes.

Following the exploration of DRL and GNNs, we introduce LLMs as central orchestrators in IoT ecosystems through the presentation of the LLMind framework. This section highlights how LLMs, with their advanced natural language processing and complex task planning capabilities, can enhance user interaction, system flexibility and network control. By engaging in natural language conversations with users, LLMs can understand complex tasks and generate control scripts using language-code transformation techniques based on finite-state machines. We discuss how the integration of LLMs with domain-specific AI modules fosters an intelligent and evolving IoT ecosystem that learns from continuous interactions between humans and machines.

Roadmap

The remainder of this chapter is articulated in the following manner. Section 5.2 shows how DRL can be used in dynamic Networks for adaptive monitoring. Section 5.3 presents a new decentralized network control scheme using graph neural networks. Section 5.4 presents a new framework dubbed LLMind to achieve autonomous control in IoT systems using LLMs as central orchestrators. Section 5.5 concludes this chapter.

5.2 Adaptive Monitoring in Dynamic Networks via DRL

We are witnessing the dawn of a new technological era where new IoT applications and devices have revolutionized every facet of our daily lives. Many of these IoT applications produce and consume data at unprecedented rates with strict reliability and latency requirements [4, 8, 9]. Not surprisingly, the bursty and unpredictable nature of their traffic can induce highly dynamic network environments that threaten their own viability. Unencumbered operation of these IoT applications requires rapid (re)actions by the Network Management and Control (NMC) system which itself depends on timely collection of network state information (NSI) [10, 11].

In highly dynamic IoT networks, one key problem is identifying the shortest path between a pair of communicating nodes. Ref. [12], for example, introduced a centralized routing protocol in which an NMC system monitors the network state (e.g., link/queuing delays) to decide which paths should be used to connect different origin-destination (OD) pairs. The monitoring process can be interpreted as a sampling process involving three steps: (a) updating: the nodes in the network report their state (i.e. their most recent queuing delay over a period) to the NMC system; (b) decision: the NMC system identifies the shortest path between each

OD pair, and computes the next report time; (c) dissemination: the NMC system disseminates the updated routing information and the next report time to all the nodes in the network.

A critical issue for this routing protocol is to determine the sampling frequency. Most network monitoring systems considers only uniform sampling (see the related work). However, uniform sampling at high frequency can be burdensome and costly due to the excessive use of network transport and computational resources, while sparse sampling can lead to misconfiguration and suboptimal routing decisions [11]. Hence, judicious sampling of network states should be an essential feature of any pragmatic NMC system. In this context, the notion of significant sampling was introduced in [12], where the next report time is computed by the NMC from its cognitive understanding of the network states and short-term behavior of exogenous offered traffic. In a nutshell, the NMC will decide the next sampling time based on the likelihood that the optimal routes should be recomputed. The problem of discovering the optimal sampling policy for the cognitive NMC is referred to as the significant sampling problem.

In an effort to get an analytical solution to significant sampling, [12] proposed the Ornstein-Uhlenbeck (OU) process as the underlying traffic/delay model and derived the optimal sampling policy when the parameters of the OU process are known to the NMC system. Despite its advantageous analytical nature, the scheme proposed in [12] has a few shortcomings. First, the derived policy ϕ' suffers from the "never-sample" problem: when the cost of sampling is large, the policy ϕ' can instruct the NMC system to "never sample", which can be detrimental in the long run. Second, considering the bursty and unpredictable nature of traffic in dynamic networks, the assumed OU delay model is narrow in scope. A realistic sampling policy should be able handle and smoothly transition through various operating regimes. In this section, we demonstrate how DRL can be utilized to solve the above problems, yielding an automated and model-free NMC system.

Related Work

Existing network monitoring systems considers only uniform sampling. Some examples are given in the following:

1. Traditional network monitoring. Simple Network Management Protocol (SNMP) [13] is a widely deployed network monitoring protocol. The controller regularly polls the agent (e.g., switches) making requests for information, and the agents respond when asked with the information requested. NetFlow [14] is a flow-based monitoring protocols built in Cisco routers. The routers update the flow statistics to the centralized Collector for aggregation and analysis on a regular basis. The time interval for updating is pre-configured.
2. Network monitoring in SDN. With the emergence of software-defined networking (SDN), new network monitoring approaches were developed for this new network architecture. OpenTM [15] is a traffic matrix estimation system design for OpenFlow networks. When sampling the network for traffic matrix estimation, the authors highlight two factors that directly affect the accuracy and overhead of estimation: (1) which switches to sample, (2) at which frequency

to sample. OpenTM focuses on the first factor, but the sampling intervals are fixed (5 seconds in the experiment). Motivated by OpenTM, an improved online monitoring system, OpenNetMon, is developed in [16], wherein the controller periodically samples the last switch on each path. In addition to passive sampling, active probing is another scheme to monitor the network state. SLAM [17] is a latency monitoring framework that sends probe packets to trigger updating from switches to the controller. The probe frequency, however, is still uniform (one probe per second in the experiment).

3. Other fields. The problem of deriving the optimal sampling frequency to tradeoff performance and overhead also exists in other fields such as sensor networks [18] and runtime verification [19]. However, the system setups are often limited to uniform sampling, and the problem is to derive the (piecewise) optimal uniform sampling.

In contrast, significant sampling is a judicious sampling strategy. The sampling intervals are computed by the NMC from its cognitive understanding of the network states. This non-uniform sampling idea is completely different from prior works wherein the NMC samples the network at uniform rate.

5.2.1 Significant Sampling

Let us consider an OD pair in a dynamic IoT network, and assume that the OD pair is connected via N paths $\{P_n : n = [N]\}$, where $[N] = 1, 2, \ldots, N$. The stochastically evolving weights[1] of the N paths are denoted by $\{X_n(t) : n = [N]\}$. In shortest path routing, an NMC system will sample the weight of the N paths by orchestrating three distinct actions: (a) the nodes (i.e., IoT devices) in each path report their weights to the NMC system; (b) the NMC system identifies the path with the smallest weight and computes the next sampling time; (c) the NMC disseminates the routing information and next report/sampling time to each nodes.

As an example, suppose that the nodes on the N paths report[2] their weights to the NMC at a sampling epoch t_{i-1}. Given observation $\{X_n(t) : n = [N], t \leq t_{i-1}\}$, the NMC will decide to route through $P_{n'}$, where $n' = \arg\min_n X_n(t_{i-1})$. This routing information, along with the next sampling epoch, will be disseminated to all the nodes on each path. The same route will be used until the next sampling epoch t_i, at which point the NMC will use $\{X_n(t) : n = [N], t \leq t_i\}$ to make new routing

[1] Weights could be delays or other cost measures.

[2] At each sampling epoch, we assume that the nodes will report the trajectory of their evolving weights since the last sampling report to the NMC. For example, at t_i, the full realizations of $\{X_n(t) : n = [N], t_{i-1} \leq t \leq t_i\}$ will be provided to the NMC system, where t_{i-1} is the last sampling time.

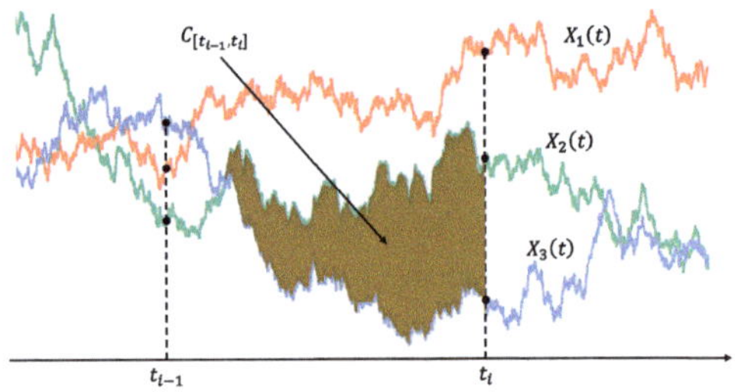

Fig. 5.1 Cost of error, $C_{[t_{i-1},t_i]}$, between two consecutive sampling epochs t_{i-1} and t_i

decisions. A fixed cost η is associated with one sampling operation, which captures the efforts required to collect and disseminate the NSI throughout the network.[3]

As illustrated in Fig. 5.1, if the shortest path changes between two sampling epochs t_{i-1} and t_i, we will incur an excess cost $C_{[t_{i-1},t_i]}$ given by

$$C_{[t_{i-1},t_i]} = \int_{t_{i-1}}^{t_i} X_{n'}(t) - \min_n X_n(t)\, dt,$$

where $n' = \arg\min_n X_n(t_{i-1})$ is the shortest path observed at epoch t_{i-1}, and $\min_n X_n(t)$ denotes the weight of the "true" shortest path at time t.

Not surprisingly, continuous sampling of the stochastic process can reduce the cost of error to zero, but will result in an extremely high sampling cost. On the other hand, sparse sampling incurs a low sampling cost, at the expense of higher cost of error. The optimal tradeoff between sampling cost and cost of error can be described through a policy that specifies the best future sampling times based on the previous observations of the weight processes. Specifically, we write the policy at epoch t_{i-1} as a deterministic function that maps the history of weight processes $s(t_{i-1}) = \{X_n(t) : n = [N], t \leq t_{i-1}\}$ to the next sampling interval T_i:

$$T_i = \phi(s(t_{i-1})).$$

In particular, if the weight processes are Markovian (i.e., all the information is captured by the most recent observation, and prior history of the process is inconsequential for optimal decision making), we can design $s(t_{i-1})$ to be $s(t_{i-1}) = \{X_n(t_{i-1}) : n = [N]\}$.

The optimal policy ϕ^*, which specifies the optimal T_i^* at epoch t_{i-1}, minimizes the cost rate (i.e., cost per unit time) over an infinite time horizon, giving

[3] To be more specific, the cost of one sampling operation is composed of communication cost and cost of action. The communication cost is introduced in steps (a) and (c), and can be considered as constant. For example, the communication cost in the step (a) can be fixed to one IP packet/node (considering a typical IP packet containing 1K Bytes, a node can send 500 real numbers to the NMC per update if we use 2 bytes to represent a real number). The cost of action is introduced by the computation in step (b), computational cost of updating routing-tables, and other side-effects. We use η as a catch-all quantity to capture the overall cost of one sampling operation.

$$\phi^* = \arg\min_{\phi} \lim_{L\to\infty} \frac{\sum_{l=i}^{L} \left(\eta + C_{[t_{l-1}, t_{l-1}+\phi(s(t_{l-1}))]}\right)}{\sum_{l=i}^{L} \phi(s(t_{l-1}))}, \tag{5.1}$$

where we assume the sampling cost η is normalized with respect to the cost of error such that they have the same scale and unit. We refer to the problem of discovering the optimal policy ϕ^* as the "significant sampling" problem for shortest path routing.

5.2.2 *Prior Solution and Limitations*

Reference [12] derived the analytical solutions for the significant sampling problem under three assumptions:

1. $N = 2$, i.e., the OD pair is connected via only two paths with weights $X_1(t)$ and $X_2(t)$. Since there are only two paths, identifying the shorter path is equivalent to identifying the sign of the difference process $X(t) = X_1(t) - X_2(t) : X_1(t) \lesseqgtr X_2(t)$ if and only if $X(t) \lesseqgtr 0$.
2. Weight processes $X_1(t)$ and $X_2(t)$ are independent Ornstein-Uhlenbeck (OU) processes with the same mean value. This means $X(t)$ is a zero-mean OU process.

   The OU process assumption follows the heavy-traffic results of Halfin and Whitt [20], which show that when service times are exponentially distributed, the sequence of appropriately normalized queue lengths will converge to the OU process. Mathematically, OU process is the only non-trivial continuous random process that is simultaneously Gaussian, Markovian, and stationary; properties that are often necessary in obtaining analytical solutions. An OU process OU(t) parameterized by $\{\mu_0, \theta_0, \sigma_0\}$ is governed by the following stochastic differential equation:

$$d\mathrm{OU}(t) = \theta_0(\mu_0 - \mathrm{OU}(t))\,dt + \sigma_0\,dW(t),$$

   where μ_0 is the long-term mean of the process, $\theta_0 > 0$ is the mean-reversion speed, and σ_0 is the volatility. This expression consists of two parts. The first part, $\theta_0(\mu_0 - \mathrm{OU}(t))\,dt$, captures the mean reverting behavior, which is akin to a force that pulls the process towards its long-term mean. The second part, $\sigma_0\,dW(t)$, is a standard Wiener process scaled by volatility factor σ_0. This second part acts as additive noise and counteracts the mean reversion. Hence, the OU process can be roughly described as noisy oscillations around μ_0.
3. Direct derivation of policy ϕ^* is quite tricky, as both numerator and denominator in (5.1) can increase unboundedly. We approximate the optimal policy ϕ^* by

$$\phi' = \arg\min_{\phi} \frac{\eta + \mathbb{E}\left[C_{[t_{i-1}, t_{i-1}+\phi(s(t_{i-1}))]}\right]}{\phi(s(t_{i-1}))}. \tag{5.2}$$

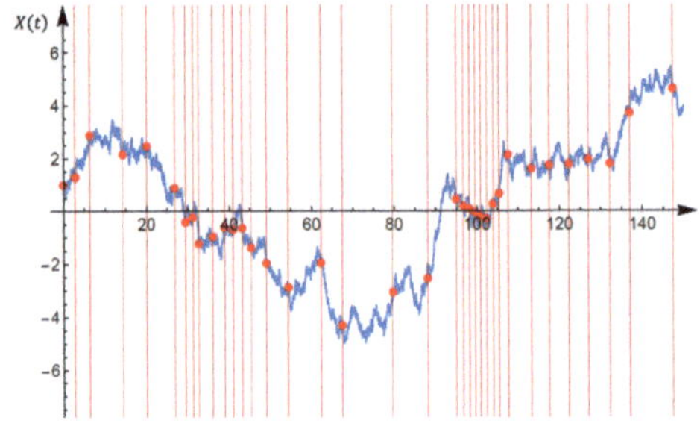

Fig. 5.2 Applying the sampling policy ϕ' on a realization of an OU process with parameters $\{\mu = 0, \theta = 0.025, \sigma = 0.5\}$. The sampling cost $\eta = 0.1$

Compared with ϕ^*, policy ϕ' is myopic and minimizes the cost rate in $[t_{i-1}, t_i]$, and takes no account of the actions after t_i. We name the policy ϕ' a "one-step look-ahead policy".

Given the above assumptions, we can focus on sampling the difference process $X(t)$, an OU process parameterized by $\{\mu = 0, \theta, \sigma\}$.[4] Since OU process is Markovian, we can simplify $s(t_{i-1})$ to $s(t_{i-1}) = X(t_{i-1})$. The policy ϕ' can then be written as

$$\begin{aligned}\phi'(X(t_{i-1}) = x) &= T_i'\Big|_{X(t_{i-1})=x} \\ &= \arg\min_{T_i>0} \frac{\eta + \mathbb{E}\left[C_{[t_{i-1}, t_{i-1}+T_i]} | X(t_{i-1}) = x\right]}{T_i},\end{aligned} \tag{5.3}$$

where [12] further derived

$$\mathbb{E}\left[C_{[t_{i-1}, t_i]} | X(t_{i-1}) = x\right] = \frac{x\left(e^{-\theta T_i} - 1\right)}{\theta} + \int_0^{e^{2\theta T_i} - 1} \frac{\sigma}{4\theta\sqrt{\theta\pi}}$$
$$\frac{\sqrt{y}}{(1+y)^{\frac{3}{2}}} e^{-\frac{\theta x^2}{y\sigma^2}} + \frac{x}{4\theta} \frac{1}{(1+y)^{\frac{3}{2}}} \operatorname{erfc}\left(-\frac{\sqrt{\theta} x}{\sqrt{y}\sigma}\right) dy. \tag{5.4}$$

The optimal sampling interval $T_i'|_{X(t_{i-1})=x}$ under this policy can then be computed numerically by solving the non-convex optimization in (5.3).

Figure 5.2 depicts a realization of an OU process with parameters $\{\mu = 0, \theta = 0.025, \sigma = 0.5\}$. The result of setting the sampling cost to $\eta = 0.1$ and applying the sampling policy ϕ' is depicted in Fig. 5.2. As shown, the sampling frequency increases when $X(t)$ is close to zero (corresponding to times when $X_1(t) \approx X_2(t)$). On the other hand, when $X(t)$ is far from zero (i.e when $X_1(t)$ and $X_2(t)$ are far apart) the algorithm will adopt a sparse sampling strategy.

[4] If both $X_1(t)$ and $X_2(t)$ are OU processes parameterized by $\{\mu_0, \theta_0, \sigma_0\}$, then $X(t)$ is an OU process parameterized by $\{\mu = 0, \theta = \theta_0, \sigma = \sqrt{2}\sigma_0\}$.

Next, we point out and elaborates on two limitations of the analytical solution in [12].

Hard to Generalize The assumptions are meant to simplify the significant sampling problem so that analytical solutions can be derived. However, the simplified problem still leads to complicated expressions that are hard to interpret, e.g., (5.4). Furthermore, extensions to more general cases (e.g., the N-path case, or non-OU processes) can be even more complex.

The Never-Sample Problem Policy ϕ' performs well when the sampling cost is small. However, when the sampling cost is large, policy ϕ' instructs the NMC system to "never sample". The details of this never-sample problem can be understood through the following Proposition 5.1.

Proposition 5.1 *Suppose $X(t)$ is a zero-mean OU process parameterized by $\{\mu = 0, \theta, \sigma\}$. At epoch t_{i-1}, we sample $X(t)$ and obtain $X(t_{i-1})$. Then, there exists a real number $\varepsilon \geq 0$ such that $\forall\ |X(t_{i-1})| \leq \varepsilon$, policy ϕ' will instruct the NMC to never sample the process if*

$$\eta > \int_0^\infty \sqrt{\frac{\sigma^2}{4\pi\theta}} - g(t)\, dt, \tag{5.5}$$

where

$$g(t) = \sqrt{\frac{\sigma^2(1 - e^{-2\theta t})}{4\pi\theta}} e^{-\frac{\theta\varepsilon^2}{\sigma^2(e^{2\theta t}-1)}} - \frac{\varepsilon e^{-\theta t}}{2} erfc\left(\sqrt{\frac{\theta\varepsilon^2}{\sigma^2(e^{2\theta t} - 1)}}\right). \tag{5.6}$$

Proof Assuming at some epoch t', we sample $X(t)$ and obtain $X(t') = \varepsilon$. The decision rule of policy ϕ' for the next sampling interval T^* can be described visually by Fig. 5.3 (see Appendix D for detailed derivations and analyses).

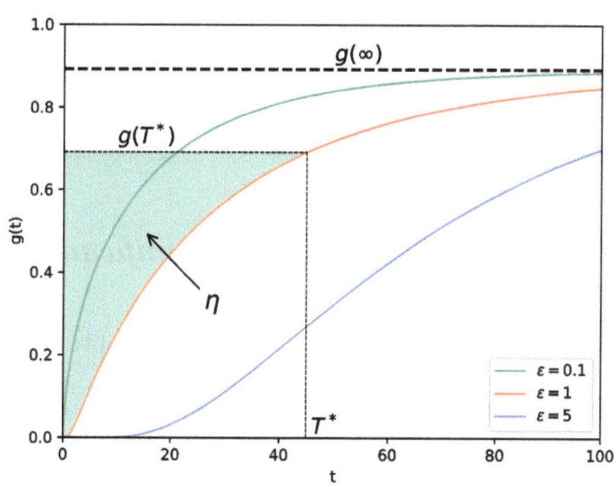

Fig. 5.3 An intuitive explanation of policy ϕ'. The difference process $X(t)$ is a zero-mean OU process parameterized by $\{\mu = 0, \theta, \sigma\}$; $X(t') = \varepsilon$ is the sampled value at epoch t'; and $g(t)$ in Eq. (5.6) is the instantaneous cost rate for $t > t'$ conditioned on $X(t') = \varepsilon$

As shown in Fig. 5.3, the shaded area increases with T; the optimal sampling interval T^* indicated by policy ϕ' is the T that makes the shaded area equal to the sampling cost η. As a result, the policy ϕ' can instruct us to "never sample" if η is greater than the shaded area even when $T \to \infty$. This gives us the lower bound for η in (5.5). When (5.5) is satisfied, if the absolute value of any sample $X(t_{i-1})$ is less than or equal to ε, policy ϕ' will instruct the NMC to never sample, and the cost rate will converge to $g(\infty) = \sqrt{\frac{\sigma^2}{4\pi\theta}}$ as time passes. □

It should be pointed out that "never sample" is not always suboptimal if our objective is to minimize the cost rate over an infinite time horizon, as in (5.1). A simple example is when $\eta \to \infty$, for which the optimal policy ϕ^* is to never sample. However, the following Proposition 5.2 further shows that policy ϕ' is guaranteed to be suboptimal if η is smaller than an upper bound.

Proposition 5.2 *Policy ϕ' is suboptimal in terms of minimizing the cost rate over an infinite time horizon if*

$$\int_0^\infty \sqrt{\frac{\sigma^2}{4\pi\theta}} - g(t)\,dt < \eta < \sqrt{\frac{\sigma^2}{4\pi\theta}\frac{1}{\theta}}. \tag{5.7}$$

In particular, sampling the weight process once at the equilibrium state can outperform ϕ'.

Proof See Appendix D. □

To overcome the two limitations above, this section puts forth a DRL solution to solve the significant sampling problem. A DRL solution has the advantage that it makes no assumption about the weight processes (often referred to as the "model-free" property of DRL), and is thus more robust to various network conditions that give rise to different stochastic variations of $X_n(t)$. This enables us to remove the assumptions made in our prior work, and provide solutions to more general settings that go beyond the 2-path limitation and traffic modeling assumption (OU process). Moreover, we leverage this DRL framework to design a better policy that minimizes the cost rate over a long time horizon. This new policy, by planning farther beyond the immediate future, can effectively address the never-sample problem faced by policy ϕ'.

5.2.3 A DRL Solution to Significant Sampling

Let us start from the formulation of a RL problem.

5.2.3.1 Significant Sampling as an RL Problem

We first transform the original significant sampling problem to a RL problem. The basic idea is that we can view the NMC system as an agent, and the underlying weight processes $\{X_n(t)\}$ as the environment. In particular, the agent will maintain a ledger that contains the state of the environment, i.e., $s(t_{i-1}) = \{X_n(t) : n = [N], t \leq t_{i-1}\}$. Given state $s(t_{i-1})$, the agent's policy ϕ will output an action $T_i = \phi(s(t_{i-1}))$, which effectively specifies the next epoch for sampling the environment and thus the time for updating the ledger. At $t_i = t_{i-1} + T_i$, the agent will sample the environment, update the ledger with $s(t_i) = \{X_n(t) : n = [N], t \leq t_i\}$, and the cycle continues.

The agent's ability to improve its actions depends on a feedback system that can assign values to various actions so that the agent can distinguish good actions from bad ones. This feedback is often expressed in terms of a *reward* function which indicates the reward associated with taking a particular action when the environment is in state $s(t_{i-1})$. For significant sampling, the environment will provide $\{X_n(t) : n = [N], t_{i-1} \leq t \leq t_i\}$ as feedback at the end of a time step, from which the agent computes a reward r_i (to be defined later) to evaluate the action T_i in this time step. We shall refer to a complete cycle from t_{i-1} to t_i as one "time step". The learning process will traverse many time steps, and each time step provides the agent with a new experience denoted by the quaternion $\{s(t_{i-1}), T_i, r_i, s(t_i)\}$.

Action-Value Function

In standard RL algorithms, the agent's objective at epoch t_{i-1} is to learn a policy that maximizes the future *return* R_i, a common definition of which is the sum of discounted future rewards

$$R_i = \sum_{l=i}^{\infty} \gamma^{l-i} r_l, \tag{5.8}$$

where $\gamma \in [0, 1]$ is a discounting factor. An action-value function (also known as the Q function) describes the maximum achievable return if the agent takes action T_i in state $s(t_{i-1})$, and then follows the optimal policy from then on. That is,

$$\begin{aligned} Q^*(s(t_{i-1}), T_i) &= \max_{\phi} \mathbb{E}_{\text{env}} \left[R_i | s(t_{i-1}), T_i, \phi \right] \\ &= \mathbb{E}_{\text{env}} \left[r_i + \gamma \max_{\phi} Q^*(s(t_i), \phi(s(t_i))) \right], \end{aligned} \tag{5.9}$$

where $Q^*(s(t_{i-1}), T_i)$ is often referred to as the Q value of action T_i in state $s(t_{i-1})$. This recursive form is known as the Bellman equation [21].

The action-value function is important in RL algorithms because it is a direct reflection of the optimal policy. That is, the action with the maximal Q value should be the action chosen by the optimal policy. On the other hand, if we know the action-value function, the optimal policy can be easily extracted by acting greedily (i.e., choose the action with the maximal Q value) [3, 21].

Learn Policy ϕ'

As an example, we can make use of the DRL framework to learn the policy ϕ'. To this end, the reward and return can be defined as

$$\begin{aligned} r_i &= -\frac{\eta + C_{[t_{i-1},t_i]}}{T_i}, \\ R_i &= r_i. \end{aligned} \tag{5.10}$$

The action-value function is then given by

$$Q^*(s(t_{i-1}), T_i) = \max_{\phi} \mathbb{E}_{\text{env}} \left[r_i \mid \phi \right]. \tag{5.11}$$

As shown in Eq. (5.11), the policy ϕ' is myopic and only maximizes the one-step reward. Inspired by (5.8), we can design a farsighted significant sampling policy that considers future rewards, well beyond the immediate one-step future.

5.2.3.2 Action-Value Function Design

In standard RL problems, the agent interacts with the environment in uniform time steps, and rewards are received at the end of time steps. The algorithm can then use the (discrete) γ-discounting to define the return as in (5.8). On the other hand, the durations of time-steps in significant sampling vary from time step to time step, and the dynamics of the environment are continuous. To address this issue, we propose a more appropriate exponentially discounting method for the computation of the return, instead of using the discrete γ-discounting.

As illustrated in Fig. 5.4, the agent takes action T_i in state $s(t_{i-1})$ and receives the feedback $\{X_n(t) : n = [N], t_{i-1} \le t \le t_i\}$ from the environment at t_i. Let us define

$$\varphi_i(t) = X_{n'_{i-1}}(t) - \min_n X_n(t), \quad t_{i-1} \le t \le t_i,$$

where n'_{i-1} is the index of the shortest path observed at epoch t_{i-1}.

The exponentially discounted reward within $[t_{i-1}, t_i]$ can then be defined as

$$r_i = -\eta e^{-\beta T_i} - \int_{t_{i-1}}^{t_i} e^{-\beta(t-t_{i-1})} \varphi_i(t) \, dt, \tag{5.12}$$

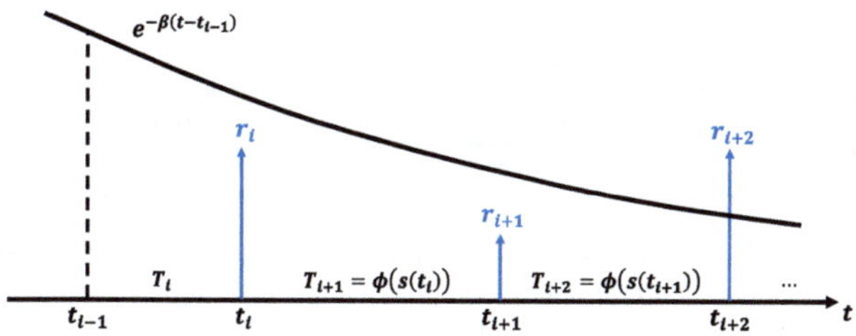

Fig. 5.4 Illustration of exponentially discounting reward for significant sampling, in which the dynamics of the environment are continuous. A reward received t time units later is worth only $e^{-\beta t}$ times what it would be worth if it were received immediately

where $\beta > 0$ is a exponential discount factor. Note that $\varphi_i(t)dt$ is the cost incurred subsequent to t_{i-1} and is therefore discounted according to the time elapsed since t_{i-1}. Likewise, the next sampling cost η is incurred at time $t_i = t_{i-1} + T_i$ and is therefore discounted corresponding to T_i, the time elapsed since t_{i-1}.

Similarly, at t_i, the agent takes action $T_{i+1} = \phi(s(t_i))$ and receives the feedback $\{X_n(t) : n = [N], t_i \leq t \leq t_{i+1}\}$ from the environment at t_{i+1}. The exponentially discounted reward within $[t_i, t_{i+1}]$ is then given by

$$r_{i+1} = -\eta e^{-\beta T_{i+1}} - \int_{t_i}^{t_{i+1}} e^{-\beta(t-t_i)} \varphi_{i+1}(t)\, dt,$$

assuming discounting begins at t_i. Future rewards within each time step can be computed in a similar fashion. Overall, the return due to the action at t_{i-1} can be defined as

$$\begin{aligned} R_i &= r_i + e^{-\beta T_i} r_{i+1} + e^{-\beta(T_i+T_{i+1})} r_{i+2} + \cdots \\ &= r_i + \sum_{l=i}^{\infty} e^{-\beta \sum_{l'=i}^{l} T_{l'}} r_{l+1}. \end{aligned} \tag{5.13}$$

Note that in the above, the discounting of the constituent rewards in R_i begins at time t_{i-1}, hence the additional discounting factors for $r_{i+1}, r_{i+2}, \cdots$. Essentially, return R_i is the "present value" of all future rewards, where a reward received t time units later is worth only $e^{-\beta t}$ times what it would be worth if it were received immediately. The discounting is introduced to prevent the return from going to infinity, as in (5.8).

The action-value function can then be rewritten in the following recursive form:

$$\begin{aligned} Q^*(s(t_{i-1}), T_i) &= \max_{\phi} \mathbb{E}_{\text{env}} \left[R_i | s(t_{i-1}), T_i, \phi\right] \qquad (5.14) \\ &= \mathbb{E}_{\text{env}} \left[r_i + e^{-\beta T_i} \max_{\phi} Q^*(s(t_i), \phi(s(t_i))) \right], \end{aligned}$$

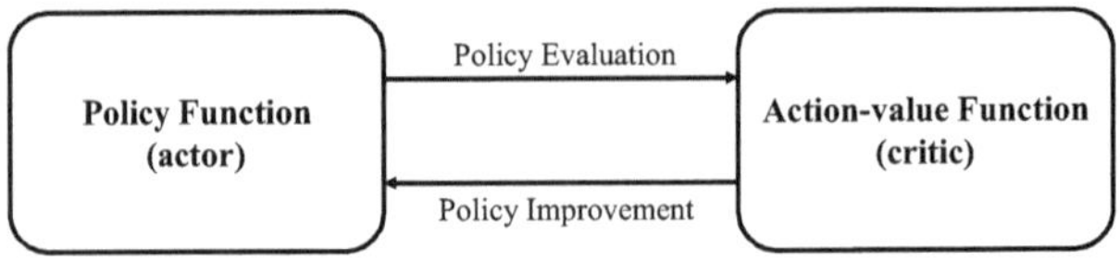

Fig. 5.5 RL algorithms as a generalized policy iteration. Policy evaluation makes the value function consistent with the current policy, and policy improvement makes the policy greedy (for higher reward) with respect to the current value function

Given the Bellman-style equation in (5.14), we can leverage DRL algorithms to learn the action-value function, and extract the corresponding optimal policy in a greedy manner. We shall refer to this new policy, which maximizes the return in (5.13), as the "multi-step look-ahead policy".

Compared with the optimal policy ϕ^* in (5.1), the multi-step look-ahead policy maximizes the sum of the exponentially discounted rewards the agent receives over an infinite time horizon. This is a better approximation to ϕ^* than the one-step look-ahead policy ϕ', especially when the discount exponent β is made to be small.

5.2.3.3 An Actor-Critic Solution

In RL, the policy function is often referred to as the "actor", as it controls the behavior of the agent; the action-value function is often referred to as the "critic", as it measures the quality of an action. An actor-critic solution for RL problems can be described as a generalized policy iteration (GPI) [21] between the actor and the critic, as illustrated in Fig. 5.5.

The policy function (actor) and action-value function (critic) will be initialized randomly, and their evolution will progress through a series of policy evaluations and policy improvements. Policy evaluation makes the action-value function more consistent with the current policy in that it reflects better the returns produced by the current policy (using Bellman equation). Policy improvement, on the other hand, makes the policy greedier with respect to the current action-value function (using policy gradient [22]). These two processes alternate and progress until the actor converges to the optimal policy, and the critic converges to the optimal Q function.

Deep Deterministic Policy Gradient (DDPG) [23, 24] is an actor-critic approach for MDP with continuous action space. This section adapts DDPG to solve the RL problem associated with significant sampling. Specifically, we use two deep neural networks (DNNs), an actor network and a critic network, to approximate the policy function and the action-value function, respectively. Figure 5.6 depicts the logical interaction of the actor and critic networks [23]. The actor, parameterized by λ_a, outputs an action T_i when the network state $s(t_{i-1})$ is used as its input, i.e., $T_i = \Psi_a(s(t_{i-1}); \lambda_a)$. The critic network, parameterized by λ_c, outputs a Q-value estimation $Q(s(t_{i-1}), T_i)$ when the state-action pair, $(s(t_{i-1}), T_i)$, is used as its input, i.e., $Q(s(t_{i-1}), T_i) = \Psi_c(s(t_{i-1}), T_i; \lambda_c)$.

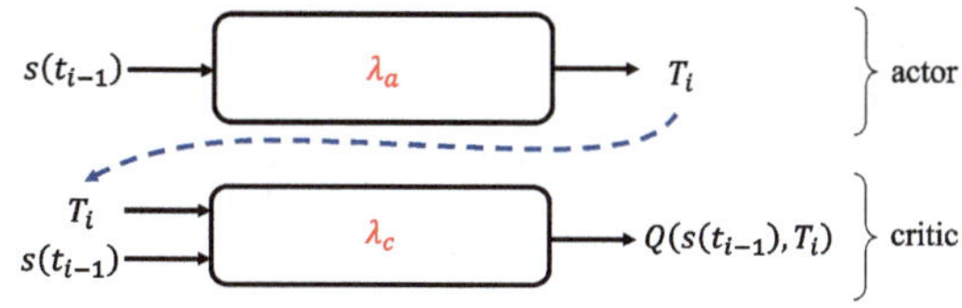

Fig. 5.6 The neural networks in actor-critic RL. The actor approximates the policy function, and the critic approximates the optimal action-value function

These two DNNs are initialized randomly. As learning continues, we use the accumulated experiences to train both DNNs, as follows [23].

Suppose we have an experience $\{s(t_{i-1}), T_i, r_i, s(t_i)\}$.

The critic is updated using the Bellman-style equation in (5.14). Specifically, given the state-action pair $\{s(t_{i-1}), T_i\}$, we update parameters λ_c in the direction that minimizes the mean square error (MSE) loss between the target Q value $r_i + e^{-\beta T_i}\Psi_c(s(t_i), \Psi_a(s(t_i); \lambda_a); \lambda_c)$ and the Q value $\Psi_c(s(t_{i-1}), T_i; \lambda_c)$ estimated by the current critic network:

$$-\nabla_{\lambda_c}\mathcal{L}_c = -\nabla_{\lambda_c}\big[r_i + e^{-\beta T_i}\Psi_c(s(t_i), \Psi_a(s(t_i); \lambda_a); \lambda_c) - \Psi_c(s(t_{i-1}), T_i; \lambda_c)\big]^2. \tag{5.15}$$

Updating λ_c in the negative gradient direction given by (5.15) makes the output of the critic network closely match the target Q value. This allows the critic network to approximate the Q function evermore closely, resulting in more accurate policy evaluations.

In practice, directly using (5.15) to update the critic usually leads to divergence of the Q function approximation. This is because the target Q value and the estimated Q value in (5.15) are highly correlated (as they are computed using the same critic network—note: the second term in the target Q is computed using the current critic network as well). To tackle this instability, people often use separate actor and critic networks (as opposed to using the current actor and critic networks in the second component of the target Q above) to compute the target Q value (see (5.17) below). This is known as "fixed Q-target" [5].

The actor is trained using policy gradient. That is, for state $s(t_{i-1})$, we update parameters λ_a in the direction that maximizes the action-value function (i.e., the output of the critic):

$$\nabla_{\lambda_a} Q^{\lambda_c} = \nabla_{\lambda_a}\Psi_c(s(t_{i-1}), \Psi_a(s(t_{i-1}); \lambda_a); \lambda_c). \tag{5.16}$$

Updating λ_a in the gradient direction given by (5.16) makes the actor greedier with respect to the current critic. This is a process of policy improvement.

5.2.3.4 Algorithm

The pseudocode for the actor-critic solution to significant sampling is given in Algorithm 10.

Algorithm 10 Significant sampling with DRL

Initialization:
Initialize a FIFO of size M for experience replay. Initialize the actor and critic networks randomly with parameters λ_a and λ_c; the target actor and target critic networks with parameters $\lambda_{a'}$ and $\lambda_{c'}$. Let $\lambda_{a'} = \lambda_a$ and $\lambda_{c'} = \lambda_c$.
Set mini-batch size to K, evaluation cycle to B, time step $i = 1$, training step $j = 1$. Set resampling window to t_w, the number of artificial experiences to V, and G.
In epoch t_0, all the nodes report weights, obtain $s(t_{i-1})$.
while 1 **do**
    New experience:
    Feed $s(t_{i-1})$ into the actor network for action T_i.
    In epoch $t_i = t_{i-1} + T_i$, all the nodes report weights.
    The NMC system computes reward r_i by Eq. (5.12).
    Store experience $\{s(t_{i-1}), T_i, r_i, s(t_i)\}$ into FIFO.
    $i = i + 1$
    Artificial experiences:
    $v = 1$.
    while $v \leq V$ **do**
        Select a random starting epoch t'_{v-1} in $[t_i - t_w, t_i]$,
        Feed state $s(t'_{v-1})$ into the actor network for T'_v.
        if $t'_{v-1} + T'_v < t_i$ **then**
            Compute reward r'_i by Eq. (5.12).
            Store experience $\{s(t'_{i-1}), T'_i, r'_i, s(t'_i)\}$.
            $v = v + 1$.
    DNN training:
    if $i * (V + 1) > M/5$ **then**
        Randomly sample G mini-batches from FIFO.
        for each mini-batch **do**
            Update λ^c in the direction given by Eq. (5.17).
            Update λ^a in the direction given by Eq. (5.18).
            Update the target networks by Eq. (5.19).
        if mod$(j, B) == 0$ **then**
            Evaluate the performance of current actor network.
            (on the evaluation trajectory)
            $j = j + G$.

As shown, we start by initializing a FIFO buffer of size M for experience replay. The goal of experience replay is to remove the temporal correlations between successive experiences, such that the experiences used for training are less dependent [5, 23]. To this end, instead of learning experiences online, we put the accumulated experiences in the FIFO (experiences will be placed into and removed from this FIFO buffer continuously), and randomly sample experiences from the FIFO to train the DNNs.

We then initialize the actor and critic networks with parameters λ_a and λ_c; and the target actor and critic networks with parameters $\lambda_{a'}$ and $\lambda_{c'}$ (for calculating target Q value). In particular, the parameters of target networks are initialized to $\lambda_{a'} = \lambda_a$ and $\lambda_{c'} = \lambda_c$. The initial sampling time is set to t_0, where $s(t_0)$ is given.

The next phase is "experience collection". At the beginning of a time step (i.e., at t_{i-1}), the agent feeds the current observation $s(t_{i-1})$ into the actor network, which outputs action T_i. Note that this action is bad early on, but will improve as learning progresses. To guarantee exploration, Gaussian noise is added to the raw output of the actor. In general, the exploration noise can also be added to the output of the critic, or the computed reward. The goal is "mislead" the agent to some extent so that the agent would explore "the paths less travelled" to discover new actions that are more optimal. After appropriate tuning, different exploration strategies can converge to the same optimum, but we found that adding noise directly to the action is more fruitful than other strategies as far as the convergence rate is concerned.

At the end of the time step (i.e., at t_i), the agent computes a reward r_i from the environment's feedback (see (5.12)). Overall, one time step provides the agent with one experience $\{s(t_{i-1}), T_i, r_i, s(t_i)\}$. All experiences will be stored in the FIFO for experience replay.

A challenge in learning the optimal policy for significant sampling is the inefficiency with which experiences are collected. Note that, one time step lasts for T_i time units, yet it gives the agent only one experience, hence is quite inefficient. This raises the following question: besides the collected experience, is there a way for the agent to generate more experiences artificially? We answer this question affirmatively by noting that in each time step, the continuously evolving weights $\{X_n(t) : n = [N]\}$ in interval $[t_{i-1}, t_i]$ are collected and reported to the NMC. As a result, at epoch t_i, all past weights $\{X_n(t) : n = [N], t \leq t_i\}$ are in fact known to the agent.

In this context, besides this new experience, the agent will resample the past and generate *artificial experiences*. Specifically, each time a new experience is collected, the agent will randomly select V starting epochs in period $[t_i - t_w, t_i]$, where t_w is a resampling window.[5] For each starting epoch $\{t'_{v-1} : v = [V]\}$, the agent will feed $s(t'_{v-1})$ into the current actor network for action T'_v. It can then compute reward r'_v which constitutes an artificial experience $\{s(t'_{v-1}), T'_v, r'_v, s(t'_v)\}$. This enables the agent to collects $V + 1$ experiences during each time step, V of which are artificial.

When the number of experiences is larger than $M/5$ (M is the buffer size), the agent starts to train the DNNs by mini-batch gradient descent [21]. Specifically, we will randomly sample G mini-batches from the FIFO, with each mini-batch containing K experiences.

For each mini-batch, we have $\{s(t_{k-1}), T_k, r_k, s(t_k)\}, k \in \{i_1, i_2, \ldots, i_K\}$. The parameters λ_c are updated in the direction that minimizes the mean square error (MSE) loss

$$
\begin{aligned}
- \nabla_{\lambda_c} \frac{1}{K} \sum_{k \in \{i_1, \ldots, i_K\}} \Big[r_k + \; e^{-\beta T_k} \Psi_{c'}(s(t_k), \Psi_{a'}(s(t_k); \lambda_{a'}); \lambda_{c'}) \\
- \Psi_c(s(t_{k-1}), T_k; \lambda_c) \Big]^2, \qquad (5.17)
\end{aligned}
$$

[5] In this section, t_w is set to be a large value (e.g., 2000 time units, much larger than the duration of one time step) because we consider stationary environment. For non-stationary environments, the choice of t_w depends on the speed at which the environment varies, because new environment requires the agent to "forget" about the past and "relearn" from the latest observations.

Unlike (5.15), the target Q value in (5.17) is computed using a target actor network and a target critic network.

The parameters λ_a are updated in the direction that maximizes the output of the critic (i.e., the Q value) by sampled policy gradient, giving

$$\nabla_{\lambda_a} Q^{\lambda_c} \approx \frac{1}{K} \sum_{k\in\{i_1,\ldots,i_K\}} \nabla_{\lambda_a} \Psi_c \left(s(t_{k-1}), T; \lambda_c\right) \Big|_{T=\Psi_a(s(t_{k-1});\lambda_a)}$$

$$= \frac{1}{K} \sum_{k\in\{i_1,\ldots,i_K\}} \nabla_T \Psi_c \left(s(t_{k-1}), T; \lambda_c\right) \nabla_{\lambda_a} \Psi_a \left(s(t_{k-1}); \lambda_a\right). \tag{5.18}$$

Each time the actor and critic networks are updated, the parameters of target networks will be updated using soft replacement [23]:

$$\lambda_{a'} = \rho\lambda_a + (1-\rho)\lambda_{a'}, \lambda_{c'} = \rho\lambda_c + (1-\rho)\lambda_{c'}. \tag{5.19}$$

Finally, to monitor and illustrate the continuous improvements of the learned policy, we evaluate the performance of the policy/actor network every B training steps. The evaluation is performed on an independent trajectory of $X_n(t)$ (this evaluation set is completely independent from the sample path of the process that were observed during the training process).

5.2.4 *Numerical Experiments*

This section presents experimental results of the actor-critic DRL solution to significant sampling. Our main goals in this study are threefold: (1) to confirm that our DRL approach can learn the one-step policy ϕ' by adopting a corresponding one-step action-value function; (2) to show that, by adopting a multi-step look-ahead action-value function, our DRL approach can learn the multi-step policy that addresses the "never-sample" problem associated with the policy ϕ', and this policy is more optimal than the policy ϕ' in terms of minimizing long-term cost rate; (3) to show that our DRL approach can provide effective solutions for more general environment with more than two paths, and with weight-evolution processes beyond the OU process.

5.2.4.1 Learn the One-Step Look-Ahead Policy

This experiment makes use of the DRL approach to learn the one-step look-ahead policy ϕ'. The importance of the analytical solutions in (5.3) is that they allow us to benchmark the performance of our DRL solution against the theoretical result. A comparable performance will provide us with the foundation to extend our objective and setup beyond what was presented in [12] with a reasonable degree of confidence in its ability to perform close to optimal.

Let us begin by assuming the same system setup as in [12]. Specifically, the OD pair is connected via two paths, and the evolving weights of both paths are represented by independent OU processes with the same mean value. Thus, the difference process $X(t)$ is a zero-mean OU process, and we can focus on sampling $X(t)$. The analytically derived policy ϕ' in (5.3) will serve as the benchmark.

To learn the one-step look-ahead policy, we can set the action-value function to be (5.11). That is, the objective of the agent is simply to maximize the one-step reward rate. The MSE loss in (5.17) can then be simplified to

$$-\nabla_{\lambda_c}\mathcal{L}_c = -\nabla_{\lambda_c}\frac{1}{K}\sum_{k\in\{i_1\ldots i_K\}}\left[\frac{r_k}{T_k} - \Psi_c(s(t_{k-1}), T_k; \lambda_c)\right]^2, \tag{5.20}$$

Compared with (5.17), the target value in (5.20) is simply r_k/T_k, and hence the target networks can be removed in this experiment.

Since the OU process is Markovian, we can set $s(t_{i-1}) = X(t_{i-1})$, i.e., the state of the environment is a real number. The actor and critic networks are designed to be fully-connected (feedforward) neural networks. The actor network consists of five fully connected layers, and the numbers of neurons in the five layers are $\{1, 32, 64, 32, 1\}$. The three hidden layers use rectifier non-linearity as activation functions, and the output layer uses hyperbolic tangent (tanh) non-linearity as the activation function. The output of the actor network u takes value in $(-1, 1)$ and is mapped to the sampling interval $T_i = T_{\max} * (u+1)/2$, where $T_{\max}$ is the maximum sampling interval. The architecture of the critic network is the same as the actor network except that a) the input layer has two neurons instead of one; b) the output layer uses no activation function, and directly produces the Q value given a state-action pair (See Fig. 5.6 for additional reference).

Gaussian noise $\mathcal{N}(0, \sigma_j)$ is added at the output of the actor to guarantee exploration. Its variance σ_j will decay as training goes on, that is, $\sigma_j = \alpha^{\lceil j/100\rceil}$ where j is the training step and $\alpha = 0.99$ is a constant.

The hyper-parameter are listed in Table 5.1.

We generate two independent trajectories of $X(t)$ for training and evaluation purposes respectively. As in [12], the sampling cost is set to $\eta = 0.1$ and the OU parameters for $X(t)$ are chosen to be $\{\mu = 0, \theta = 0.025, \sigma = 0.5\}$. The duration of the training trajectory is 2×10^5 time units, and the duration of the evaluation trajectory is 2×10^4 time units.

Over the course of training, for every $B = 100$ training steps, we assess the performance of the actor network on the evaluation trajectory. Specifically, let us denote by ϕ_j the sampling policy learned by the actor network after j training steps. We will apply the policy ϕ_j on the whole evaluation trajectory, and compute the cost rate by

$$\overline{c}_j = \frac{\sum_{l=1}^{L_j}(\eta + C_{[t_{l-1},t_l]})}{\sum_{l=1}^{L_j} T_l}, \tag{5.21}$$

Table 5.1 Hyper-parameter settings

Description	Symbol	Value
FIFO size	M	32,768
Mini-batch size	K	128
Resampling window	t_w	2000
# artificial experiences	V	31
# mini-batches per training	G	5
Maximum sampling interval	$T_{\max}$	40
Exponential discount factor	β	0.01
Exploration noise decay	α	0.99
Target model update	ρ	0.1
Learning rate (actor)	–	0.0005
Learning rate (critic)	–	0.001
Optimizer	–	Adam

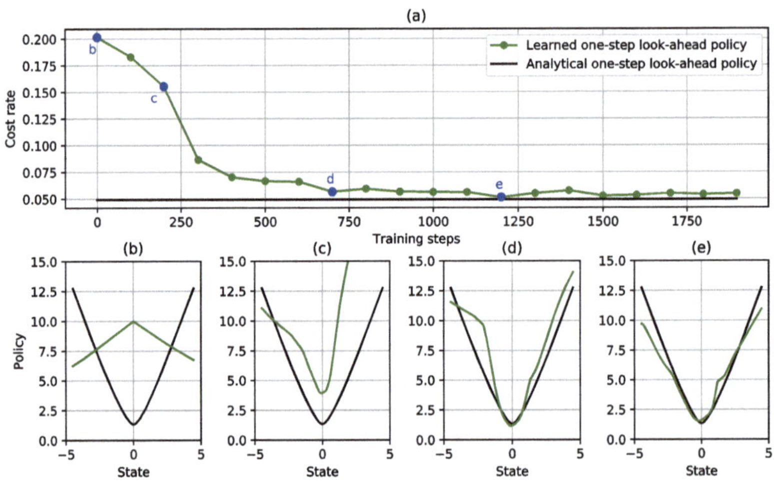

Fig. 5.7 The cost rate achieved on the evaluation trajectory, and the policy improvements in the RL process: (**a**) cost rate achieved on the evaluation trajectory versus the number of training steps; (**b**) the learned policy of actor at point b; (**c**) the learned policy of actor at point c; (**d**) the learned policy of actor at point c; (**e**) the learned policy of actor at point e

where L_j is the number of samples of policy ϕ_j when applied on the whole evaluation trajectory, and the sampling epochs on the evaluation trajectory are $\{t_i : i = [L_j]\}$.

Figure 5.7a presents the cost rate as a function of training steps in the RL process. As a benchmark, we also plot the minimum achievable cost rate (0.0492) which is obtained by applying the analytically derived one-step look-ahead policy given by (5.3) on the evaluation trajectory. We can see that the cost rate achieved by the learned policy decreases as training continues and eventually approaches the optimal cost. The learned policies at different times of the learning process are plotted in Fig. 5.7b–e. The x-axis is the observed value of the weight process $X(t)$, and the y-

axis is the corresponding action T_i given by the learned policy at that training step. The benchmark, again, is the analytically derived one-step look-ahead policy. As shown, the actor network will learn the optimal policy after enough training steps.

5.2.4.2 Learn the Multi-step Look-Ahead Policy

Section 5.2.4.1 confirms that the actor-critic DRL approach can learn the one-step look-ahead policy. This is an important achievement in and of itself because the DRL system is model-free. Specifically, unlike the analysis giving rise to the analytical solution, the agent in DRL is unaware that the underlying delay model is an OU process and yet "learns" to obtain a near-optimal sampling policy.

This subsection evaluates the multi-step look-ahead policy based on the actor-critic framework. We have repeated the experiment in Sect. 5.2.4.1 assuming a small sampling cost followed by assuming a large sampling cost.

Figure 5.8 shows the performance of the one-step and multi-step look-ahead policy when $\eta = 0.1$. The performance curves for one-step look-ahead policy, both analytical and learned, are reproduced from Fig. 5.7 (the difference is that the y-axis is in log scale). As shown, with $\eta = 0.1$, the multi-step look-ahead policy achieves only minor gain over the one-step look-ahead policy. This indicates that the one-step look-ahead policy is a good policy when η is small in terms of minimizing the long-term cost rate. This is not the case, however, for more general η.

In the second experiment, we use a larger sampling cost $\eta = 15$. Notice that $\eta = 15$ falls into the region $12.2 < \eta < 35.68$ given by (5.7) if we set $\varepsilon = 0.1$. Thus, if the absolute value of one sample is less than or equal to 0.1, the one-step look-ahead policy will instruct us to never sample, and the cost rate will eventually converge to $g(\infty) = \sqrt{\frac{\sigma^2}{4\pi\theta}}$.

Figure 5.9 shows the cost rates achieved by the learned one-step and multi-step look-ahead policies. The analytical cost rate for one-step look-ahead policy is $g(\infty) \approx 0.89$ (i.e., the dashed straight line in Fig. 5.9). We note that the learned one-step look-ahead policy is better than the analytical results. This is because an

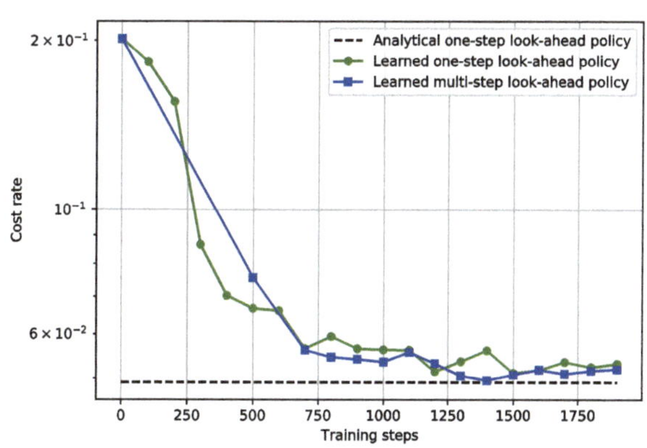

Fig. 5.8 The cost rate achieved on the evaluation trajectory when the sampling cost $\eta = 0.1$

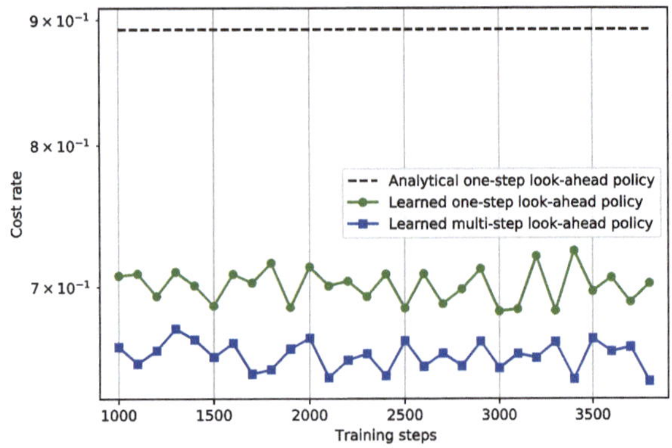

Fig. 5.9 The cost rate achieved on the evaluation trajectory when the sampling cost $\eta = 15$

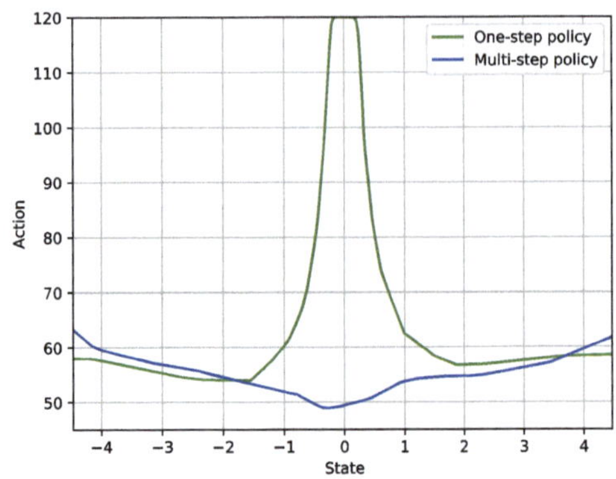

Fig. 5.10 The one-step and multi-step look-ahead policies. The sampling cost $\eta = 15$

"artificial" maximum sampling interval $T_{\max}$ is set in the implementation (recall that the output of the actor DNN is $[0, T_{\max}]$, where $T_{\max} = 120$ in this experiment). Thus, if the absolute value of one sample is less or equal to 0.1, the action chosen by the one-step look-ahead policy is 120 rather than infinity. This is confirmed in Fig. 5.10, where the learned policies are plotted. As shown, for the one-step look-ahead policy, the learned policy is consistent with our prediction. When a sampled value is close to 0, the action given by the learned one-step look-ahead policy reaches the maximal sampling interval. On the other hand, for the multi-step look-ahead policy, the learned policy matches with our intuition of a good policy, i.e., when $X(t)$ is close to zero, sampling should be more frequent; when $X(t)$ is far away from zero, sampling should be more sparse. The never-sample problem is well addressed by our multi-step look-ahead policy.

Overall, the learned multi-step policy achieves the best long-term cost rate. Considering the average cost rate achieved on the evaluation trajectory, our multi-step policy outperforms the analytical one-step policy, and the learned one-step policy by 39% and 8%.

5.2.4.3 Extensions to More General Environments

Our previous discussions focused on the same system setup as in [12], i.e., the OD pair is connected via only two paths, and the weight processes of both paths are OU processes. Thanks to the model-free property of RL, an advantage of our DRL approach is that we can provide solutions for more general system setups with only slight modifications to the algorithm. This subsection will study two generalizations.

N Paths and OU Process

This experiment removes the $N = 2$ assumption. We assume there are N paths and the evolving weight of each path is an OU process. In particular, we choose $N = 6$ in the experiment.

The state of the environment is defined to be $s(t_{i-1}) = \{X_n(t_{i-1}) : n = [N]\}$. That is, the state input of the actor and critic networks is a vector of length N. We can then increase the number of neurons in the input layer of the actor and critic networks to N and $N + 1$, respectively. The hyper-parameters are the same as that in Table 5.1. The six OU processes are generated using the same parameter set $\{\mu = 0, \theta = 0.025, \sigma = 0.3\}$, and the sampling cost $\eta = 0.1$. We use the DRL framework to learn the multi-step look-ahead policy in the implementation.

To evaluate the learning results, we apply the learned policy on a realization of $X_n(t)$, and plot the sampling results in Fig. 5.11. The learned policy matches our intuition of a good policy: it samples more frequently when the values of $\{X_n(t) : n = [6]\}$ are close, and more sparsely when one of the three weights is much smaller than the other five paths.

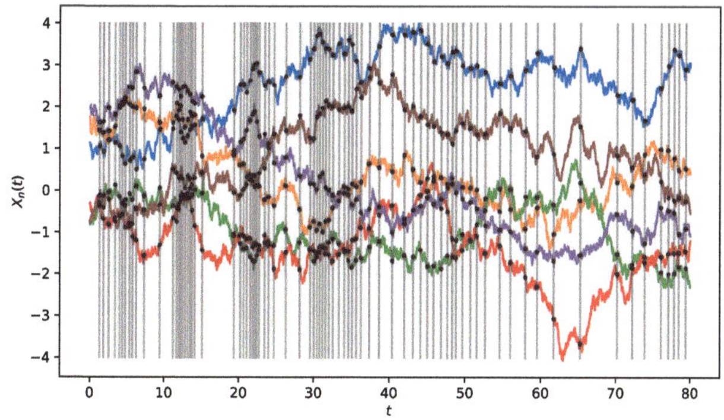

Fig. 5.11 The results of applying the learned policy on a realization of $\{X_n(t) : n = [6]\}$

Two Paths and Weight Processes with Memory

In previous sections, we showed that the DRL framework can solve the significant sampling problem when the underlying weight process is taken to be an OU process. A favorable property of the OU process is that it is Markovian. Evolution of such an environment is completely captured through the latest sample, and hence we can define the state of the environment as $s(t_{i-1}) = \{X_n(t_{i-1}) : n = [N]\}$.

Unlike the OU process, real-world weight processes may exhibit complex interdependencies over time and have memory. This suggests that optimal sampling of such processes depends not only on the latest sample, but also on the history of process. To address this issue, one may define the state of the environment as $s(t_{i-1}) = \{X_n(t) : n = [N], t \leq t_{i-1}\}$, that is, we feed the full history of the weight process into the DNNs for decision making. However, such an approach will require an incredibly large and potentially infinite state space due to the continuous nature of $X_n(t)$, and is thus impractical. Hence, we seek to reduce the state space required for the optimal sampling of general processes. This section combines the following two schemes to achieve this goal:

1. We characterize each weight process, $X_n(t)$, in the frequency domain, and use the ledger to represent $X_n(t)$ as a real vector obtained by uniformly sampling the process (as per Nyquist sampling theorem [25]). This allows us to reduce/compress the information contained in the continuous process/signal.
2. We assume the decision depends only on a finite period of recent history, and truncate early uniformly sampled points before this period.

Denote by $\boldsymbol{X}(t_{i-1}) = \{X_n(t_{i-1}) : n = [N]\}$ the N samples at a sampling epoch t_{i-1}. The state of the environment can then be defined as (5.22) following the above two schemes.

$$s(t_{i-1}) = \Big\{\boldsymbol{X}\left(t_{i-1} - \frac{L-1}{f_{\max}}\right), \ \boldsymbol{X}\left(t_{i-1} - \frac{L-2}{f_{\max}}\right), \ldots, \boldsymbol{X}\left(t_{i-1} - \frac{1}{f_{\max}}\right), \boldsymbol{X}(t_{i-1})\Big\}, \tag{5.22}$$

where $f_{\max}$ is the uniform sampling rate, and the history before $t_{i-1} - (L-1)/f_{\max}$ is considered irrelevant to decision making. In a nutshell, state $s(t_{i-1})$ is composed of the samples at L most recent sampling epochs, if we sample the history of $X_n(t)$ uniformly at rate $f_{\max}$.

The uniform sampling rate $f_{\max}$ is designed to be $f_{\max} = \max\{f_1, f_2, \ldots, f_N\}$, where f_n is the uniform sampling rate for the n-th path. To determine f_n, we first transform $X_n(t)$ to the frequency domain. If $X_n(t)$ is band-limited, f_n can be set to the Nyquist rate of $X_n(t)$ (twice its bandwidth). If $X_n(t)$ is non-band limited as most real-world signals/weight processes, we will pass $X_n(t)$ through a low-pass filter and truncate high frequency components in the stopband of the filter (empirically, the spectrum of most practical process centers around DC, and progressively gets smaller and smaller as frequency increases). Then, the filtered $X_n(t)$ is a band-limited signal, and we set f_n to be its Nyquist rate.

To confirm the proposed schemes are effective, we conduct the following experiment, in which an artificial weight process with memory is used as input. In this example, we assume the OD pair is connected via two paths, and we focus on sampling the difference process $X(t)$. In particular, we artificially generate $X(t)$ by adding a periodicity (i.e., a rectangular wave) to an OU process,

$$X(t) = \mathrm{OU}(t) + \sqrt{\frac{2\sigma^2}{\theta}} \sum_i \mathrm{Rect}(t - iD_R), \tag{5.23}$$

where OU(t) is an OU process parameterized by $\{\mu = 0, \theta = 0.025, \sigma = 0.5\}$; the rectangular wave Rect$(t) = \mathrm{sgn}(t), -D_R/2 \le t < D_R/2$ (D_R is the duration of one period, we set $D_R = 20$ in the experiment). The rectangular wave is scaled by the two standard deviations of the OU process. A realization of $X(t)$ is shown in Fig. 5.13. Compared with OU process, this new process undergoes sudden changes, and hence is challenging for the DRL framework to learn a good sampling policy.[6]

To learn a good policy for the weight process in (5.23), the first step is to determine an uniform sampling rate as in (5.22). Following scheme a), we set the uniform sampling rate $f_{\max} = 1$ sample per unit time (sampling at this rate, the dominant spectrum of $X(t)$ has already been captured). As a result, at an epoch t_{i-1}, we will uniformly sample the history of $X(t)$, and set the state of the environment at t_{i-1} to be a collection of L most recent uniform samples, i.e., $s(t_{i-1}) = \{X(t_{i-1} - l), l = L-1, L-2, \ldots, 0\}$. We will monitor the performance of the learned policy for different values of L.

In the experiment, the sampling cost is set to $\eta = 0.1$, and we let the DRL framework learn the multi-step look-ahead policy. All experimental setups are the same as previous except for the DNN architecture.[7] The shape of the actor is now $\{L, 64, 128, 64, 32, 1\}$, where the L input neurons are used to represent state $s(t_{i-1})$. The shape of the critic is $\{L + 1, 64, 128, 64, 32, 1\}$.

First, we set $L = 1$. This means the actor network need to determine the optimal sampling interval T_i from state $s(t_{i-1}) = X(t_{i-1})$. We then assess the learned policy on an evaluation trajectory every $B = 4000$ training steps, and plot the achieved cost

[6] We successfully experimented with various processes with memory, including ARMA, ARIMA, filtered OU, etc. but the process in Eq. (5.23) is a good candidate to showcase the ability of our algorithm, as the sudden changes of the rectangular wave present a challenging environment.

[7] Besides the feedforward network, other neural network architectures are also possible. For processes with memory, we recommend using RNNs [26], as they can capture the temporal relations of sequential inputs. For environment in (5.23), our additional simulations show that feedforward network and RNN achieve almost the same performance. Thus, we stick to feedforward network in the simulations.

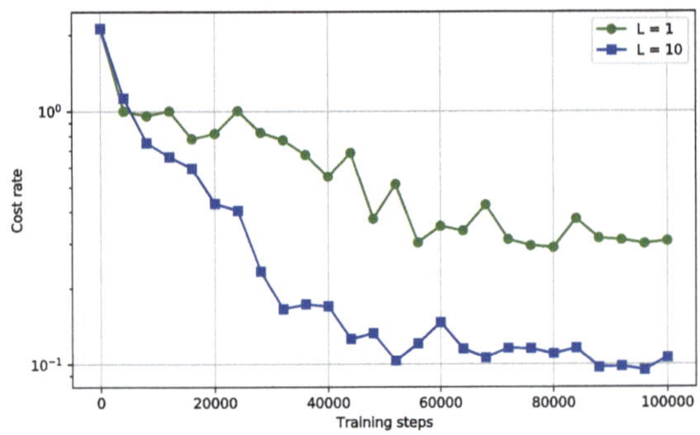

Fig. 5.12 The cost rate achieved on the evaluation trajectory given different input length L

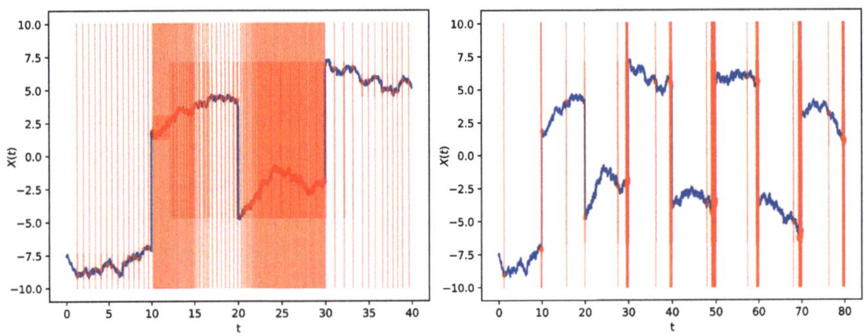

(a) $L = 1$ (b) $L = 10$

Fig. 5.13 Evaluating the learned policy on the trajectory given different input length L. (**a**) $L = 1$. (**b**) $L = 10$

rate in Fig. 5.12. Then, we increase L. The final performance of the learned policy improves as L increases until $L = 10$. No further improvements were observed for $L > 10$. Note that when $L = 10$, the history captured in state $s(t_{i-1})$ is half the period of the rectangular wave. Overall, choosing $L = 10$ helps improve the average cost rate (at convergence) by a factor of 3.2 in comparison to when $L = 1$.

To evaluate the learning results, we apply the learned policy (after 80,000 training steps) on a realization of $X(t)$, and plot the sampling results in Fig. 5.13. As can be seen, the learned policy is bad when $L = 1$. This means the actor network is unable to make a good decision given only the latest observation of $X(t)$. On the other hand, when $L = 10$, the learned policy samples the weight process mostly at the half-cycle of the rectangular wave where the sign of the rectangular wave changes.

5.3 Decentralized Network Control with Graph Neural Networks

In the previous sections, we focused on centralized control schemes for IoT network management, exploring how they leverage a global perspective to optimize network performance. While these centralized approaches offer significant advantages, they also encounter challenges such as scalability limitations, increased latency, and vulnerability to single points of failure. To overcome these issues, this section examines decentralized control schemes [27], specifically focusing on decentralized channel management in wireless local area networks (WLANs) using GNNs.

5.3.1 *Decentralized Channel Management*

Wireless local area networks (WLANs), commonly known as Wi-Fi, have become the most widely used means for IoT devices to connect to the Internet. According to Cisco [28], by 2023, the number of networked devices will be 5 billion in North America, and the number of IoT connections is expected to reach 14.7 billion globally. In particular, 72% of IoT connections will be made via Wi-Fi. The exploding number of wireless connections and advanced multimedia applications have led to an unprecedented amount of traffic over Wi-Fi networks. On the other hand, the channel resources assigned to Wi-Fi have been limited to the 2.4 GHz and 5 GHz unlicensed spectrum [29], reinforcing the need for optimizing channel allocation and network management.

Given a network of Wi-Fi APs and a fixed number of channels, channel allocation exploits spatial multiplexing and assigns the channels to the APs such that mutual interference among APs is minimized. Intuitively, a judicious channel allocation scheme assigns different channels to adjacent APs and the same channels to non-adjacent APs, thereby boosting network throughput while minimizing interference. However, the latter is not feasible for large-scale networks with a limited number of channels.

Prior solutions on WLAN channel allocation rely on a central controller to manage APs and make allocation decisions with the global network information [30–36]. The problem can be formulated as a weighted graph coloring problem, which is NP-complete [37]. To tackle the NP-hardness, various schemes, such as local-search [30], integer linear programming [31], deep reinforcement learning (DRL) [34, 35], etc., have been proposed. Among them, three notable related works are TurboCA [33], Net2Seq [34], and graph convolutional network (GCN)-based channel allocation [35].

TurboCA, the commercial solution implemented in the Cisco Meraki series, is a heuristic algorithm optimized on the data collected from millions of real-world APs. The design principle of TurboCA is client-centric: APs with higher client density and usage will be assigned more channels; APs with fewer/zero clients are more inclined to switch to new channels than those with more clients; single AP failure is avoided by a multiplicative network metric design and neighborhood clearance. Motivated by TurboCA, the authors of [34] proposed Net2Seq, a DRL-based channel allocation scheme, aiming to minimize mutual interference among APs. Net2Seq tailors a few feature-extraction schemes and utilizes deep neural networks (DNNs) for decision-making. With the data-driven nature, Net2Seq is able to adapt its solution to traffic demands, thereby outperforming TurboCA in terms of minimizing network interference. In contrast, [35] proposed GCNs to extract features directly from the interference graph of APs. In their approach, the channel allocation state and the topological information are fed into a centralized GCN, which, in turn, outputs the state-action values for all possible actions, determining the channel allocation plan [5]. Their solution, however, is limited to small-scale problems, where each AP can only occupy a single channel, the total number of APs is few (e.g., 10), and the interference graph is sparse.

In this section, we introduce a decentralized learning-based solution for channel allocation in WLANs. Our approach leverages GNNs to exploit topological insight and extract useful features from the traffic loads and channel occupations of APs. Unlike the centralized DNN in [34] and GCN in [35], we design the GNN in a way that APs are allowed to actively select channels based on the locally sensed states of their neighbors. Specifically, APs do not need to update the central controller about their states and passively wait for the allocation decisions; instead, they make local decisions about which channels to occupy, yielding an efficient, scalable, and robust channel allocation solution. The GNN is trained with the policy gradient method in a model-free manner, which needs only system observations, not system models; and thus, can be used in any application scenarios. Benefiting from the permutation equivariance of GNNs, the proposed approach is robust to AP reordering, and hence, can be implemented directly on networks with arbitrarily ordered APs. Experiment results corroborate our theoretical findings and verify that the performance of the proposed method approaches that of the centralized solution.

5.3.2 Problem Formulation

Consider a WLAN with N APs denoted by $\{n_1, \ldots, n_N\}$ (see Fig. 5.14). The APs have different traffic demands and can choose from available radio resources, i.e., channels, to satisfy these demands. There exists mutual interference between two APs if they are audible to each other and using the same channel. By adaptively allocating channels among APs, our goal is to minimize the mutual interference in the system while satisfying the traffic demands of the APs.

Suppose there are M channels available in the network, and the traffic demands of APs are represented by $\boldsymbol{d} = [d_1, \ldots, d_N]^\top$ with d_i denoting the demand of AP n_i, generated by local user devices they serve. Channels are selected by the APs based on the traffic demands $\boldsymbol{d}$ via a control policy $\pi(\boldsymbol{d}) = \boldsymbol{C} = [\boldsymbol{c}_1, \ldots, \boldsymbol{c}_N]$, where $\boldsymbol{c}_i = [c_{i1}, \ldots, c_{iM}]^\top$ is the channel selection of AP n_i with $c_{i\ell} \in \{0, 1\}$ set to 1 if

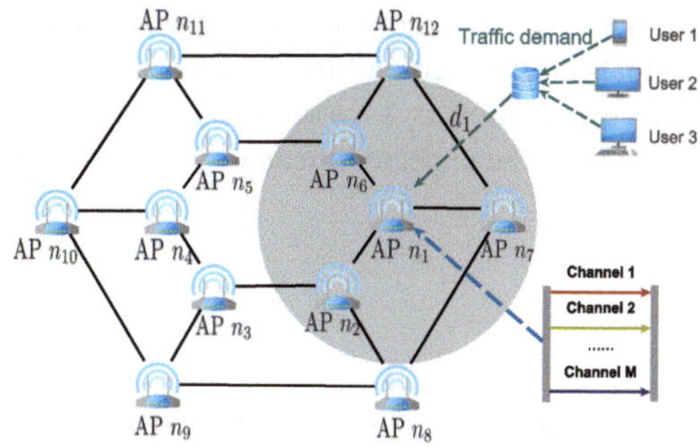

Fig. 5.14 A WLAN with $N = 12$ APs. APs are loaded with respective traffic demands of their local user devices and use different channels for data transmission. Each AP is interfered by the neighboring APs within the communication/interference range (grey circle for AP n_1)

channel ℓ is selected by AP n_i, and 0 otherwise, for $i = 1, \ldots, N$ and $\ell = 1, \ldots, M$. Let M_i denote the number of channels selected by AP n_i, i.e., $M_i = \sum_{\ell=1}^{M} c_{i\ell}$. Once the channels are selected, the traffic load of an AP is equally split on the selected channels. For example, if AP n_i is loaded with traffic demand d_i, the traffic on each channel is d_i/M_i.

The interference pattern among APs can be captured by a neighborhood map $\mathcal{N} : (i, j) \rightarrow \{0, 1\}$, $i, j = 1, \ldots, N$, where $\mathcal{N}(i, j) = 1$ means that the APs n_i and n_j are within each other's communication range.[8] AP n_i is interfered by AP n_j if and only if: (i) they are within each other's communication range, i.e., $\mathcal{N}(i, j) = 1$, (ii) AP n_j is transmitting, and (iii) their selected channels overlap with each other. Multiple channels can be selected by a single AP to alleviate the traffic burden on each channel, although this may increase the interference between APs. This formulation implies a trade-off between the transmission burden and the interference effect.

With the above preliminaries, the interference of AP n_j to AP n_i on channel ℓ can be represented by a function of the traffic demands d_i, d_j, the neighborhood map $\mathcal{N}(i, j)$, the channel selections $c_{i\ell}, c_{j\ell}$, and the total numbers of selected channels M_i, M_j by n_i, n_j, i.e.,

$$I_{i,j,\ell}\big(d_i, d_j, \mathcal{N}(i, j), c_{i\ell}, c_{j\ell}, M_i, M_j\big). \tag{5.24}$$

The explicit expression of function (5.24) depends on specific systems and environments. Our objective is to reduce the mutual interference and balance the channel utilization over the whole network. In this context, we follow [34] and define the *maximal channel utilization* as the interference of AP n_i on the worst channel

$$u_i(\boldsymbol{d}, \mathcal{N}, \boldsymbol{C}) = \max_{\ell=1,\ldots,M} \sum_{j=1}^{N} I_{i,j,\ell}\big(d_i, d_j, \mathcal{N}(i, j), c_{i\ell}, c_{j\ell}, M_i, M_j\big) \tag{5.25}$$

[8] This is commonly determined by measuring the inter-AP received signal strength and considering a threshold to establish neighborhood (e.g. − 82 dBm) in real-world applications.

with $I_{i,i,\ell} = 0$ by default. Since the traffic demands are instantaneous at APs, the mutual interference and the channel utilization are instantaneous as well. A long-term expectation is a more meaningful metric. We, thus, consider the objective function as the expected mean of $\{u_i(\boldsymbol{d}, \mathcal{N}, \boldsymbol{C})\}_{i=1}^N$ over APs

$$\mathbb{E}_{\boldsymbol{d}}\Big[\frac{1}{N}\sum_{i=1}^{N} d_i u_i(\boldsymbol{d}, \mathcal{N}, \boldsymbol{C})\Big], \tag{5.26}$$

where $u_i(\boldsymbol{d}, \mathcal{N}, \boldsymbol{C})$ is weighed by the traffic demand d_i to give more importance on busy APs and the expectation $\mathbb{E}_{\boldsymbol{d}}[\cdot]$ is w.r.t. the distribution of the traffic demands $\boldsymbol{d}$. Note that the objective function can be any combination of $\{u_i(\boldsymbol{d}, \mathcal{N}, \boldsymbol{C})\}_{i=1}^N$ that is not limited to (5.26). We formulate the channel management problem in WLANs as

$$\mathbb{P} := \min_{\boldsymbol{C}} \mathbb{E}_{\boldsymbol{d}}\Big[\frac{1}{N}\sum_{i=1}^{N} d_i u_i(\boldsymbol{d}, \mathcal{N}, \boldsymbol{C})\Big], \tag{5.27}$$

$$s.t. \quad \boldsymbol{c}_i \in \{0, 1\}^M, \ \forall\, i = 1, \ldots, N, \tag{5.28}$$

where the channel selections $\boldsymbol{C}$ are controlled by the policy π based on the traffic demands $\boldsymbol{d}$. While some theoretical models exist for the interference function $I_{i,j,\ell}(\cdot)$ in $u_i(\cdot)$ [cf. (5.25)] and the distribution of the traffic demands $\boldsymbol{d}$ in $\mathbb{E}_{\boldsymbol{d}}[\cdot]$, we stress that problem (5.27) is given without any explicit model because the latter may not be accurate for real-world systems leading to inevitable degradation of model-based approaches. While this is a centralized problem that integrates the global network information in the problem formulation, we aim to find a solution that allows for a decentralized implementation, i.e., the control policy can be shared over all APs and each AP can deploy it to allocate channels locally with only neighborhood information. These facts result in three main challenges for solving problem (5.27):

1. The problem requires a solution that can be implemented in a decentralized manner, where each AP communicates with its neighbors, aggregates the neighborhood information, and makes its own decisions.
2. The problem requires a model-free solution that does not rely on theoretical system models, which precludes the application of model-based heuristic methods.
3. The problem requires continuous relaxation of discrete actions to compute gradients for conventional learning methods, which leads to performance degradation.

In the next section, we propose a learning-based approach that allows decentralized channel allocation and solves the problem in a model-free manner.

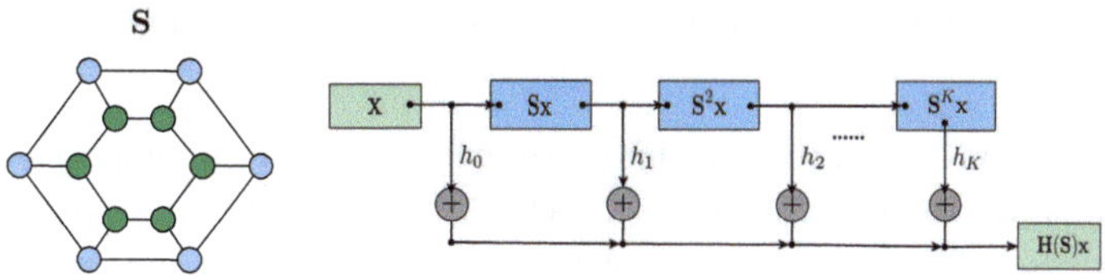

Fig. 5.15 The GCF of order K. The WLAN in Fig. 5.14 is modeled as a graph with a shift operator $\boldsymbol{S}$ (left). The GCF shifts a signal $\boldsymbol{x}$ over $\boldsymbol{S}$ and aggregates the shifted signals $\{\boldsymbol{S}^k\boldsymbol{x}\}_{k=0}^K$ with coefficients $\{h_k\}_{k=0}^K$ to generate the output $\boldsymbol{H}(\boldsymbol{S})\boldsymbol{x}$ (right)

5.3.3 A GNN-based Decentralized Solution

With the aforementioned challenges, we pursue to solve problem (5.27) with a learning procedure that combines *graph neural networks (GNNs)* and the *policy gradient* method [22]. The former allows for a decentralized implementation, while the latter yields a model-free solution. Specifically, we start by parameterizing the policy $\pi(\boldsymbol{d})$ with GNNs and re-formulate problem (5.27) as a statistical learning problem. We proceed to solve this learning problem with the policy gradient method, which needs only system observations but not system models.

5.3.3.1 Graph Neural Networks

A WLAN can be modeled by a graph $\mathcal{G} = (\mathcal{V}, \mathcal{E})$ with the node set $\mathcal{V}$ representing APs and the edge set $\mathcal{E}$ representing communication links, as shown in Fig. 5.15. The graph structure is captured by a matrix $\boldsymbol{S} \in \mathbb{R}^{N\times N}$, referred to as the *graph shift operator*, with the (i, j)th entry non-zero $[\boldsymbol{S}]_{ij} \neq 0$ if and only if there exists an edge between the nodes n_i and n_j, i.e., $(n_i, n_j) \in \mathcal{E}$. The system states are modeled by a graph signal $\boldsymbol{x} \in \mathbb{R}^N$ with the ith entry $[\boldsymbol{x}]_i$ representing the state of node n_i, such as the traffic demands of APs $\boldsymbol{d}$. The graph signal exhibits an irregular structure inherent in the graph topology and the coupling between them can be exploited for feature extraction.

Graph Convolutional Filters (GCFs) The GCF is one of the key tools in graph signal processing, which leverages the structural information to model representations from graph signals [38]. Specifically, the GCF is a polynomial of the shift operator matrix $\boldsymbol{S}$ over the graph signal $\boldsymbol{x}$ as

$$\boldsymbol{H}(\boldsymbol{S})\boldsymbol{x} = \sum_{k=0}^{K} h_k \boldsymbol{S}^k \boldsymbol{x} \tag{5.29}$$

with $\{h_k\}_{k=0}^K$ the filter coefficients. The shift operation $\boldsymbol{S}\boldsymbol{x}$ assigns to node n_i the aggregated information $[\boldsymbol{S}\boldsymbol{x}]_i$ from its neighboring nodes and $\boldsymbol{S}^k\boldsymbol{x}$ represents information exchanges up to a neighborhood of radius k – see Fig. 5.15. The GCF is a shift-and-sum operator that generalizes the convolution from the time/image domain to the graph domain and generates features from a more complete picture of the graph.

GNNs The GNN is a layered architecture, where each layer comprises a bank of GCFs and a pointwise nonlinearity [39–41]. At layer $\ell = 1, \ldots, L$, there are $F_{\ell-1}$ input signals $\{\boldsymbol{x}_{\ell-1}^{g}\}_{g=1}^{F_{\ell-1}}$ generated at layer $\ell - 1$. The latter are processed by a bank of GCFs $\{\boldsymbol{H}_{\ell}^{fg}(\boldsymbol{S})\}_{fg}$, aggregated over the input index g, and passed through a pointwise nonlinearity $\sigma(\cdot) : \mathbb{R} \to \mathbb{R}$ to generate the output signals of layer ℓ as

$$\boldsymbol{x}_{\ell}^{f} = \sigma\Big(\sum_{g=1}^{F_{\ell-1}} \boldsymbol{H}_{\ell}^{fg}(\boldsymbol{S})\boldsymbol{x}_{\ell-1}^{g}\Big), \ \forall f = 1, \ldots, F_{\ell}, \tag{5.30}$$

where common examples of $\sigma(\cdot)$ include the rectified linear unit (ReLU), absolute value, etc. Without loss of generality, we assume a single input $\boldsymbol{x}_0^1 = \boldsymbol{x}$ and a single output $\boldsymbol{x}_L^1$. The overall GNN can be represented as a nonlinear mapping $\Phi(\boldsymbol{x}, \boldsymbol{S}, \mathcal{H})$, where the inputs are the graph signal $\boldsymbol{x}$, the shift operator $\boldsymbol{S}$, and the output is the task-relevant feature. The architecture parameters $\mathcal{H}$ are the filter coefficients of all GCFs throughout L layers.

Statistical Learning Problem By parameterizing the policy $\pi(\boldsymbol{d})$ with the GNN $\Phi(\boldsymbol{d}, \boldsymbol{S}, \mathcal{H})$ of parameters $\mathcal{H}$, we can re-formulate problem (5.27) as a statistical learning problem

$$\begin{aligned} \mathbb{P}_{\mathcal{H}} := \min_{\mathcal{H}} \mathcal{L}(\mathcal{H}) &= \min_{\mathcal{H}} \mathbb{E}_{\boldsymbol{d}}\Big[\frac{1}{N}\sum_{i=1}^{N} d_i u_i(\boldsymbol{d}, \boldsymbol{S}, \mathcal{H})\Big], \\ s.t. \quad &[\Phi(\boldsymbol{d}, \boldsymbol{S}, \mathcal{H})]_i \in \{0, 1\}^M, \ \forall i = 1, \ldots, N. \end{aligned} \tag{5.31}$$

The goal is to learn the optimal parameters $\mathcal{H}^*$ that minimize the objective function. It is challenging to solve this problem by conventional unsupervised learning methods because they (i) rely on explicit theoretical models of the interference function $I_{i,j,\ell}(\cdot)$ and the traffic demand distribution $\mathbb{E}_{\boldsymbol{d}}[\cdot]$; (ii) require well-designed continuous relaxation for the discrete action space, to make gradient computations possible. Motivated by these challenges, we propose to leverage the model-free policy gradient method, which is widely used in reinforcement learning, to conduct the training procedure.

5.3.3.2 Policy Gradient

The training procedure optimizes the GNN parameters $\mathcal{H}$ with gradient descent. At each iteration t, we update the current parameters $\mathcal{H}^{(t)}$ as

$$\mathcal{H}^{(t+1)} = \mathcal{H}^{(t)} - \alpha^{(t)} \nabla_{\mathcal{H}} \mathcal{L}(\mathcal{H}^{(t)}), \tag{5.32}$$

where $\alpha^{(t)} > 0$ is the step-size. The preceding update requires the system model and the continuous relaxation to compute the gradient $\nabla_{\mathcal{H}}\mathcal{L}(\mathcal{H}^{(t)})$, where the former is not assumed available in problem (5.31) and the latter leads to performance degradation. We overcome these issues by leveraging the policy gradient method. It exploits the likelihood ratio property to provide a stochastic approximation for the gradient of policy functions that take the form of $\mathbb{E}[F(\boldsymbol{d}, \Phi(\boldsymbol{x}, \boldsymbol{S}, \mathcal{H}))]$, where $F(\cdot)$ is a general unknown function [22].

Specifically, the policy gradient method considers the GNN outputs $\Phi(\boldsymbol{d}, \boldsymbol{S}, \mathcal{H})$ as samples drawn from a probability distribution $p_{\boldsymbol{d},\boldsymbol{S},\mathcal{H}}(\Phi)$ specified by $\boldsymbol{d}$, $\boldsymbol{S}$ and $\mathcal{H}$, which has a delta density function $f_{\boldsymbol{d},\boldsymbol{S},\mathcal{H}}(\Phi) = \delta\big(\Phi - \Phi(\boldsymbol{d}, \boldsymbol{S}, \mathcal{H})\big)$. This allows to compute the Jacobian of the policy function as

$$\nabla_{\mathcal{H}}\mathbb{E}[F(\boldsymbol{d}, \Phi(\boldsymbol{d}, \boldsymbol{S}, \mathcal{H}))] = \mathbb{E}[F(\boldsymbol{d}, \Phi)\nabla_{\mathcal{H}}\log f_{\boldsymbol{d},\boldsymbol{S},\mathcal{H}}(\Phi)]. \tag{5.33}$$

It translates the computation of $\nabla_{\mathcal{H}}\mathbb{E}[F(\boldsymbol{d}, \Phi(\boldsymbol{d}, \boldsymbol{S}, \mathcal{H}))]$ to the multiplication of $F(\boldsymbol{d}, \Phi)$ and $\nabla_{\mathcal{H}}\log f_{\boldsymbol{d},\boldsymbol{S},\mathcal{H}}(\Phi)$. However, computing the gradient of the delta density function $\nabla_{\mathcal{H}}\log f_{\boldsymbol{d},\boldsymbol{S},\mathcal{H}}(\Phi)$ requires the system model $F(\cdot)$ as well. We resolve this issue by approximating the delta density function with another density function centered around $\Phi(\boldsymbol{d}, \boldsymbol{S}, \mathcal{H})$. Examples include the Gaussian distribution, binomial distribution and categorical distribution. Then, we can estimate the gradient of the loss function $\nabla_{\mathcal{H}}\mathcal{L}(\mathcal{H}^{(t)})$ as

$$\nabla_{\mathcal{H}}\mathcal{L}(\mathcal{H}^{(t)}) = \mathbb{E}_{\boldsymbol{d}}\Big[\frac{1}{N}\sum_{i=1}^{N} d_i u_i(\boldsymbol{d}, \boldsymbol{S}, \mathcal{H}^{(t)})\nabla_{\mathcal{H}}\log f_{\boldsymbol{d},\boldsymbol{S},\mathcal{H}^{(t)}}(\Phi)\Big], \tag{5.34}$$

where the expectation $\mathbb{E}_{\boldsymbol{d}}[\cdot]$ is approximated by observing $\mathcal{T}$ samples of the traffic demands $\{\boldsymbol{d}_\tau\}_{\tau=1}^{\mathcal{T}}$ and taking the average. We remark that (5.34) is model-free, i.e., it does not require any theoretical model of the interference function $u_i(\cdot)$ but only function values $u_i(\boldsymbol{d}_\tau, \boldsymbol{S}, \mathcal{H}^{(t)})$ evaluated at the sampled traffic demands $\boldsymbol{d}_\tau$, which can be observed from the system directly. Moreover, it deals with the discrete action space by sampling actions from policy distributions, e.g., binomial distributions and categorical distributions, which avoids the performance degradation induced by the continuous relaxation in conventional unsupervised learning methods.

5.3.3.3 Decentralized Implementation

The trained GNN admits a decentralized implementation, i.e., each node can compute outputs and make decisions locally with neighborhood information. To see this, let $[\boldsymbol{S}]_{ij}$ be the (i, j)th entry of $\boldsymbol{S}$ and $[\boldsymbol{x}]_i$ the ith entry of $\boldsymbol{x}$. The graph sparsity allows to compute the shift operation $\boldsymbol{S}\boldsymbol{x}$ as

$$[\boldsymbol{S}\boldsymbol{x}]_i = \sum_{j=1}^{N}[\boldsymbol{S}]_{ij}[\boldsymbol{x}]_j = \sum_{j:\,\mathcal{N}(i,j)=1}[\boldsymbol{S}]_{ij}[\boldsymbol{x}]_j, \tag{5.35}$$

for $i = 1, \ldots, N$. That is, the entry $[\boldsymbol{S}\boldsymbol{x}]_i$ at node n_i can be computed with the entries $\{[\boldsymbol{x}]_j\}_j$ at nodes $\{n_j\}_j$ that are connected to node n_i. This indicates that the entries of K shifted signals $\{[\boldsymbol{S}^k\boldsymbol{x}]_i\}_{k=0}^K$ can be computed at node n_i by exchanging information with its neighboring nodes as well. Since the aggregation of the filter coefficients $\{h_k\}_{k=0}^K$ is a linear combination of signal values at individual node, which dose not involve any feature fusion among nodes, the entry of the filter output $[\boldsymbol{H}(\boldsymbol{S})\boldsymbol{x}]_i$ at node n_i can be computed with local neighborhood information; hence, GCFs allow for a decentralized implementation.

GNNs inherit the decentralized implementation from GCFs because the nonlinearity $\sigma(\cdot)$ is pointwise and local [42–44]. This analysis shows that while GCFs and GNNs are expressed in compact forms (5.29) and (5.30), they can be implemented in a decentralized manner, i.e., each node does not need to know the entire graph topology, but only have the communication capacity to exchange information with neighbors and the computation capacity to aggregate the neighborhood information.

5.3.3.4 Permutation Equivariance

GNNs account for the graph structure to extract features and thus, are equivariant to topology permutations, i.e., the node reordering. This is a key property that allows GNNs to exploit internal symmetries of graph signals, which demonstrates the advantage of GNNs to process networked data and justifies the effectiveness of the proposed approach. Specifically, define the permutation matrix as $\boldsymbol{P} \in \{0, 1\}^{N \times N}$ with $\boldsymbol{P}\mathbf{1} = \mathbf{1}$ and $\boldsymbol{P}^\top\mathbf{1} = \mathbf{1}$. The permuted matrix $\boldsymbol{P}\boldsymbol{S}\boldsymbol{P}^\top$ reorders the columns and rows of $\boldsymbol{S}$, while the permuted vector $\boldsymbol{P}\boldsymbol{x}$ reorders the entries of $\boldsymbol{x}$. According to [45, 46], the following theorem formally characterizes the permutation equivariance of GNNs.

Theorem 5.1 (Proposition 1 [45]) *Consider the GNN $\Phi(\boldsymbol{x}, \boldsymbol{S}, \mathcal{H})$ with the graph signal $\boldsymbol{x}$ and the underlying graph $\boldsymbol{S}$. Let $\boldsymbol{P}$ be a permutation matrix. Then, it holds that*

$$\Phi(\boldsymbol{P}\boldsymbol{x}, \boldsymbol{P}\boldsymbol{S}\boldsymbol{P}^\top, \mathcal{H}) = \boldsymbol{P}\Phi(\boldsymbol{x}, \boldsymbol{S}, \mathcal{H}). \tag{5.36}$$

Theorem 5.1 states that GNNs applied on permuted graphs $\boldsymbol{P}\boldsymbol{S}\boldsymbol{P}^\top$ and permuted signals $\boldsymbol{P}\boldsymbol{x}$ generate equally permuted outputs. This result indicates that GNNs can harness the same information regardless of what permuted versions of graphs and signals are processed. The latter acts as a data augmentation technique that facilitates the training procedure, which will be validated in the next section. In the context of channel allocation, this property yields the implementation of GNN-based policies independent of the order of APs, i.e., they can be implemented directly on arbitrarily ordered APs. This follows the intuitive requirement of decentralized channel management solutions in WLANs because APs will not be labeled with a fixed order in practice and the explicit order is not the prior information available during implementation.

5.3.4 Numerical Evaluations

In this section, we present empirical results for performance evaluation. While the proposed approach is model-free, we make system observations (i.e., observations of the traffic demands and the interference) from given models. However, we do not assume any knowledge of these models during training/testing but only generate samples for observation.

Setup The network consists of N APs and the topology is generated randomly with the edge probability q, i.e., AP n_i is connected to AP n_j with probability q for $i, j = 1, \dots, N$. Traffic demands are sampled from distribution $\max(\mathbb{N}(0.8, 0.4), 0)$, where $\mathbb{N}(\mu, \rho)$ denotes the normal distribution with mean μ and standard deviation ρ, and the interference is derived from the model $I_{i,j,\ell} = \mathcal{N}(i, j) c_{i\ell} c_{j\ell} d_j / M_j$. The GNN has 4 layers, each containing 32, 64, 64, 32 filters of order 3 and the ReLU nonlinearity. The policy distribution is the categorical distribution and the batch size is 64. Our results are averaged over 10 randomly generated graphs.

Small-Scale Scenario First, we consider a small-scale scenario with 10 APs, 4 channels and $q = 0.25$ edge probability, where zero interference can be theoretically achieved by allocating different channels to adjacent APs and the same channels to nonadjacent APs. We compare the proposed approach with two baselines: (i) GNN with conventional unsupervised learning [32], and (ii) uniformly random channel selection. The former relaxes discrete actions as continuous variables for training and conducts discrete quantification for testing, which is a model-based solution that depends on closed-form system models to compute gradients, while the latter is model-free. Figure 5.16 shows the results. The proposed approach exhibits the optimal performance, i.e., zero interference among APs. The GNN with unsupervised learning performs worse despite the model-based property. This is because it approximates discrete actions via a continuous relaxation, and hence, suffers from performance degradation. Moreover, we see from the figure that it does not show learning behaviour. This is because (i) the maximal operation in (5.25)

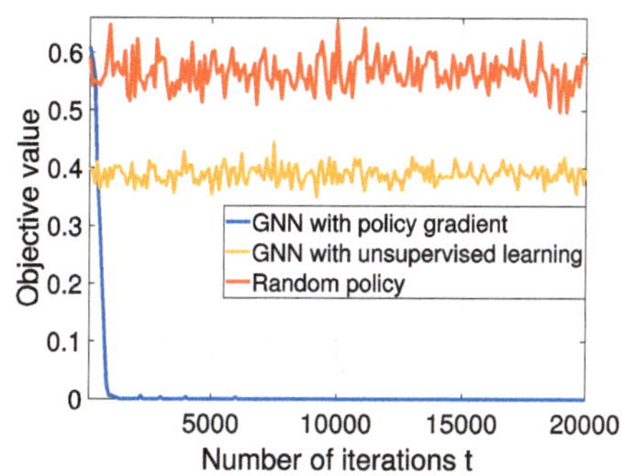

Fig. 5.16 Performance of GNN with policy gradient compared to GNN with conventional unsupervised learning and the random channel selection policy

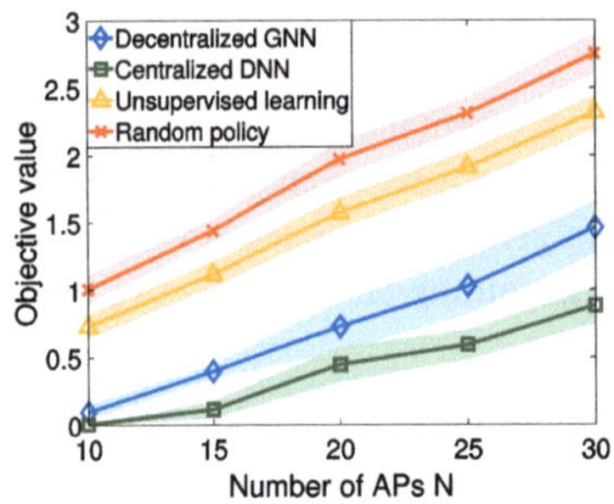

(a)

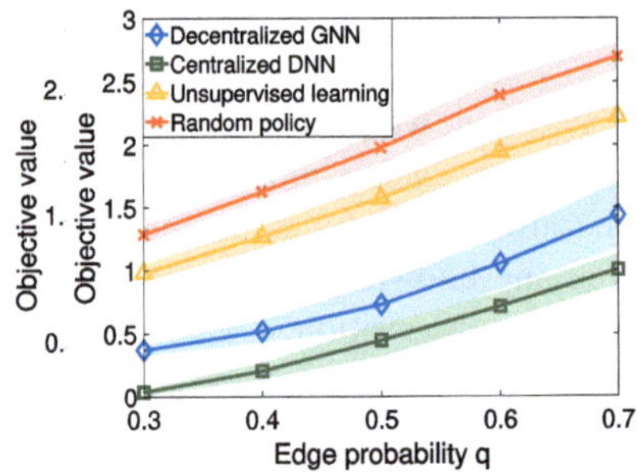

(b)

Fig. 5.17 Performance of decentralized GNN compared with the baselines. The results are averaged over 10 random networks and the shaded area shows the std. dev. (**a**) Different numbers of APs, i.e., different network sizes, with $q = 0.5$ edge probability. (**b**) Different edge probabilities q, i.e., different network densities, with 20 APs

makes the gradient propagation difficult even for continuous variables, and (ii) the discrete quantification during testing creates further difficulty.

Large-Scale Scenario Next, we evaluate the proposed approach on larger networks with various numbers of APs in Fig. 5.17a and various numbers of edges in Fig. 5.17b. The former expands the network size, while the latter increases the network density. We consider three baselines for performance comparison: (i) deep neural network (DNN), (ii) GNN with unsupervised learning, and (iii) random channel selection. The first follows [34] to parameterize the control policy with a DNN, which is a centralized solution that requires the global network information during implementation, while the latter two are decentralized solutions. The centralized DNN performs best but its execution requires access to the states, i.e., the traffic demands, of all the APs, which may not be practical for large-scale scenarios. The proposed approach achieves a performance close to the one achieved by the centralized DNN, but requires only local neighborhood information for implementation, yielding a satisfactory trade-off between performance and scalability. Moreover, the proposed approach generalizes well for large and dense networks, in which decentralization plays an important role.

GNN vs. DNN Next, we compare the proposed approach with the centralized DNN in terms of the training efficiency and the graph connection. In Fig. 5.18a, we see that the proposed approach converges much faster than the centralized DNN and shows better training efficiency. The result corroborates the fact that the GNN is permutation equivariant and the latter acts as a data augmentation technique to facilitate the training procedure, as indicated in Theorem 5.1. In Fig. 5.18b, we see that the performance degradation of the decentralized GNN w.r.t. the centralized DNN first increases with the graph connectivity, i.e., from 0.355 at $q = 0.6$ to 0.433 at $q = 0.7$. This is because the interference becomes more severe and the global

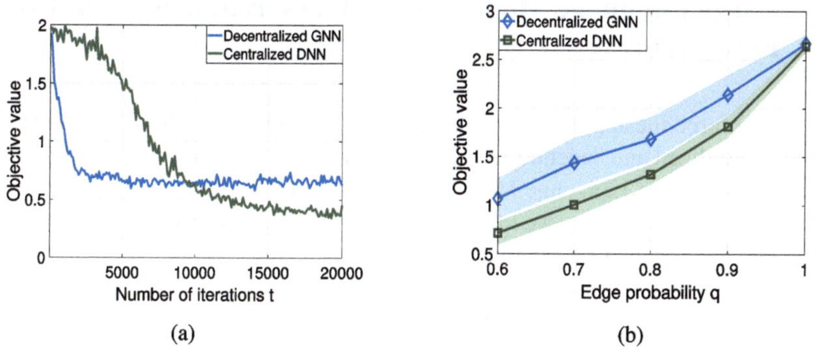

Fig. 5.18 (**a**) Training performance of decentralized GNN and centralized DNN for a network with 20 APs and $q = 0.5$ edge probability. (**b**) Performance of decentralized GNN and centralized DNN for a network with 20 APs and different graph connections, i.e., different edge probabilities q

Fig. 5.19 Performance of decentralized GNN compared with the baselines w.r.t. the mean and max objectives. The network has 20 APs and $q = 0.5$ edge probability, and the results are averaged over 10 random networks

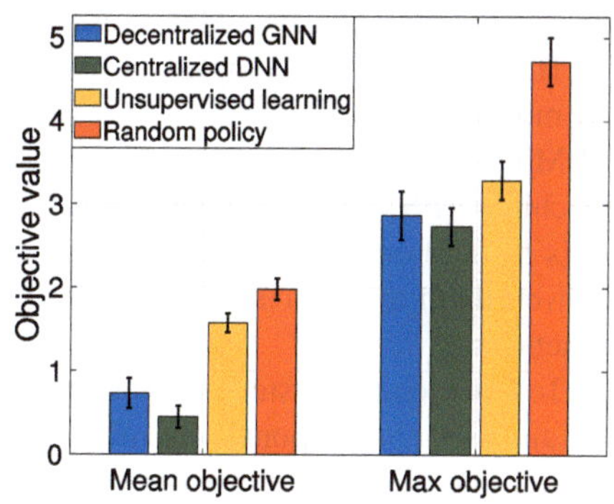

information becomes more important for channel allocation. The performance degradation then decreases from $q = 0.7$ to $q = 1$ and converges to almost zero for fully-connected graphs. We attribute this behaviour to the fact that the neighborhood information becomes richer as the graph connectivity increases, and the proposed approach performs comparably well with the centralized DNN.

Optimization Criteria Lastly, we show the adaptivity of the proposed approach by testing a different objective function. We consider the objective function as the expected maximum of the maximal channel utilization over APs, i.e., the maximal channel utilization on the worst AP. The results are shown in Fig. 5.19. The proposed approach performs close to the centralized DNN and outperforms the other baselines on both objective functions. This indicates that it is capable of handling different optimization criteria specified by different application requirements and can be used in different decentralized tasks of channel allocation.

5.4 Large Language Models as Central Orchestrators in IoT Ecosystems

Having explored the applications of advanced AI techniques such as DRL and GNN in both centralized and decentralized IoT network control, we now turn our attention to the role of Large Language Models (LLMs) as central orchestrators within IoT ecosystems. In this section, we introduce LLMind, an innovative AI framework that harnesses the power of LLMs to coordinate and enhance collaboration among IoT devices and domain-specific AI modules.

LLMind positions LLMs at the core of the IoT ecosystem, leveraging their advanced task planning, language understanding, and generation capabilities to facilitate seamless interaction and control between human users and the network of devices. Through natural conversations conducted via user-friendly social media platforms, users can communicate complex tasks to the system in everyday language. The LLM engages in these dialogues, interpreting user intentions and devising comprehensive plans to execute the desired tasks.

The execution of complex tasks—often requiring the cooperation of multiple AI modules and IoT devices—is realized through the generation of control scripts. LLMind employs a Language-Code transformation approach based on finite-state machines (FSMs) to translate high-level user instructions into executable code. This method enables the LLM to parse natural language inputs and systematically convert them into precise commands that orchestrate device actions and AI module functions.

To enhance the system's speed and effectiveness, LLMind incorporates semantic analysis and response optimization techniques. Semantic analysis allows the framework to comprehend the contextual nuances of user instructions, ensuring accurate interpretation and execution. Response optimization techniques streamline the communication process between the LLM, AI modules, and IoT devices, reducing latency and improving overall system performance.

Ultimately, LLMind is designed not only to innovate IoT device control and enrich user experiences but also to foster an intelligent and integrated IoT ecosystem. By continuously learning from interactions between users and machines, the framework evolves and becomes more sophisticated over time. This ongoing development enhances the ecosystem's ability to handle increasingly complex tasks, adapt to new devices and technologies, and provide more intuitive and efficient user interactions.

In the following, we delve deeper into the architecture and components of the LLMind framework. We explore how LLMs integrate with domain-specific AI modules, the mechanisms behind natural language interaction and planning, and the methodologies used for control script generation. We also discuss the system's ability to learn from user interactions, its potential applications, and the challenges and future directions for integrating LLMs into IoT ecosystems.

5.4.1 Background

Large Language Model (LLM) has attracted tremendous attention from both academia and industry. It is an advanced generative artificial intelligence designed to comprehend and generate human-like text based on extensive training data. In addition to excellent language skills, LLM also exhibits a high level of reasoning and planning capability [47, 48]. Moreover, the remarkable feature of in-context learning substantially broadens the range of potential application scenarios for LLM, inspiring the design of diverse general-purpose AI agents.

Our daily lives are inundated with an abundance of IoT devices, ranging from smart desk lamps to bulky smart home appliances such as refrigerators and washing machines. The emergence of LLM presents a transformative opportunity to revolutionize the development of IoT devices by empowering them with advanced intelligence. Imagine an LLM-based AI agent coordinating diverse IoT devices to collaboratively accomplish intricate tasks. LLM's remarkable planning and scheduling flexibility surpasses scripted intelligence. Its in-context learning ability enables swift adaptation and optimal resource utilization for efficient task solutions. Consequently, users can effortlessly express their needs without concerns about operational details.

Integrating LLMs into IoT device control presents a number of challenges despite its promising potential. First, the LLM cannot handle specialized tasks like object detection or facial recognition, while there are already AI models excelling in these domains. The first challenge lies in combining the power of fragmented AI modules with the LLM to handle various complex tasks. Furthermore, due to the LLM's slow inference speed and high computational costs, it is crucial to investigate methods that can enhance the system's response speed and overall efficiency. Another significant challenge involves the effective control of IoT devices by the AI agent. The LLM must accurately learn the diverse functionalities and operational characteristics of various IoT devices. Additionally, the design of efficient interaction mechanisms that enable users to naturally and intuitively express their needs to the LLM-based AI agent presents a further challenge.

In this work, we propose LLMind, a framework that transforms conventional IoT device control by integrating LLM as the orchestrator for accomplishing complex tasks. This integration leads to the construction of an LLM-empowered IoT ecosystem. Through novel approaches, our framework effectively overcomes the aforementioned challenges.

First, the agent integrates domain-specific AI modules, with the LLM acting as the task allocation governor. This incorporation enables the general-purpose LLM to effectively manage specific tasks by invoking specialized AI modules, enhancing system flexibility and practicality. Moreover, by incorporating new AI modules, the AI agent acquires new capabilities and continues to evolve.

Second, to enhance the response speed and overall efficiency of the agent, we utilize a semantic analysis NLP model to classify and locally store LLM-generated plans (control scripts) for specific user commands. By identifying similar

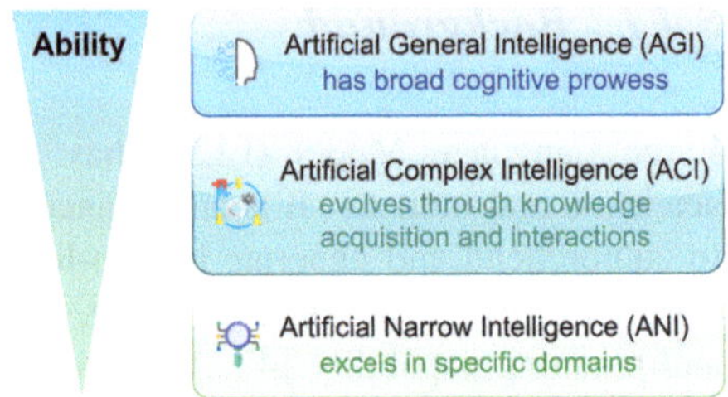

Fig. 5.20 Artificial Complex Intelligence (ACI)

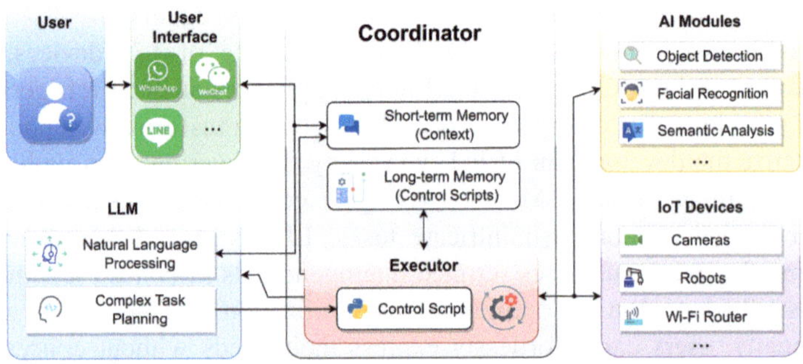

Fig. 5.21 System diagram

instructions in historical records, even with varying tones or expressions, the agent can retrieve validated solutions, thus avoiding complete replanning. This approach enables the system to be more and more skillful through interactions.

Third, the agent leverages the LLM for complex task planning and corresponding control script generation using a novel finite-state machine approach, enabling efficient and precise invocation of AI modules and IoT devices. This methodology significantly improves accuracy and success rates.

Last but not least, we connect our demo system to the widely used social media platform, WeChat, allowing users to conveniently interact with the LLM agent.

We refer to the intelligence level attained by our framework as Artificial Complex Intelligence (ACI), which serves as a bridge between Artificial Narrow Intelligence (ANI) [49] and Artificial General Intelligence (AGI) [50], as illustrated in Fig. 5.20. ACI integrates the capabilities of various ANI modules and undergoes continuous evolution towards AGI. Our framework goes beyond mere device control, providing a comprehensive blueprint for a future intelligent, integrated, and collaborative IoT device ecosystem empowered by LLM.

5.4.2 *The LLMind Framework*

This section presents the framework design. Our framework, depicted in Fig. 5.21, consists of five components: user interface, LLM, coordinator, AI modules, and IoT devices. The operational workflow of a demo system is illustrated in the flowchart presented in Fig. 5.22.

Users can interact with the system by sending text messages through social media software. The LLM promptly responds to confirm message receipt. It is important to note that the system filters user messages. For casual chats and greetings, the system provides text replies without further processing. However, for specific requests and commands, the system formulates plans and strives to fulfill the user's instructions using integrated AI modules and connected IoT devices.

After receiving a user instruction, the system prioritizes enhancing efficiency by searching for validated and feasible historical scripts stored locally in the coordinator's long-term memory. This process utilizes NLP techniques to locate previously issued user commands with similar semantic meanings. If a matching control script is found, it is executed accordingly. However, in the absence of a matching control script, the LLM undertakes the planning process and generates a new control script to invoke AI modules and IoT devices.

The system performs preliminary feasibility validation on the newly generated script, checking for syntax errors or hardware-related issues. If the script fails to execute correctly, the system updates the context (coordinator's short-term memory), gathering error information and triggering the LLM to regenerate the script based on the updated context. If the script continues to fail after multiple retries (more than three attempts), the system collects error information and reports it to the user. Conversely, if the script executes successfully, upon completion, the system summarizes the intermediate results gathered during script execution and generates an appropriate response based on the user's instructions.

5.4.2.1 User Interface

Our proposed framework is open to various interaction methods such as text, pictures, and voices. It depends on valid LLM input or the availability of methods to convert user input into a suitable format for LLM processing. Social media software is a good choice since it offers a seamless and natural interaction experience that aligns with our daily lives.

5.4.2.2 LLM

Within our framework, the LLM performs two key functionalities [51, 52]. First, it engages in conversations with users, generating contextually appropriate responses. Second, it does the planning and generates control scripts to address the tasks raised by the user, utilizing specialized AI modules and IoT devices as part of the available resources and methods. Python was selected as the programming language for our demo system.

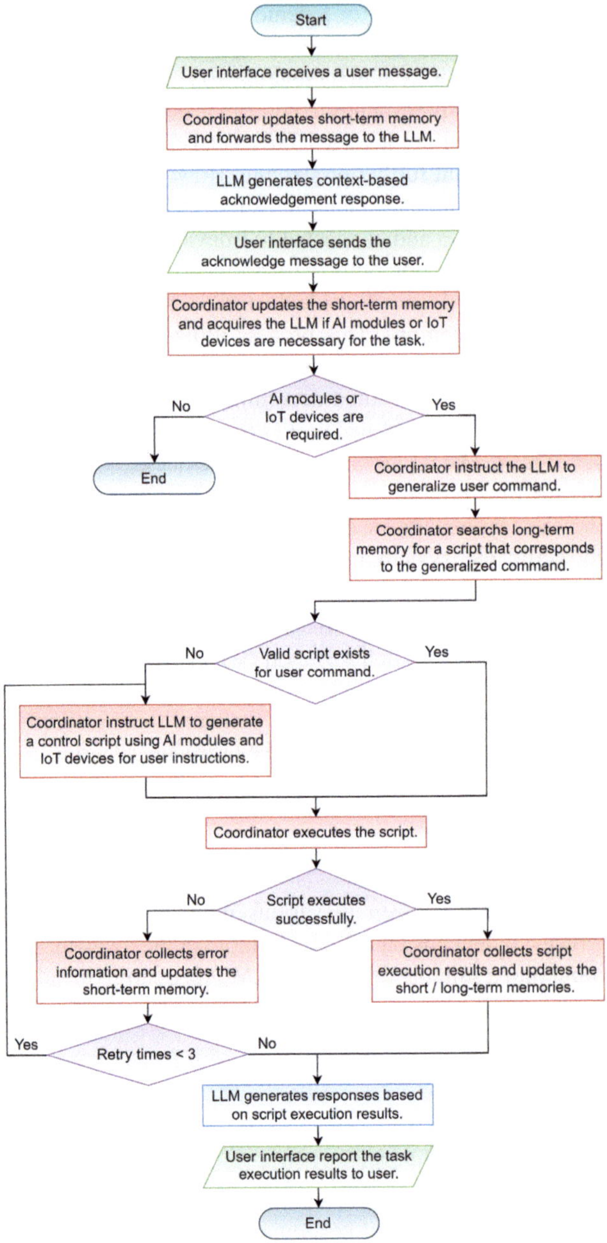

Fig. 5.22 System operational process flowchart

Role-Playing Prompts

1: You are a serious housekeeper managing the employer's daily life.
2: You are working remotely and unable to perform tasks personally. However, fortunately, you have numerous assistants available at your employer's house to provide support.
3: You do not need to give an exact answer to a question if you lack the necessary information, but assure your employer that you will try to figure it out.
4: You must not ask your assistants to perform tasks that are not part of your employer's instructions.
5: Your reply should be simple and concise.
6: You should use the same language as your employer.

User Information Prompts

1: Your employer is male.
2: The name of your employer is Eason.

Conversation

To enhance the lifelike and immersive response of the LLM, we adopt a role-playing technique in which the LLM takes on the persona of a housekeeper. This approach is supported by carefully crafted system-level prompts, shown as follows. It is important to note that in order to provide an optimal user experience, specific user-related details may be required.

Script Generation

The LLM itself does not directly interact with AI modules and IoT devices. Instead, it generates Python control scripts to invoke AI modules and send control commands to IoT devices through network connections. Within this framework, we have proposed a Language-Code transformation scheme based on a finite-state machine (FSM) for generating scripts. Our Language-Code transformation scheme tackles two critical issues.

First, a user may describe the task he wants the AI agent to do with simple (sometimes ambiguous) human language, but the successful execution of a task relies on complex code descriptions that precisely define the operation of AI modules and IoT devices. The mismatch between high-level human command and precise code description necessitates a reliable Language-Code transformation that can be refined iteratively until the human command is realized to produce outcomes satisfactory to the human.

To address the first issue, the scheme enriches the information in the human command by leveraging knowledge about the environment. For instance, when a user instructs the agent to fetch a bottle of iced Coke, the agent should understand that iced Cokes are stored in the refrigerator, and the refrigerator is located in the kitchen corner, even if not explicitly mentioned. In our system, we enrich the command's information with pre-stored environment knowledge using in-context learning.

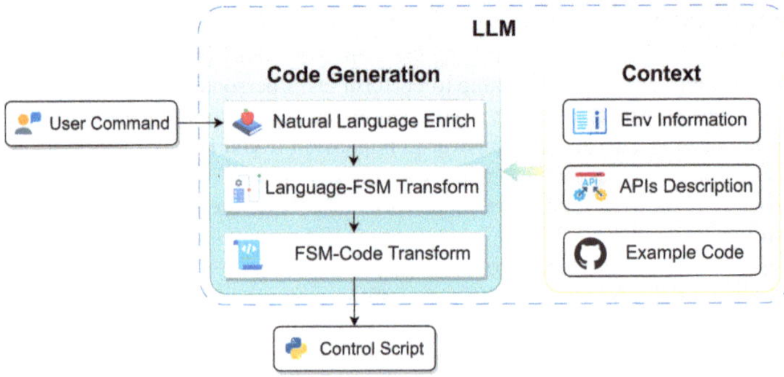

Fig. 5.23 Language-Code transformation procedure

Second, the LLM must precisely transform the enriched information (expressed in natural language) into executable code. However, the language description can become lengthy after enrichment to enhance clarity, and as a result, the generated code may become excessively complex, posing a challenge to error-free execution. Directly using the LLM to generate complex code from long language descriptions is unreliable due to its limited ability to handle intricate code with advanced functionality, as observed in [53]. This may result in the loss of crucial information from the human command or render the generated code unusable.

To address this problem, we use FSM to bridge the gap between language and code descriptions. First, we utilize the task decomposition abilities of the LLM to break down the overall task into sub-tasks, representing them as states in the FSM. The interconnection between these sub-tasks is captured through transitions between FSM states. Next, we utilize the LLM to convert each sub-task within the FSM states into executable Python code. Finally, by combining the code descriptions of each state and incorporating state-transition conditions, we obtain the final code for the overall task.

Compared to the "Language-Code" transformation, our novel "Language-FSM-Code" transformation scheme maintains the same level of precision in conveying information since an FSM can precisely and comprehensively describe a task [54]. Additionally, by providing APIs that enable the LLM to reference and interact with AI modules and IoT devices, the complexity of sub-task transformation within each FSM state is manageable for current LLMs. Notably, our scheme enhances reliability without causing information loss.

Figure 5.23 depicts the Language-Code transformation procedure, which consists of three phases: First, natural language enrichment; next, Language-FSM transformation; and finally FSM-Code transformation. Our scheme utilizes pre-stored knowledge, including environment information, API documentation of AI modules and IoT devices, and example codes, to aid in the transformation.

In phase one, the system enriches the task description using the environment information stored in the short-term memory of the coordinator. This information encompasses a list of available AI modules and IoT devices, their functional descriptions, as well as IoT device locations. The environment information enables a precise and executable task description.

Phase two leverages the enriched task description and API information to generate the FSM representation of the task. The provision of API information helps the LLM to gain a better understanding of the capabilities of AI modules and IoT devices, thereby facilitating the development of a reasonable and efficient FSM design. Without API information, the FSM representation may include overly complex sub-tasks or redundant processes, causing inefficiencies.

In phase three, the FSM-Code transformation is divided into three sub-phases. An FSM has two components: state operation and state transition. We generate code descriptions for each component in separate sub-phases, merging them in the final sub-phase. Our method simplifies the LLM's task and improves reliability compared to a single-round interaction.

In the first sub-phase, we provide the LLM with the FSM derived in phase two and API information. This enables the LLM to generate state operation code that incorporates the desired functionality. In the second sub-phase, example codes are provided to assist the LLM in understanding Python-based state transitions. Finally, the LLM is provided with the FSM, code generated in the previous sub-phases, and supplementary reference codes to facilitate the production of the final code.

5.4.2.3 Coordinator

In our framework, the coordinator serves as the central hub, facilitating the interconnection and seamless operation of various system components. With its pivotal role in managing the flow of information and tasks, the coordinator ensures the smooth functioning of complex AI systems. It comprises three essential parts: short-term memory, long-term memory, and script executor.

The short-term memory contains the contextual information provided to the LLM for both conversation and code-generation purposes. This information serves as a crucial resource for the AI system, providing the necessary knowledge and understanding of user interactions, system states, and task requirements. It comprises four main components:

1. Environment Information: The component includes a list of available AI modules and IoT devices with their function descriptions, as well as the installation locations of IoT devices. It enables the LLM to be aware of the available AI modules, IoT devices, and their capabilities.
2. API Descriptions: This component offers a comprehensive description of the API functions used to invoke AI modules and control IoT devices within the script, encompassing details such as inputs, outputs, specifications, expected time costs, and other relevant information. These details enable the LLM to effectively leverage the available functionality.
3. Chat History: It retains the record of previous conversations between the system and the user, allowing the LLM to refer back to previous interactions.

4. Execution Results and Error Reports: The short-term memory stores the execution results of generated scripts and error reports returned by the script executor. This information helps the LLM assess the outcome of script execution and identify issues or errors that may have occurred.

The context is updated in real-time based on user interactions and the execution status of control scripts, ensuring that it is readily available for intelligent decision-making and adaptive behavior within the AI system.

The long-term memory functions as a database for previously generated scripts. Valid scripts are stored locally and linked to their corresponding user instructions. This enables the system to reuse preexisting scripts when a similar instruction is given by the user in the future, resulting in improved efficiency and faster response speed. Importantly, the agent's ability continues to grow through the accumulation and generalization of experience. However, there is an issue with how the system assesses the similarities between user instructions. Our system implements a semantic analysis NLP model that maps words to a vector space. The semantic similarity is then determined by computing the distance between these vectors. To enhance the search speed, we vectorize the instruction associated with the script before storing it in the database.

The control script generated by the LLM is passed to the executor, which provides the necessary runtime environment and offers API functions that can be called within the scripts. Furthermore, the executor monitors hardware status, handles runtime data, generates execution reports, and updates both short and long-term memories.

5.4.2.4 AI Modules

Currently, LLM cannot effectively handle many domain-specific tasks. To address this limitation, we integrated domain-specific AI modules into the framework, aiming to enhance overall performance by assigning specialized tasks to dedicated modules, resulting in improved capabilities, accuracy, and performance beyond LLM's general knowledge.

AI modules can encompass various AI techniques, such as computer vision, natural language processing, speech recognition, sentiment analysis, recommendation systems, and more. Each module is trained and optimized for its specific task, enabling it to provide highly accurate and specialized results.

Furthermore, AI modules enable the AI system to adapt and evolve rapidly in response to emerging challenges and advancements in specific domains. New modules can be developed and integrated into the system as needed, expanding its capabilities and keeping pace with the evolving landscape of AI technologies.

In the context of IoT device control, AI modules play a crucial role in handling specialized tasks that may be beyond the scope of a general-purpose AI model like an LLM. For example, tasks like object detection, facial recognition, anomaly detection, or localization require specialized training data and algorithms. By incorporating dedicated AI modules for these tasks, the AI system can provide more accurate and reliable results, enhancing the overall system performance and user experience.

5.4.2.5 IoT Devices

IoT devices encompass a wide range of physical objects embedded with sensors, actuators, and connectivity features that enable internet connectivity and data exchange. Examples include smart appliances, wearables, industrial equipment, and environmental sensors.

To ensure efficient control of diverse IoT devices within our proposed framework, manufacturers need to expose a set of well-documented API functions that encapsulate the necessary functionality for interacting with the devices. These functions should offer a clear and standardized interface, enabling the AI agent to initiate specific actions or retrieve information from the IoT devices.

It is important to emphasize that manufacturers should prioritize security considerations when exposing API functions for remote control. By implementing robust authentication mechanisms, access controls, and encryption protocols within the API functions, they can ensure that only authorized entities have control over the devices. This approach guarantees secure and protected communications, safeguarding against unauthorized access or tampering.

5.4.2.6 Related Works

Despite the existing research on LLM-based intelligent control systems [55–60], this remains a new area that requires further investigation to enhance user experiences and facilitate device collaboration. This subsection reviews several recent works in the context of our framework.

Auto-GPT [60], an open-source application built on GPT-4, enables autonomous tasks through chained LLM thoughts. It provides GPT-based models with memory and autonomy for independent task execution and experiential learning. However, Auto-GPT has limitations [61], such as high costs due to rapid token consumption, inability to retain learning across tasks or generalize reusable functions, and limited access to resources, constraining its capabilities. Our framework is an approach that allows the LLM to access AI modules and IoT devices, enriching the available resources for the agent. Additionally, we introduce a novel historical script retrieval mechanism, which enhances system efficiency and enables the system to become more experienced and skillful over time.

LLM-Planner [58] introduces a few-shot planning method using LLMs for versatile embodied agents that rapidly comprehend and execute natural language instructions. It generates high-level plans adapting to the environment. Simulation results on the ALFRED dataset showcase LLM-Planner's competitive few-shot performance with minimal training data, enabling sample-efficient embodied agents.

Our work makes further advances by providing technical implementation details on how to use the LLM for real-world task planning and scheduling, along with a new FSM-based approach that significantly improves task execution success rate.

HuggingGPT [57] is a framework that connects and coordinates AI models, leveraging LLMs' language comprehension and reasoning. It includes stages such as task planning, model selection, execution, and response generation. However, as mentioned in [57], HuggingGPT still has limitations in planning feasibility, efficiency, token lengths, and stability. Our work extends HuggingGPT by integrating AI modules into the system to assist in task execution. The LLM invokes AI modules as necessary to enhance its capabilities.

The framework proposed in [55] combines LLMs, digital twin systems, and industrial automation for intelligent production planning and control. It uses prompts to provide context for LLM agents, enabling contextually appropriate autonomous responses. These agents interpret descriptive information in the digital twins and control the physical system via service interfaces. Integrating LLMs with digital twin systems enhances production system scalability and adaptability. Our proposed framework offers an alternative approach by utilizing predefined APIs to control IoT devices, avoiding the complexity and cost of building a digital twin, while still achieving a satisfactory effect.

5.4.3 *Experiments and Demos*

We have verified the feasibility of the framework through experiments. In the experiments, the user sends a text message to the AI agent. It then formulates a solution by examining the user's command and the system's available resources and methods, including AI modules and connected IoT devices. Following that, the AI agent generates an executable script and responds to the user's request based on the execution outcome.

We deployed the coordinator and AI modules of the system on an edge server. The edge server is connected to the same local network as the IoT devices via WiFi and can interact with ChatGPT using OpenAI's APIs.

The available IoT devices in the system are as follows:

1. Security cameras: The cameras are positioned within the room to capture images of the surrounding environment. These images are transmitted to the edge server via a local network. This experiment employs three cameras: two placed at the corners of the room, approximately 2 meters above the ground, and the third mounted on top of a TurtleBot mobile robot.
2. TurtleBot mobile robot: The TurtleBot mobile robot can receive target position coordinates through a network connection and autonomously plan paths to navigate within the room, intelligently avoiding obstacles.
3. WiFi router: The WiFi router can manage bandwidth allocation for each device and enforce restrictions on their maximum network speed.

Table 5.2 Dialog contents between different roles

Num.	Conversations	Contents
1	User → Housekeeper	Can you help me to count the number of people in the room and identify who they are?
2	Housekeeper → User	I can help you count the number of people in the room and identify them. Please give me a moment to gather the information.
3	Assistant → Housekeeper	Do you require any assistance?
4	Housekeeper → Assistant	Yes, please. I need assistance with counting the number of people in the room and identifying them.
5	Assistant → Housekeeper	Please repeat the instruction using as few words as possible.
6	Housekeeper → Assistant	Count people in room, identify them.
In the absence of a valid script in the long-term memory, the LLM generates a script, which is then executed in the executor.		
7	Assistant → Housekeeper	We have found 5 people in the room. The list of identified 3 people are ['Mike', 'Ada', 'Joe']. There are also 2 people that can not be identified. Their locations are shown in the following list in a format of [x, y]: [[10, 1], [12, 5]].
8	Housekeeper → User	There are 5 people in the room. Three of them have been identified as Mike, Ada, and Joe. The other 2 people cannot be identified and their locations are at [10, 1] and [12, 5].

Additionally, the system integrates the following AI modules:

1. Object detection: We employ the YOLO v8 model [62] to detect various objects in the scene, including people.
2. Face recognition: We use an open-source project called Face Recognition [63] to identify individuals in the scene.

5.4.3.1 Scenario 1: Check-in and Security

Guests arrive at a party. The user sends a message to the AI agent: "Count the number of people in the room and identify who they are."[9] Table 5.2 provides the dialogue content between different roles in the experiment. Among them, the LLM acts as the housekeeper, while the coordinator serves as the assistant.

In step 2, the LLM, acting as the housekeeper, responds to the message sent by the user.

[9] Demo video 1: https://youtu.be/k7Y8gKV5lt8.

In steps 3 and 4, following each Q&A exchange between the LLM and the user, as shown in steps 1 and 2, the coordinator questions the LLM about the need for assistance. In the scenario where the LLM assumes the role of a remote housekeeper who relies on assistants in the user's residence to carry out tasks. This inquiry serves to verify if the user has given specific instructions. Ambiguous instructions, as well as unrelated casual conversations, will prompt the user to provide additional information or be filtered out. When the LLM indicates a need for help, it signifies that the user has assigned a task that could potentially be handled using AI modules and IoT devices. We employ a role-playing methodology that allows for the implementation of various roles and role-playing settings to facilitate the context-filtering process.

To ensure effective communication between the LLM and the coordinator, the LLM's response in step 4 must incorporate specific keywords that facilitate word matching. These rules are already embedded in the context and are not shown in Table 5.2. Specifically, when the LLM's response is affirmative and assistance is needed, it should include the keyword "Yes". Conversely, if the response is negative, it must include the keyword "No". In this experiment, the LLM's response is affirmative, based on user's instruction.

In steps 5 and 6, the coordinator prompts the LLM to generate a summary of the user's instructions to facilitate the search for relevant historical scripts stored in long-term memory. This search process involves comparing the semantic content of the user's instructions. However, the original task description may suffer from low quality and be mixed with various linguistic habits. Additionally, the LLM may require multiple interactions with the user to fully grasp the details of specific tasks. To handle these conditions, in step 5, we request the LLM to provide a concise summary of the user's instructions. Subsequently, we conduct a search for relevant scripts based on the LLM's summary to enhance accuracy.

In this experiment, the coordinator could not find a suitable control script in long-term memory for the task. Consequently, the LLM is instructed to generate a new script. The code is generated using an FSM approach, as depicted in Fig. 5.24a. In addition to relying on the security camera installed on the ceiling, if the person cannot be identified due to low photo resolution or incomplete facial information, the AI agent will instruct the TurtleBot to move to the position of the unknown person and take a photo for further assurance. The code generates two lists that store the execution results, one for known persons and another for unknown persons.

In step 7, once the script execution is complete, the LLM receives the execution results of the API functions. Throughout the script execution process, API functions may generate log information, encompassing intermediate results, warnings, and errors. This log information is collected and sent to the LLM as context (short-term memory) for analysis.

Finally, in step 8, the LLM generates a report based on the user's question and the script's log information and sends it to the user.

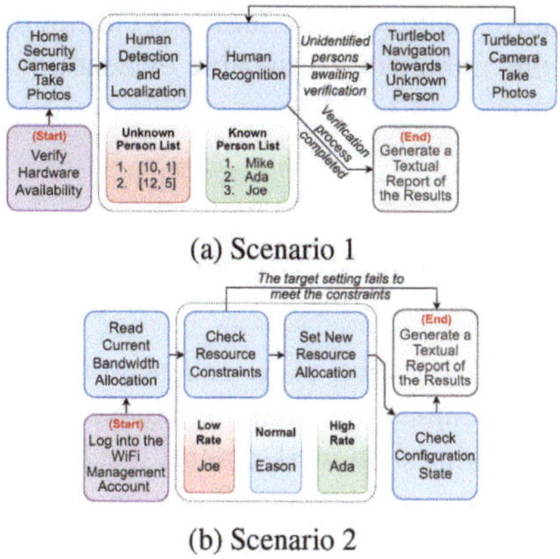

Fig. 5.24 FSM of generated code. (**a**) Scenario 1. (**b**) Scenario 2

5.4.3.2 Scenario 2: Network Management

Our framework is designed to handle both one-off complex tasks and scenarios that involve multiple interactions with users. Next, we consider a scenario where a family of three individuals shares the same home WiFi network.[10] The interaction process between the AI agent and the user in this scenario is similar to the one described in Scenario 1.

In this scenario, Eason, the user, is watching a movie, and asks the AI agent, "Can you improve my internet speed? My movie has a slight lag." When the AI agent, encounters this task for the first time, it generates an FSM based on the available APIs of the WiFi router, as depicted in Fig. 5.24b. Following the transition rules defined in this FSM, the LLM generates executable scripts to address the request.

It is important to note that in Scenario 2, we limit the total bandwidth of the shared network to 100 MB/s for demonstration purposes. Additionally, the speed limit for each user is fixed in a stepped manner, categorized as Low Rate (20 MB/s), Normal (30 MB/s), and High Rate (50 MB/s). In the first attempt, the AI agent successfully uses the network management API to elevate the user's bandwidth from the Low Rate to the Normal level without exceeding the total bandwidth of the shared network. It then informs the user about the adjustment result.

However, if Eason remains dissatisfied with the network speed, he can ask, "Can you increase my internet speed once more?" Since the script for "improving internet speed for Eason" is already stored in the long-term memory, the previously generated script is executed directly. However, in this experiment, during the execution of the script, it is determined that further increasing the network speed for Eason is not possible. This limitation arises because another user, Ada, is already in the High Rate list. Increasing the network speed for Eason again would exceed the total bandwidth limit of the shared network. Therefore, this execution is

[10] Demo video 2: https://youtu.be/GCc2cdb8nS0.

terminated, and the reason is communicated to Eason. Scenario 2 demonstrates that our framework is capable of handling tasks that involve multiple interactions with users.

5.5 Concluding Remarks

The integration of AI-enhanced network control represents a paradigm shift in IoT communication, enabling adaptive, scalable, and intelligent network management. Traditional network control mechanisms, which rely on fixed rules and centralized decision-making, are ill-equipped to handle the ever-growing complexity and scale of modern IoT environments. This chapter has explored how AI techniques such as Deep Reinforcement Learning (DRL), Graph Neural Networks (GNNs), and Large Language Models (LLMs) can transform IoT network control, ensuring autonomous, context-aware, and resilient network operations.

By examining DRL-based adaptive monitoring, we demonstrated how AI can optimize network state information collection, reducing energy consumption and balancing performance trade-offs. The use of GNNs for decentralized network control highlighted the importance of scalable and collaborative intelligence in managing large-scale IoT networks. Additionally, the introduction of LLMs in IoT orchestration emphasized the potential of natural language processing for human-machine interaction in network control.

As IoT continues to evolve, future research should focus on real-time AI inference at the edge, the development of self-optimizing network control models, and the integration of hybrid AI-driven architectures that combine centralized efficiency with decentralized intelligence. Additionally, advancements in federated learning and adaptive reinforcement learning will further enhance the robustness of AI-driven network control.

Ultimately, AI-powered IoT network management will enable seamless, energy-efficient, and resilient connectivity, paving the way for autonomous, self-learning, and highly scalable IoT ecosystems. By bridging the gap between data-driven intelligence and real-world IoT challenges, AI-driven network control is set to redefine the future of IoT communications.

References

1. Y. Shao, Q. Cao, S.C. Liew, H. Chen, Partially observable minimum-age scheduling: the greedy policy. IEEE Trans. Commun. **70**(1), 404–418 (2021)
2. G. Chen, S.C. Liew, Y. Shao, Uncertainty-of-information scheduling: a restless multiarmed bandit framework. IEEE Trans. Inf. Theory **68**(9), 6151–6173 (2022)
3. P. Shen, Y. Shao, Q. Cao, L. Lu, Dynamic gnodeb sleep control for energy-conserving radio access network. IEEE Trans. Cogn. Commun. Netw. **10**(4), 1371–1385 (2024)

4. Y. Shao, S.C. Liew, H. Chen, Y. Du, Flow sampling: network monitoring in large-scale software-defined IoT networks. IEEE Trans. Commun. **69**(9), 6120–6133 (2021)
5. V. Mnih, K. Kavukcuoglu, D. Silver, A.A. Rusu, J. Veness, M.G. Bellemare, A. Graves, M. Riedmiller, A.K. Fidjeland, G. Ostrovski, et al., Human-level control through deep reinforcement learning. Nature **518**(7540), 529 (2015)
6. Z. Wu, S. Pan, F. Chen, G. Long, C. Zhang, S.Y. Philip, A comprehensive survey on graph neural networks. IEEE Trans. Neural Netw. Learn. Syst. **32**(1), 4–24 (2020)
7. W.X. Zhao, K. Zhou, J. Li, T. Tang, X. Wang, Y. Hou, Y. Min, B. Zhang, J. Zhang, Z. Dong, et al., A survey of large language models, vol. 1, no. 2 (2023). arXiv preprint arXiv:2303.18223
8. I. Lee, K. Lee, The internet of things (IoT): applications, investments, and challenges for enterprises. Bus. Horiz. **58**(4), 431–440 (2015)
9. M. Bennis, M. Debbah, H.V. Poor, Ultra reliable and low-latency wireless communication: tail, risk, and scale. Proc. IEEE **106**(10), 1834–1853 (2018)
10. V.W. Chan, Optical flow switching networks. Proc. IEEE **100**(5), 1079–1091 (2012)
11. V.W. Chan, Cognitive optical networks, in *IEEE Int. Conf. Commun. (ICC)* (IEEE, 2018)
12. A. Rezaee, V.W. Chan, Cognitive network management and control with significantly reduced state sensing. IEEE Trans. Cogn. Commun. Netw. **5**(3), 783–794 (2019)
13. J. Case, K. McCloghrie, M. Rose, S. Waldbusser, Simple network management protocol (snmp) (1990)
14. B. Claise, G. Sadasivan, V. Valluri, M. Djernaes, Cisco systems netflow services export version 9. RFC 3954 (2004)
15. A. Tootoonchian, M. Ghobadi, Y. Ganjali, Opentm: traffic matrix estimator for openflow networks, in *Int. Conf. Passive Active Netw. Meas.* (Springer, Berlin, 2010), pp. 201–210
16. N.L. Van Adrichem, C. Doerr, F.A. Kuipers, Opennetmon: network monitoring in openflow software-defined networks, in *IEEE Netw. Operations Manag. Symp. (NOMS)* (IEEE, 2014), pp. 1–8
17. C. Yu, C. Lumezanu, A. Sharma, Q. Xu, G. Jiang, H.V. Madhyastha, Software-defined latency monitoring in data center networks, in *Int. Conf. Passive Active Netw. Meas.* (Springer, Berlin, 2015), pp. 360–372
18. M. Martínez-Rey, F. Espinosa, A. Gardel, Analysis of the optimal sampling rate for state estimation in sensor networks with delays. ISA Trans. **68**, 293–301 (2017)
19. B. Bonakdarpour, S. Navabpour, S. Fischmeister, Sampling-based runtime verification, in *Int. Symp. Formal Methods* (Springer, Berlin, 2011), pp. 88–102
20. S. Halfin, W. Whitt, Heavy-traffic limits for queues with many exponential servers. Oper. Res. **29**(3), 567–588 (1981)
21. R. Sutton, A. Barto, F. Bach, *Reinforcement Learning: An Introduction* (MIT Press, Cambridge, 2018)
22. R.S. Sutton, D.A. McAllester, S.P. Singh, Y. Mansour, Policy gradient methods for reinforcement learning with function approximation, in *Advances in Neural Information Processing Systems* (2000), pp. 1057–1063
23. T.P. Lillicrap, J.J. Hunt, A. Pritzel, N. Heess, T. Erez, Y. Tassa, D. Silver, D. Wierstra, Continuous control with deep reinforcement learning (2015). arXiv:1509.02971
24. D. Silver, G. Lever, N. Heess, T. Degris, D. Wierstra, M. Riedmiller, Deterministic policy gradient algorithms, in *Int. Conf. Mach. Learning (ICML)* (2014)
25. C.E. Shannon, Communication in the presence of noise. Proc. Inst. Radio Eng., vol. 371 (1934)
26. M. Schuster, K.K. Paliwal, Bidirectional recurrent neural networks. IEEE Trans. Signal Process. **45**(11), 2673–2681 (1997)
27. Z. Gao, Y. Shao, D. Gündüz, A. Prorok, Decentralized channel management in WLANs with graph neural networks, in *IEEE International Conference on Communications* (IEEE, 2023), pp. 3072–3077
28. Cisco, Cisco annual Internet report: 2018–2023. Cisco Executive Perspectives White paper (2020)

29. E.H. Ong, J. Kneckt, O. Alanen, Z. Chang, T. Huovinen, T. Nihtilä, IEEE 802.11ac: enhancements for very high throughput WLANs, in *IEEE International Symposium on Personal, Indoor and Mobile Radio Communications (PIMRC)* (2011)
30. L. Abdelwedoud, A. Busson, I.G. Lassous, Use of a weighted conflict graph in the channel selection operation for Wi-Fi networks, in *IEEE/IFIP Annual Conference on Wireless On-demand Network Systems and Services Conference (WONS)* (2021)
31. M. Seyedebrahimi, F. Bouhafs, A. Raschellà, M. Mackay, Q. Shi, SDN-based channel assignment algorithm for interference management in dense Wi-Fi networks, in *European Conference on Networks and Communications (EuCNC)* (2016)
32. A. Kaushik, M. Alizadeh, O. Waqar, H. Tabassum, Deep unsupervised learning for generalized assignment problems: a case-study of user-association in wireless networks, in IEEE International Conference on Communications Workshops (ICC Workshops) (2021)
33. A. Bhartia, B. Chen, F. Wang, D. Pallas, R. Musaloiu-E, T.T.T. Lai, H. Ma, Measurement-based, practical techniques to improve 802.11ac performance, in *ACM Internet Measurement Conference (IMC)* (2017)
34. O. Iacoboaiea, J. Krolikowski, Z.B. Houidi, D. Rossi, Real-time channel management in WLANs: deep reinforcement learning versus heuristics, in *IFIP Networking Conference (NETWORKING)* (2021)
35. K. Nakashima, S. Kamiya, K. Ohtsu, K. Yamamoto, T. Nishio, M. Morikura, Deep reinforcement learning-based channel allocation for wireless LANs with graph convolutional networks. IEEE Access **8**, 31823–31834 (2020)
36. Y. Shao, Y. Cai, T. Wang, Z. Guo, P. Liu, J. Luo, D. Gündüz, Learning-based autonomous channel access in the presence of hidden terminals. IEEE Trans. Mob. Comput. **23**(5), 3680–3695 (2023)
37. P. Formanowicz, K. Tanaś, A survey of graph coloring-its types, methods and applications. Found. Comput. Decis. Sci. **37**(3), 223–238 (2012)
38. A. Gavili, X. Zhang, On the shift operator, graph frequency, and optimal filtering in graph signal processing. IEEE Trans. Signal Process. **65**(23), 6303–6318 (2017)
39. M. Henaff, J. Bruna, Y. LeCun, Deep convolutional networks on graph-structured data (2015). arXiv preprint arXiv:1506.05163
40. M. Defferrard, X. Bresson, P. Vandergheynst, Convolutional neural networks on graphs with fast localized spectral filtering, in *Advances in Neural Information Processing Systems*, vol. 29 (2016)
41. Z. Gao, E. Isufi, A. Ribeiro, Stochastic graph neural networks. IEEE Trans. Signal Process. **69**, 4428–4443 (2021)
42. Q. Li, F. Gama, A. Ribeiro, A. Prorok, Graph neural networks for decentralized multi-robot path planning, in *IEEE/RSJ International Conference on Intelligent Robots and Systems (IROS)* (2020)
43. Z. Gao, F. Gama, A. Ribeiro, Wide and deep graph neural network with distributed online learning. IEEE Trans. Signal Process. **70**, 3862–3877 (2022)
44. J.B. Hansen, S.N. Anfinsen, F.M. Bianchi, Power flow balancing with decentralized graph neural networks. IEEE Trans. Power Syst. **38**(3), 2423–2433 (2022)
45. F. Gama, J. Bruna, A. Ribeiro, Stability properties of graph neural networks. IEEE Trans. Signal Process. **68**, 5680–5695 (2020)
46. Z. Gao, E. Isufi, A. Ribeiro, Stability of graph convolutional neural networks to stochastic perturbations. Signal Process. **188**, 108216 (2021)
47. W. Huang, F. Xia, T. Xiao, H. Chan, J. Liang, P. Florence, A. Zeng, J. Tompson, I. Mordatch, Y. Chebotar, et al., Inner monologue: embodied reasoning through planning with language models (2022). arXiv preprint arXiv:2207.05608
48. J. Mao, Y. Qian, H. Zhao, Y. Wang, GPT-driver: learning to drive with GPT (2023). arXiv preprint arXiv:2310.01415
49. A. Kaplan, M. Haenlein, Siri, Siri, in my hand: Who's the fairest in the land? On the interpretations, illustrations, and implications of artificial intelligence. Bus. Horiz. **62**(1), 15–25 (2019)

50. B. Goertzel, C. Pennachin, *Artificial General Intelligence*, vol. 2 (Springer, Berlin, 2007)
51. H. Cui, Y. Du, Q. Yang, Y. Shao, S.C. Liew, LLMind: orchestrating AI and IoT with LLMs for complex task execution. IEEE Commun. Mag. **63**(4), 214–220 (2024)
52. Y. Du, H. Deng, S.C. Liew, K. Chen, Y. Shao, H. Chen, The power of large language models for wireless communication system development: a case study on FPGA platforms, in *IEEE Vehicular Technology Conference* (2023)
53. M. Asay, Are large language models wrong for coding? (2023)
54. P. Gladyshev, Finite state machine approach to digital event reconstruction. Digital Investig. **1**, 130–149 (2004)
55. Y. Xia, M. Shenoy, N. Jazdi, M. Weyrich, Towards autonomous system: flexible modular production system enhanced with large language model agents (2023). arXiv preprint arXiv:2304.14721
56. W. Huang, C. Wang, R. Zhang, Y. Li, J. Wu, L. Fei-Fei, Voxposer: composable 3D value maps for robotic manipulation with language models (2023). arXiv preprint arXiv:2307.05973
57. Y. Shen, K. Song, X. Tan, D. Li, W. Lu, Y. Zhuang, HuggingGPT: solving AI tasks with ChatGPT and its friends in HuggingFace (2023). arXiv preprint arXiv:2303.17580
58. C.H. Song, J. Wu, C. Washington, B.M. Sadler, W.-L. Chao, Y. Su, LLM-planner: few-shot grounded planning for embodied agents with large language models, in *Proceedings of the IEEE/CVF International Conference on Computer Vision* (2023), pp. 2998–3009
59. L. Wang, C. Ma, X. Feng, Z. Zhang, H. Yang, J. Zhang, Z. Chen, J. Tang, X. Chen, Y. Lin, et al., A survey on large language model based autonomous agents (2023). arXiv preprint arXiv:2308.11432
60. A. Official, AutoGPT official (2023)
61. Jina.ai, Auto-GPT unmasked: The hype and hard truths of its production pitfalls
62. G. Jocher, YOLOv8 documentation (2020)
63. ageitgey, ageitgey/face_recognition (2019)

Chapter 6
Future Trends and Challenges

The future of AI in IoT is ours to shape – wisely, ethically, and inclusively.

6.1 Emerging Applications of AI in IoT

Following the detailed discussions of AI-enhanced IoT communication in the preceding chapters, it becomes clear that advanced techniques – ranging from real-time feedback-enhanced transmissions (Chap. 2) to semantic-level data exchange (Chap. 3), multi-agent coordination of resources (Chap. 4), and holistic network control (Chap. 5) – are unlocking new, transformative applications across diverse industries. These methods serve as the backbone of a new wave of intelligent, data-driven solutions that can respond to real-world conditions in milliseconds, automatically optimize performance, and integrate seamlessly with broader ecosystems of connected devices [1, 2].

A prominent example of this convergence is found in the domain of deep and wide coverage IoT. By drawing on the real-time feedback introduced in Chap. 2, these systems can continuously adjust coding schemes to match the changing decoding status of the receiver. In the application of collaborative robotics and autonomous drones [3], when combined with semantic communication concepts from Chap. 3, devices can exchange only the most relevant contextual information, minimizing bandwidth usage and latency. This becomes crucial when multiple autonomous agents must coordinate actions, avoid collisions, and handle sudden obstacles in real time. Equipping drones or robots with the multi-agent resource allocation capabilities discussed in Chap. 4 further refines their ability to share scarce wireless spectrum and computational resources, mitigating interference and ensuring stable connections. Through the network control frameworks described in Chap. 5, fleets of robots or drones can be managed holistically, allowing tasks to be dynamically assigned or rerouted based on real-time sensor inputs and system priorities [4].

The healthcare sector also stands to benefit immensely from these advancements [5]. Wearable IoT devices and hospital-wide sensor networks have already shown

Y. Shao, *AI-Empowered IoT Communications*, Wireless Networks,
https://doi.org/10.1007/978-3-031-98816-5_6

promise for remote patient monitoring and predictive diagnostics. Yet, when augmented by AI techniques outlined in earlier chapters, especially semantic communication (Chap. 3) and flexible network control (Chap. 5), these devices can refine their transmissions to highlight critical signals or anomalies, ensuring that clinical staff receives actionable data promptly. At the same time, multiple access optimization (Chap. 4) helps allocate wireless resources so that high-priority data, such as alerts from at-risk patients, travels swiftly through congested networks. Over time, such AI-empowered IoT solutions have the potential to shift healthcare from reactive treatment to proactive and preventive care [6], substantially improving patient outcomes and hospital resource management.

In industrial environments, the principles set forth in previous chapters come together to deliver unprecedented levels of efficiency and agility [7]. Real-time feedback-enhanced transmissions (Chap. 2) ensure robust connectivity in remote conditions, while semantic communication (Chap. 3) alleviates communication overhead for massive sensor arrays monitoring everything from temperature fluctuations to mechanical vibrations. When entire production lines or supply chains must coordinate, AI-driven multiple access solutions (Chap. 4) and advanced network control methods (Chap. 5) can orchestrate hundreds or thousands of devices, assigning resources automatically to match current needs. The result is an industry capable of rapid reconfiguration, where processes are dynamically tuned to reduce downtime, energy use, and operational costs. In some cases, factories can move toward "lights-out" operations, running autonomously for extended periods with minimal human intervention [8].

Elsewhere, environmental and civic applications illustrate how AI transforms IoT from a simple connectivity framework into an intelligent infrastructure for sustainable development [9]. Sensor networks measuring soil conditions or crop health can leverage semantic communication and adaptive feedback loops to deliver data-driven insights with high precision and low energy consumption. AI-based control mechanisms then adjust irrigation schedules, pesticide usage, or harvesting timelines in real time, improving yields while conserving resources. In smart cities, the same suite of AI for IoT techniques introduced throughout Chaps. 2 to 5 can be harnessed for traffic management, pollution monitoring, and energy optimization, enabling urban systems to anticipate congestion, respond to outages, and balance electrical loads proactively [10].

Collectively, these scenarios underscore the enormous potential of AI-enhanced IoT architectures to reshape industries, services, and daily life. Rather than merely connecting more devices to the internet, the focus has shifted toward embedding intelligence at every layer, from sensor nodes capable of on-device inference, to gateways that distill essential information, to cloud platforms that orchestrate vast networks of autonomous agents. As we explore the remaining sections of this final chapter, it is equally important to recognize that these emerging applications also pose significant challenges in terms of scalability, ethical guidelines, and social impact – topics to which we now turn in the pages ahead.

6.2 Challenges in Scalability and Management

As AI-enabled IoT systems continue to expand in both scope and complexity, organizations face an array of difficulties in coordinating the ever-growing number of connected devices and ensuring reliable, efficient performance. These challenges become particularly pronounced when emerging applications, such as those discussed in Sect. 6.1, scale from pilot projects to large-scale commercial deployments. While the communication strategies outlined in Chaps. 2 through 5 provide powerful tools for intelligent resource allocation, semantic data exchange, and network coordination, practitioners must still contend with operational bottlenecks, performance degradation, and evolving management requirements. Addressing these issues demands not only technical innovations but also organizational and strategic foresight that can adapt to rapidly changing conditions.

One immediate issue is the exponential increase in data generation that occurs when hundreds, thousands, or even millions of IoT devices come online simultaneously. Although semantic communication (Chap. 3) can help manage bandwidth and limit unnecessary transmissions, the aggregate data load may still grow at a pace that outstrips available network or computational capacity. This surge in traffic puts pressure on gateways, edge servers, and cloud platforms, leading to potential congestion or latency spikes that undermine the real-time decision-making capabilities of AI models. As a result, system designers must balance the fine-grained optimizations discussed in earlier chapters with top-level network planning that anticipates both current and future demands, often involving dynamic network slicing, scalable cloud-edge infrastructures, and periodic reconfiguration of the network topology [11].

Compounding these issues, many AI-empowered IoT deployments operate in heterogeneous environments where devices differ widely in terms of computational power, memory, battery life, and communication protocols. Heterogeneity complicates not only how the devices connect – whether via Wi-Fi, LoRaWAN, NB-IoT, or other standards – but also how tasks are distributed among them. Techniques such as multi-agent deep reinforcement learning (Chap. 4) can coordinate resource allocation across diverse endpoints, yet it remains non-trivial to ensure each device executes the correct AI algorithms and maintains consistent local models, especially when frequent updates or model retraining are required. When some devices become overloaded or fall offline, the system must adapt without compromising overall functionality. These dynamic reconfigurations are a hallmark of AI-enhanced network control (Chap. 5), but the complexity escalates substantially with scale, necessitating robust fault tolerance and resilient orchestration mechanisms.

Security and privacy considerations also intensify with larger deployments [12]. While encryption, authentication, and secure boot processes can protect individual devices, the cumulative attack surface expands as the number of connected nodes increases. Coordinated attacks or malicious exploits pose greater risks, whether through denial-of-service campaigns or the introduction of misleading data that confuses AI algorithms. Managing security certificates and access controls for each node becomes a formidable task, especially when devices are widely dispersed and possibly located in harsh environments with limited physical security. In such cases, even straightforward management procedures, like distributing security patches or updating firmware, can represent a complex logistical undertaking. The synergy between AI and IoT magnifies these stakes, since compromised data flows may directly affect how critical systems are controlled and adjusted in real time.

Beyond technical considerations, the organizational aspects of scaling up AI-empowered IoT solutions can be equally challenging. Enterprises may need to retrain staff and cultivate specialized skill sets in data science, networking, and systems engineering. Decision-makers often confront the difficulty of integrating legacy systems with new AI-driven capabilities, a process that can require large investments in both technology and personnel. Moreover, as IoT networks expand globally, compliance with diverse regulations – ranging from data sovereignty laws to environmental standards – adds layers of complexity. This combination of technical, human, and regulatory factors underscores the need for well-designed governance models that can sustain the growth of AI-infused IoT ecosystems while preserving the reliability, security, and ethical standards outlined in earlier chapters.

Despite these obstacles, the prospects for effectively scaling AI-empowered IoT remain strong. The approaches to semantic and feedback-based communication from Chaps. 2 and 3, when coupled with multi-agent coordination (Chap. 4) and robust network control strategies (Chap. 5), provide a solid foundation for building intelligent systems capable of automatically adapting to fluctuating conditions. In many instances, adopting a hierarchical or federated structure can help distribute the computational load, with edge devices handling time-sensitive tasks and the cloud taking responsibility for more complex analyses. As the next section will explore, however, this scaling process is intertwined with broader ethical and social considerations – an increasingly pressing dimension of AI and IoT deployments that requires careful planning and transparent stakeholder engagement.

6.3 Ethical and Social Implications

As AI-enhanced IoT systems expand in complexity and scale, they inevitably intersect with broader ethical and social considerations that extend beyond purely technical challenges. The automation and connectivity strategies described throughout this book hold immense promise for optimizing processes, conserving resources, and improving quality of life. Yet the same technologies, when deployed at large scale, can also carry risks that impact individuals, communities, and society at large. Addressing these risks in a transparent, equitable, and forward-looking manner is therefore essential to fostering trust and ensuring that AI and IoT remain forces for positive transformation rather than sources of unintended harm.

One primary concern is the potential erosion of privacy [12]. As billions of devices capture environmental data, along with personal details about our health, behavior, and interactions, organizations amass unprecedented amounts of information. While techniques such as semantic communication can help reduce redundant or non-essential data transmissions, it does not fully resolve questions about how much data is necessary or appropriate to collect. Furthermore, the integration of AI can blur the line between raw sensor readings and derived insights, allowing inferences about an individual's preferences or vulnerabilities in ways that may not have been intended. This asymmetry between data collectors and data subjects can undermine individual autonomy, especially when users lack clear visibility into how their data is processed or used to shape automated decisions.

Bias in AI algorithms poses another ethical challenge [13], particularly for large-scale IoT networks that influence public services, healthcare, or critical infrastructure. If the training data or the model design inadvertently embeds biased assumptions, the resulting predictions and actions may systematically disadvantage certain groups. For example, an AI-enabled traffic control system might route congestion away from affluent areas at the expense of lower-income neighborhoods, or a predictive healthcare model might yield less accurate results for historically underrepresented populations. The risk of amplifying social inequalities grows as AI is woven into the underlying fabric of IoT communication, since negative feedback loops can silently perpetuate unfair outcomes over time. Counteracting such tendencies demands careful model validation, diverse and representative data sets, and ongoing monitoring to identify disparities in performance.

Security vulnerabilities, already a major topic in IoT, take on new gravity when AI systems are involved. A compromised network that influences AI-based decisions could trigger far-reaching consequences, from the malfunction of industrial processes to the manipulation of environmental control systems in smart cities. Even more subtly, adversaries might inject tainted data that skews an AI model's understanding of reality or systematically degrade semantic communication by altering key contextual information. As with biases, the large-scale deployment of AI-empowered IoT heightens the likelihood and potential damage of security breaches, making robust encryption, identity management, and anomaly detection indispensable components of a broader governance framework.

Beyond these immediate concerns, the rise of AI in IoT has broader socioeconomic and cultural ramifications [14]. On one hand, AI-empowered IoT applications can create substantial efficiencies that drive productivity, reduce operational costs, and spur new industries, generating employment opportunities in data science, IoT engineering, and maintenance. On the other hand, the increased automation of tasks may displace certain segments of the workforce, particularly in roles that involve repetitive or easily digitized processes. While it is beyond the scope of this volume to prescribe specific policy interventions, it is crucial to recognize that the adoption of AI-enhanced IoT solutions can deepen societal divides if not accompanied by inclusive economic planning, worker retraining, and social safety nets.

Finally, the ethical challenges of AI-empowered IoT also speak to broader questions of accountability and governance. As devices and AI models gain the capacity to make autonomous decisions, attributing responsibility for errors, failures, or harmful outcomes becomes more complex. Traditional regulatory frameworks may not be equipped to handle the shared accountability between device manufacturers, network operators, AI developers, and end users. In this context, decision-makers must grapple with how best to define transparent guidelines, oversight mechanisms, and liability structures that protect public welfare without stifling innovation. This balancing act is further complicated by global variations in legal standards, cultural norms, and public expectations about data ownership and personal rights.

Collectively, these ethical and social implications underscore the multifaceted nature of AI-empowered IoT. While it holds tremendous potential to benefit humanity – improving efficiency, enhancing health outcomes, reducing resource consumption, and unlocking new forms of creativity – its success depends on careful design, responsible data stewardship, and meaningful engagement with stakeholders. The next section will expand upon how organizations and individuals can prepare for the future of AI-empowered IoT, bridging the gap between technological capabilities and the need for sustainable, ethical, and resilient deployments.

6.4 Preparing for the Future

The rapid evolution of AI-powered IoT, as explored throughout this book, makes it clear that the innovations we witness today are only a precursor to more complex, interconnected systems poised to transform industries and daily life. From adaptive feedback loops and semantic communication (Chaps. 2 and 3) to multi-agent coordination and advanced network control (Chaps. 4 and 5), recent advances offer glimpses of an IoT ecosystem that learns, adapts, and collaborates in ways that were scarcely imaginable just a decade ago. Yet the drive toward automation, scalability, and pervasive intelligence also reveals significant organizational and societal challenges (Sects. 6.2 and 6.3) that must be confronted for these technologies to serve the common good. In preparing for the future of AI in IoT, individuals, institutions, and communities must actively shape the course of development rather than simply reacting to it.

One of the most immediate priorities lies in creating frameworks that support sustainable growth and governance. As IoT deployments scale, so too must the architectural principles and management strategies that govern them. In particular, it will be vital to design systems that can seamlessly accommodate an ever-increasing number of heterogeneous devices while preserving stability, security, and real-time responsiveness. Emerging standards and protocols will continue to play a central role, offering developers and vendors common ground on which to build interoperable solutions. When combined with robust network slicing and intelligent resource allocation, these standards can enable flexible, scalable IoT deployments that adapt to changing requirements without compromising on performance or reliability.

Parallel to these technological developments, organizations must develop human capital capable of steering AI and IoT innovation responsibly. This includes not only data scientists and engineers but also ethicists, social scientists, and legal experts who can foresee and address the broader societal impacts outlined in Sect. 6.3.

Fostering interdisciplinary collaboration can help bridge technical expertise with nuanced understanding of user needs, ethical considerations, and policy constraints. By integrating these perspectives at an early stage – whether through dedicated task forces, university-industry partnerships, or cross-functional project teams – stakeholders can guide AI-empowered IoT solutions toward outcomes that reflect shared values and inclusive growth [15].

Equally essential is the commitment to continual learning and innovation. The AI algorithms and communication protocols described in this book are, by their nature, iterative. Deep learning architectures evolve swiftly, and communication technologies undergo frequent revisions in response to new demands and emerging use cases. Consequently, organizations aspiring to thrive in this environment must embrace a culture of experimentation, rapid prototyping, and ongoing education. In such a landscape, competitiveness is tied to the ability to pivot quickly, adopt cutting-edge methods, and integrate novel insights from a global network of research institutions, startups, and open-source communities.

A significant part of preparing for the future also involves acknowledging potential disruptions and proactively building resilience. Economic factors, unforeseen technological breakthroughs, cybersecurity threats, and shifts in regulatory frameworks can all reshape the terrain on which AI-empowered IoT operates. Investing in redundancy, robust security measures, and contingency planning mitigates the risks posed by these uncertainties. Mechanisms such as distributed ledger technologies, zero-trust architectures, and decentralized edge computing platforms can limit the impact of localized failures or targeted cyberattacks [16]. Likewise, careful monitoring of regulatory trends, including data protection laws and emerging guidelines for ethical AI, helps organizations remain nimble and compliant as policies continue to evolve.

Ultimately, the trajectory of AI in IoT will hinge not just on technical prowess but on a collective willingness to engage with the moral, economic, and social dimensions that accompany rapid innovation. Those who embrace this complexity, by designing scalable infrastructures, nurturing diverse talent, and balancing the interests of multiple stakeholders, will be well-positioned to guide AI-empowered IoT toward solutions that are both transformative and equitable. In doing so, they will help ensure that this technology, with all its capacity for reshaping industries and daily life, remains firmly rooted in human values and broader societal goals.

With these considerations in mind, this final section offers a call to action: to view AI-enhanced IoT not merely as a set of advanced tools but as a catalyst for smarter, more empathetic ways of living and working. By uniting technical innovation with robust ethical frameworks and a commitment to collaboration, we can harness the extraordinary promise of AI-empowered IoT to address global challenges, foster economic growth, and improve individual well-being. The future, in this sense, is a shared project – one that demands ongoing dialogue, experimentation, and a willingness to evolve as new insights and responsibilities emerge.

References

1. D.C. Nguyen, M. Ding, P.N. Pathirana, A. Seneviratne, J. Li, D. Niyato, O. Dobre, H.V. Poor, 6G Internet of Things: A comprehensive survey. IEEE Internet Things J. **9**(1), 359–383 (2021)
2. M. Marjani, F. Nasaruddin, A. Gani, A. Karim, I.A. T. Hashem, A. Siddiqa, I. Yaqoob, Big IoT data analytics: architecture, opportunities, and open research challenges. IEEE Access **5**, 5247–5261 (2017)
3. F. Vicentini, Collaborative robotics: a survey. J. Mech. Des. **143**(4), 040802 (2021)
4. R. Galin, R. Meshcheryakov, Automation and robotics in the context of Industry 4.0: the shift to collaborative robots, in *IOP Conference Series: Materials Science and Engineering*, vol. 537 (IOP Publishing, Bristol, 2019), p. 032073
5. M.H. Kashani, M. Madanipour, M. Nikravan, P. Asghari, E. Mahdipour, A systematic review of IoT in healthcare: Applications, techniques, and trends. J. Netw. Comput. Appl. **192**, 103164 (2021)
6. R. Shah, A. Chircu, IoT and AI in healthcare: A systematic literature review. Issues Inf. Syst. **19**(3), 33–41 (2018)
7. P. Bhambri, S. Rani, Challenges, opportunities, and the future of industrial engineering with IoT and AI, in *Integration of AI-Based Manufacturing and Industrial Engineering Systems with the Internet of Things* (2024), pp. 1–18
8. B. Chander, S. Pal, D. De, R. Buyya, Artificial intelligence-based Internet of Things for Industry 5.0, in *Artificial Intelligence-based Internet of Things Systems* (2022), pp. 3–45
9. M.W. Khan, S. Saad, S. Ammad, K. Rasheed, Q. Jamal, Smart infrastructure and AI, in *AI in Material Science* (CRC Press, Boca Raton, 2024), pp. 193–215
10. M.E.E. Alahi, A. Sukkuea, F.W. Tina, A. Nag, W. Kurdthongmee, K. Suwannarat, S.C. Mukhopadhyay, Integration of IoT-enabled technologies and AI for smart city scenario: recent advancements and future trends. Sensors **23**(11), 5206 (2023)
11. S. Duan, D. Wang, J. Ren, F. Lyu, Y. Zhang, H. Wu, X. Shen, Distributed artificial intelligence empowered by end-edge-cloud computing: A survey. IEEE Commun. Surv. Tutorials **25**(1), 591–624 (2022)
12. O.E.L. Castro, X. Deng, J.H. Park, Comprehensive survey on AI-based technologies for enhancing IoT privacy and security: Trends, challenges, and solutions. Human-centric Comput. Inf. Sci. **13**(1), 1–34 (2023)
13. H. Bangui, B. Buhnova, M. Ge, Social Internet of Things: ethical AI principles in trust management. Procedia Comput. Sci. **220**, 553–560 (2023)
14. L. Sineviciene, L. Hens, O. Kubatko, L. Melnyk, I. Dehtyarova, S. Fedyna, Socio-economic and cultural effects of disruptive industrial technologies for sustainable development. Int. J. Glob. Energy Issues **43**(2–3), 284–305 (2021)
15. L. Xiao, X. Wan, X. Lu, Y. Zhang, D. Wu, IoT security techniques based on machine learning: How do IoT devices use AI to enhance security? IEEE Signal Process. Mag. **35**(5), 41–49 (2018)
16. M. Abdullahi, Y. Baashar, H. Alhussian, A. Alwadain, N. Aziz, L.F. Capretz, S.J. Abdulkadir, Detecting cybersecurity attacks in Internet of Things using artificial intelligence methods: A systematic literature review. Electronics **11**(2), 198 (2022)

Appendix A
Semantic Encoding, Decoding, and Combined Semantic Encoding and Decoding

This appendix provides supplementary materials for Chap. 3.

A.1 Proof of Proposition 3.1

We first prove the "if" part: for any stochastic encoding scheme $\boldsymbol{U}$, there are a series of deterministic encoding schemes $\boldsymbol{U}_1, \boldsymbol{U}_2, \ldots$ such that

$$L_{\boldsymbol{U}} = \sum_i \lambda_i L_{\boldsymbol{U}_i}, \quad D_{\boldsymbol{U},\boldsymbol{Q}} = \sum_i \lambda_i D_{\boldsymbol{U}_i,\boldsymbol{Q}},$$

where $\lambda_i \in [0, 1]$ and $\sum_i \lambda_i = 1$. We rewrite $\boldsymbol{U}$ as

$$\boldsymbol{U} = \sum_{i_1,i_2,\ldots,i_N} u(s_{i_1}|w_1)u(s_{i_2}|w_2)\ldots u(s_{i_N}|w_N)\boldsymbol{\Delta}_{i_1,i_2,\ldots,i_N}.$$

Let $\delta_{i_1,i_2,\ldots,i_N}(s_m|w_n)$ be the element of $\boldsymbol{\Delta}_{i_1,i_2,\ldots,i_N}$ on the n-th row and m-th column. We have

$$\begin{aligned} D_{\boldsymbol{U},\boldsymbol{Q}} &= \sum_n p(w_n) \sum_{\hat{w}} d(w_n, \hat{w}) \sum_m u(s_m|w_n) q(\hat{w}|s_m) \\ &= \sum_{i_1,i_2,\ldots,i_N} u(s_{i_1}|w_1)u(s_{i_2}|w_2)\ldots u(s_{i_N}|w_N) \\ &\quad \times \sum_n p(w_n) \sum_{\hat{w}} d(w_n, \hat{w}) \sum_m \delta_{i_1,i_2,\ldots,i_N}(s_m|w_n) q(\hat{w}|s_m) \end{aligned}$$

Y. Shao, *AI-Empowered IoT Communications*, Wireless Networks,
https://doi.org/10.1007/978-3-031-98816-5

$$= \sum_{i_1,i_2,\ldots,i_N} u(s_{i_1}|w_1)u(s_{i_2}|w_2)\ldots u(s_{i_N}|w_N)D_{\mathbf{\Delta}_{i_1,i_2,\ldots,i_N},\mathbf{Q}}.$$

and

$$\begin{aligned} L_U &= \sum_n p(w_n)\sum_m u(s_m|w_n)\ell(s_m) \\ &= \sum_{i_1,i_2,\ldots,i_N} u(s_{i_1}|w_1)u(s_{i_2}|w_2)\ldots u(s_{i_N}|w_N)\sum_n p(w_n)\sum_m \delta_{i_1,i_2,\ldots,i_N}(s_m|w_n)\ell(s_m) \\ &= \sum_{i_1,i_2,\ldots,i_N} u(s_{i_1}|w_1)u(s_{i_2}|w_2)\ldots u(s_{i_N}|w_N)L_{\mathbf{\Delta}_{i_1,i_2,\ldots,i_N}}. \end{aligned}$$

Thus, the distortion-cost pair $(L_U, D_{U,Q})$ can be achieved by deterministic encoding schemes in $\left\{\mathbf{\Delta}_{i_1,i_2,\ldots,i_N} : \forall i_1, i_2, \ldots, i_N \in [M]\right\}$ via time sharing.

Now we prove the "only if" part. Let

$$\mathbf{A} = \{\mathbf{\Delta}_1, \mathbf{\Delta}_2, \ldots, \mathbf{\Delta}_{|\mathbf{A}|}\} \subseteq \{\mathbf{\Delta}_{i_1,i_2,\ldots,i_N} : \forall i_1, i_2, \ldots, i_N \in [M]]\}$$

be an arbitrary but fixed set. We define a distribution $(\lambda_1, \lambda_2, \ldots, \lambda_{|\mathbf{A}|})$, where $\lambda_i \in [0, 1]$ and $\sum_{i=1}^{|\mathbf{A}|} \lambda_i = 1$. Denote

$$L \triangleq \sum_{i=1}^{|\mathbf{A}|} \lambda_i L_{\mathbf{\Delta}_i}, \quad D \triangleq \sum_{i=1}^{|\mathbf{A}|} \lambda_i D_{\mathbf{\Delta}_i,\mathbf{Q}}.$$

Now we construct a stochastic encoding scheme $\mathbf{U}$ such that $L = L_U$ and $D = D_{U,Q}$. By letting

$$u(s|w) \triangleq \sum_{i=1}^{|\mathbf{A}|} \lambda_i \mathbf{\Delta}_i(s|w), \tag{A.1}$$

for any w, s, we have $u(s|w) \geq 0$ and

$$\sum_s u(s|w) = \sum_s \sum_{i=1}^{|\mathbf{A}|} \lambda_i \mathbf{\Delta}_i(s|w) = \sum_{i=1}^{|\mathbf{A}|} \lambda_i \sum_s \mathbf{\Delta}_i(s|w) = \sum_{i=1}^{|\mathbf{A}|} \lambda_i \times 1 = 1.$$

Therefore, $\mathbf{U}$ is a legitimate encoding scheme. Further,

$$L_U = \sum_n p(w_n) \sum_m u(s_m|w_n)\ell(s_m) = \sum_n p(w_n) \sum_m \sum_{i=1}^{|\boldsymbol{A}|} \lambda_i \boldsymbol{\Delta}_i(s|w)\ell(s_m)$$

$$= \sum_{i=1}^{|\boldsymbol{A}|} \lambda_i \sum_n p(w_n) \sum_m \boldsymbol{\Delta}_i(s|w)\ell(s_m) = \sum_{i=1}^{|\boldsymbol{A}|} \lambda_i L_{\boldsymbol{\Delta}_i} = L,$$

and

$$D_{U,Q} = \sum_n p(w_n) \sum_{\hat{w}} d(w_n, \hat{w}) \sum_m u(s_m|w_n) q(\hat{w}|s_m)$$

$$= \sum_n p(w_n) \sum_{\hat{w}} d(w_n, \hat{w}) \sum_m \sum_{i=1}^{|\boldsymbol{A}|} \lambda_i \boldsymbol{\Delta}_i(s|w) q(\hat{w}|s_m)$$

$$= \sum_{i=1}^{|\boldsymbol{A}|} \lambda_i \sum_n p(w_n) \sum_{\hat{w}} d(w_n, \hat{w}) \sum_m \boldsymbol{\Delta}_i(s|w) q(\hat{w}|s_m)$$

$$= \sum_{i=1}^{|\boldsymbol{A}|} \lambda_i D_{\boldsymbol{\Delta}_i, Q} = D.$$

The proposition is proved.

A.2 Properties of the Six Subsets in Definition 3.18

This appendix summarizes some properties of the six subsets in Definition 3.18.

Proposition A.1 (Orders of $\underline{S}'_w$, $\underline{S}''_w$ and $\overline{S}'_w$, $\overline{S}''_w$) *For any meaning $w \in \mathcal{W}$, the sets $\underline{S}'_w$, $\underline{S}''_w$ and $\overline{S}'_w$, $\overline{S}''_w$ satisfy*

1. *$\forall\, s_{m_1}, s_{m_2} \in \underline{S}'_w$, $m_1 < m_2$, we have*

$$\ell\left(s_{m_1}\right) < \ell\left(s_{m_2}\right), \varphi\left(w, s_{m_1}\right) > \varphi\left(w, s_{m_2}\right).$$

2. *$\forall\, s_{m_1}, s_{m_2} \in \underline{S}''_w$, $m_1 < m_2$, we have*

$$\ell\left(s_{m_1}\right) < \ell\left(s_{m_2}\right), \varphi\left(w, s_{m_1}\right) < \varphi\left(w, s_{m_2}\right).$$

3. *$\forall\, s_{m_1}, s_{m_2} \in \overline{S}'_w$, $m_1 < m_2$, we have*

$$\ell\left(s_{m_1}\right) < \ell\left(s_{m_2}\right), \varphi\left(w, s_{m_1}\right) < \varphi\left(w, s_{m_2}\right).$$

4. $\forall\, s_{m_1}, s_{m_2} \in \overline{S}''_w$, $m_1 < m_2$, *we have*

$$\ell\left(s_{m_1}\right) < \ell\left(s_{m_2}\right), \varphi\left(w, s_{m_1}\right) > \varphi\left(w, s_{m_2}\right).$$

Proof We prove (1) by contradiction. For a give $w \in \mathcal{W}$, suppose that there exist $s_{m_1}, s_{m_2} \in \underline{S}'_w$, $m_1 < m_2$ such that one of the following two conditions holds.

- $\varphi\left(w, s_{m_1}\right) \le \varphi\left(w, s_{m_2}\right)$.
- $\ell\left(s\right) \ge \ell\left(s'\right)$, $\varphi\left(w, s_{m_1}\right) > \varphi\left(w, s_{m_2}\right)$.

Note that $m_1 < m_2$ implies that $\ell\left(s_{m_1}\right) \le \ell\left(s_{m_2}\right)$. The two conditions above are respectively equivalent to the two conditions below.

1. $\ell\left(s\right) \le \ell\left(s'\right), \varphi\left(w, s_{m_1}\right) \le \varphi\left(w, s_{m_2}\right)$.
2. $\ell\left(s_{m_1}\right) = \ell\left(s_{m_2}\right), \varphi\left(w, s_{m_1}\right) > \varphi\left(w, s_{m_2}\right)$.

If (a) holds, s_{m_2} should be removed from $\underline{S}'_w$. If (b) holds, s_{m_1} should be removed from $\underline{S}'_w$. Therefore, S'_w is not the smallest subset of S satisfying the conditions in Definition 4.4. The proofs of (2) to (4) are similar to (1) and are omitted. □

Proposition A.2 *For a given* w, *and the associated* $\underline{S}'_w$, $\underline{S}''_w$ *and* $\overline{S}'_w$, $\overline{S}''_w$. *Define*

$$\underline{m}_{w,1} = \min\left\{m : s_m \in \underline{S}'_w\right\}, \quad \underline{m}_{w,2} = \max\left\{m : s_m \in \underline{S}'_w\right\},$$
$$\underline{m}_{w,3} = \min\left\{m : s_m \in \underline{S}''_w\right\}, \quad \underline{m}_{w,4} = \max\left\{m : s_m \in \underline{S}''_w\right\},$$
$$\overline{m}_{w,1} = \min\left\{m : s_m \in \overline{S}'_w\right\}, \quad \overline{m}_{w,2} = \max\left\{m : s_m \in \overline{S}'_w\right\},$$
$$\overline{m}_{w,3} = \min\left\{m : s_m \in \overline{S}''_w\right\}, \quad \overline{m}_{w,4} = \max\left\{m : s_m \in \overline{S}''_w\right\}.$$

It holds that

1. $\ell\left(s_{\underline{m}_{w,1}}\right) = L_{\min}$ *and* $\varphi\left(w, s_{\underline{m}_{w,1}}\right) = \min_{s':s'\in S, \ell(s')=L_{\min}} \varphi(w, s')$.
2. $\varphi\left(w, s_{\underline{m}_{w,2}}\right) = \min_{s'\in S} \varphi\left(w, s'\right)$ *and*
$$\ell\left(s_{\underline{m}_{w,2}}\right) = \min_{s':s'\in S, \varphi(w,s')=\varphi\left(w, s_{\underline{m}_{w,2}}\right)} \ell\left(s'\right).$$
3. $\varphi\left(w, s_{\underline{m}_{w,3}}\right) = \min_{s'\in S} \varphi\left(w, s'\right)$ *and*
$$\ell\left(s_{\underline{m}_{w,3}}\right) = \max_{s':s'\in S, \varphi(w,s')=\varphi\left(w, s_{\underline{m}_{w,3}}\right)} \ell\left(s'\right).$$
4. $\ell\left(s_{\underline{m}_{w,4}}\right) = L_{\max}$ *and* $\varphi\left(w, s_{\underline{m}_{w,4}}\right) = \min_{s':s'\in S, \ell(s')=L_{\max}} \varphi(w, s')$.

5. $\ell\left(s_{\overline{m}_{w,1}}\right) = L_{\min}$ *and* $\varphi\left(w, s_{\overline{m}_{w,1}}\right) = \max_{s':s'\in\mathcal{S},\ell(s')=L_{\min}} \varphi(w, s')$.
6. $\varphi\left(w, s_{\overline{m}_{w,2}}\right) = \max_{s'\in\mathcal{S}} \varphi\left(w, s'\right)$ *and*

$$\ell\left(s_{\overline{m}_{w,2}}\right) = \min_{s':s'\in\mathcal{S},\varphi(w,s')=\varphi\left(w,s_{\overline{m}_{w,2}}\right)} \ell\left(s'\right).$$

7. $\varphi\left(w, s_{\overline{m}_{w,3}}\right) = \max_{s'\in\mathcal{S}} \varphi\left(w, s'\right)$ *and*

$$\ell\left(s_{\overline{m}_{w,3}}\right) = \max_{s':s'\in\mathcal{S},\varphi(w,s')=\varphi\left(w,s_{\overline{m}_{w,3}}\right)} \ell\left(s'\right).$$

8. $\ell\left(s_{\overline{m}_{w,4}}\right) = L_{\max}$ *and* $\varphi\left(w, s_{\overline{m}_{w,4}}\right) = \max_{s':s'\in\mathcal{S},\ell(s')=L_{\max}} \varphi(w, s')$.

Proof The proofs of the eight conditions are similar and thus we only prove (1) in the following. We first prove that $\ell\left(s_{\underline{m}_{w,1}}\right) = L_{\min}$. It is sufficient to prove that

$$\min\left\{\ell\left(m\right) : s_m \in \underline{\mathcal{S}}'_w\right\} = L_{\min}.$$

Let $\hat{\mathcal{S}}_w \triangleq \left\{\hat{s} \in \mathcal{S} : \ell\left(\hat{s}\right) = L_{\min}\right\}$. Suppose $\hat{\mathcal{S}}_w \cap \underline{\mathcal{S}}'_w = \emptyset$. By Definition 3.18, for any $s' \in \hat{\mathcal{S}}_w \subseteq \mathcal{S} \setminus \underline{\mathcal{S}}'_w$, there exists an $s \in \mathcal{S}'_w$ such that

$$\ell\left(s\right) \leq \ell\left(s'\right), \varphi\left(w, s\right) \leq \varphi\left(w, s'\right).$$

Thus, $L_{\min} \leq \ell\left(s\right) \leq \ell\left(s'\right) = L_{\min}$ which implies that $s \in \hat{\mathcal{S}}_w$. Thus, $\hat{\mathcal{S}}_w \cap \underline{\mathcal{S}}'_w \neq \emptyset$. On the other hand, $s \in \underline{\mathcal{S}}'_w$, i.e., $s \in \hat{\mathcal{S}}_w \cap \underline{\mathcal{S}}'_w \subseteq \hat{\mathcal{S}}_w \cap \left(\mathcal{S} \setminus \underline{\mathcal{S}}'_w\right) = \emptyset$, which does not hold. Therefore,

$$\ell\left(s_{\underline{m}_{w,1}}\right) = \min\left\{\ell\left(m\right) : s_m \in \underline{\mathcal{S}}'_w\right\} = L_{\min}.$$

Now we prove that $\varphi\left(w, s_{\underline{m}_{w,1}}\right) = \min_{s':s'\in\mathcal{S},\ell(s')=L_{\min}} \varphi(w, s')$. Proposition A.1 shows that for any $s_{m_1}, s_{m_2} \in \underline{\mathcal{S}}'_w$, $m_1 < m_2$, we have $\ell\left(s_{m_1}\right) < \ell\left(s_{m_2}\right)$. Thus, $\left|\hat{\mathcal{S}}_w \cap \underline{\mathcal{S}}'_w\right| = 1$, i.e., $\hat{\mathcal{S}}_w \cap \underline{\mathcal{S}}'_w = \left\{s_{\underline{m}_{w,1}}\right\}$.

If $|\hat{\mathcal{S}}_w| = 1$, then clearly (1) holds. Otherwise, by Definition 3.18, $\forall\, s' \in \hat{\mathcal{S}}_w \setminus \left\{s_{m_{w,1}}\right\} \subseteq \mathcal{S} \setminus \underline{\mathcal{S}}'_w$, there exists an $s \in \underline{\mathcal{S}}'_w$ such that

$$\ell\left(s\right) \leq \ell\left(s'\right), \varphi\left(w, s\right) \leq \varphi\left(w, s'\right).$$

Since $\ell\left(s'\right) = L_{\min}$, we have $\ell\left(s\right) = L_{\min}$, i.e. $s = s_{m_{w,1}}$. Thus, $\varphi\left(w, s_{m_{w,1}}\right) \leq \varphi\left(w, s'\right)$ for any $s' \in \hat{\mathcal{S}}_w$. □

Proposition A.3 *For any* $i = 1, 2, 3, 4$, *it holds that*

- $\underline{\mathrm{P}i} = (L_{\underline{U}_i}, D_{\underline{U}_i,Q})$, *where* $\underline{\boldsymbol{U}}_i = \boldsymbol{\Delta}_{\underline{m}_{w_1,i},\underline{m}_{w_2,i},\ldots,\underline{m}_{w_N,i}}$
- $\overline{\mathrm{P}i} = (L_{\overline{U}_i}, D_{\overline{U}_i,Q})$, *where* $\overline{\boldsymbol{U}}_i = \boldsymbol{\Delta}_{\overline{m}_{w_1,i},\overline{m}_{w_2,i},\ldots,\overline{m}_{w_N,i}}$.

Definition A.1 We define

$$\underline{\Psi}' \triangleq \left\{\boldsymbol{\Delta}_{i_1,i_2\ldots,i_N} : \forall n, s_{i_n} \in \underline{S}'_{w_n}\right\}, \underline{\Psi}'' \triangleq \left\{\boldsymbol{\Delta}_{i_1,i_2\ldots,i_N} : \forall n, s_{i_n} \in \underline{S}''_{w_n}\right\},$$

$$\underline{\Psi} \triangleq \left\{\boldsymbol{\Delta}_{i_1,i_2,\ldots,i_N} : \forall n, s_{i_n} \in \underline{S}_{w_n}\right\}, \overline{\Psi}' \triangleq \left\{\boldsymbol{\Delta}_{i_1,i_2\ldots,i_N} : \forall n, s_{i_n} \in \overline{S}'_{w_n}\right\},$$

$$\overline{\Psi}'' \triangleq \left\{\boldsymbol{\Delta}_{i_1,i_2\ldots,i_N} : \forall n, s_{i_n} \in \overline{S}''_{w_n}\right\}, \overline{\Psi} \triangleq \left\{\boldsymbol{\Delta}_{i_1,i_2,\ldots,i_N} : \forall n, s_{i_n} \in \overline{S}_{w_n}\right\},$$

$$\Psi \triangleq \left\{\boldsymbol{\Delta}_{i_1,i_2,\ldots,i_N} : \forall n, s_{i_n} \in \overline{S}_{w_n} \cup \underline{S}_{w_n}\right\}.$$

It is easy to see that $\underline{\Psi}' \cup \underline{\Psi}'' \subseteq \underline{\Psi}$ and $\overline{\Psi}' \cup \overline{\Psi}'' \subseteq \overline{\Psi}$.

Lemma A.1 *For any* $\boldsymbol{U} = \boldsymbol{\Delta}_{i_1,i_2\ldots,i_N}$, *the following four conditions hold.*

1. $\exists\, \boldsymbol{U}_{\mathrm{l,d}} \in \underline{\Psi}'$ *such that* $D_{U,Q} \geq D_{U_{\mathrm{l,d}},Q}$ *and* $L_U \geq L_{U_{\mathrm{l,d}}}$.
2. $\exists\, \boldsymbol{U}_{\mathrm{r,d}} \in \underline{\Psi}''$ *such that* $D_{U,Q} \geq D_{U_{\mathrm{r,d}},Q}$ *and* $L_U \leq L_{U_{\mathrm{r,d}}}$.
3. $\exists\, \boldsymbol{U}_{\mathrm{l,u}} \in \overline{\Psi}'$ *such that* $D_{U,Q} \leq D_{U_{\mathrm{l,u}},Q}$ *and* $L_U \geq L_{U_{\mathrm{l,u}}}$.
4. $\exists\, \boldsymbol{U}_{\mathrm{r,u}} \in \overline{\Psi}''$ *such that* $D_{U,Q} \leq D_{U_{\mathrm{r,u}},Q}$ *and* $L_U \leq L_{U_{\mathrm{r,u}}}$.

Proof The proofs of the four conditions are similar. We only prove (1). By Definition 3.18, for any n, we can find an i'_n such that $s_{i'_n} \in \underline{S}'_{w_n}$ and

$$\ell\left(s_{i'_n}\right) \leq \ell\left(s_{i_n}\right), \varphi\left(w, s_{i'_n}\right) \leq \varphi\left(w, s_{i_n}\right).$$

Let $\boldsymbol{U}_{\mathrm{l,d}} = \boldsymbol{\Delta}_{i'_1,i'_2\ldots,i'_N}$, we have $D_U \geq D_{U_{\mathrm{l,d}},Q}$ and $L_U \geq L_{U_{\mathrm{l,d}}}$. □

A.3 Proof of Theorem 3.1

Before proving Theorem 3.1, we first give some properties of the G function in Definition 3.19.

Lemma A.2 *For any* $a_1, a_2, b_1, b_2 \in \mathbb{R}$, *where* $b_1, b_2 > 0$, *the following three arguments are equivalent:*

1. $\frac{a_1}{b_1} \geq \frac{a_2}{b_2}$.
2. $\frac{a_1+a_2}{b_1+b_2} \geq \frac{a_2}{b_2}$.
3. $\frac{a_1}{b_1} \geq \frac{a_1+a_2}{b_1+b_2}$.

Define three vectors $[a_1, b_1]^\top$, $[a_2, b_2]^\top$ and $[a_1 + a_2, b_1 + b_2]^\top$. Lemma A.2 follows directly from the relative magnitudes of the angles of the three vectors.

Corollary A.1 *For any* $n = 1, 2, \ldots, N$ *and* $s', s'', s''' \in S_{w_n}$, *where* $\ell(s') < \ell(s'') < \ell(s''')$, *the following three arguments are equivalent.*

- $G(w_n, s', s'') \le G(w_n, s', s''')$.
- $G(w_n, s', s'') \le G(w_n, s'', s''')$.
- $G(w_n, s', s''') \le G(w_n, s'', s''')$.

The following three arguments are equivalent:

- $G(w_n, s', s'') \ge G(w_n, s', s''')$.
- $G(w_n, s', s'') \ge G(w_n, s'', s''')$.
- $G(w_n, s', s''') \ge G(w_n, s'', s''')$.

Lemma A.3 *For any* $a_1, a_2, \ldots, a_K, b_1, b_2, \ldots, b_K, c \in \mathbb{R}$ *and* $b_1, b_2, \ldots, b_K > 0$, *where* $K > 0$, *we have*

1. $\max\limits_{k\in[1,K]} \frac{a_k}{b_k} \ge \frac{\sum_k a_k}{\sum_k b_k} \ge \min\limits_{k\in[1,K]} \frac{a_k}{b_k}$.
2. $\frac{a_k}{b_k} \ge c, \forall k$ *implies that* $\frac{\sum_k a_k}{\sum_k b_k} \ge c$.
3. $\frac{a_k}{b_k} \le c, \forall k$ *implies that* $\frac{\sum_k a_k}{\sum_k b_k} \le c$.

Algorithm 4 constructs a sequence of deterministic encoding schemes that shape R_{enc}. In this process, two key steps are lines 2 and 20, which indicate the next encoding scheme at a time step t. Next, we analyze these two steps.

Definition A.2 For $t = 1, 2, 3, \ldots, \underline{T}$, we define

$$\underline{G}^{(t)} \triangleq \min_{n,m:s_m \in \underline{S}_{w_n}, m > \underline{i}_n^{(t-1)}} G\left(w_n, s_{\underline{i}_n^{(t-1)}}, s_m\right).$$

For $t = 1, 2, 3, \ldots, \overline{T}$, we define

$$\overline{G}^{(t)} \triangleq \max_{n,m:s_m \in \overline{S}_{w_n}, m > \overline{i}_n^{(t-1)}} G\left(w_n, s_{\overline{i}_n^{(t-1)}}, s_m\right).$$

Lemma A.4 $\underline{G}^{(t)}$ *is a non-decreasing function of* $t = 1, 2, \ldots, \underline{T}$ *and* $\overline{G}^{(t)}$ *is a non-increasing function of* $t = 1, 2, \ldots, \overline{T}$.

Proof We first prove that $\underline{G}^{(t)}$ is a non-decreasing function. It is sufficient to prove that $\underline{G}^{(t)} \le \underline{G}^{(t+1)}$ if for any $t = 1, 2, \ldots, \underline{T} - 1$. Recall that $s_{\underline{m}^{(t)}} \in \underline{S}_{w_{\underline{n}^{(t)}}}$ and $s_{\underline{m}^{(t+1)}} \in \underline{S}_{w_{\underline{n}^{(t+1)}}}$. If $\underline{n}^{(t+1)} \ne \underline{n}^{(t)}$, then by Algorithm 4, $\underline{i}_{\underline{n}^{(t+1)}}^{(t)} = \underline{i}_{\underline{n}^{(t+1)}}^{(t-1)}$. Therefore,

$$\underline{G}^{(t)} = \min_{n,m:s_m \in \underline{S}_{w_n}, m > \underline{i}_n^{(t-1)}} G\left(w_n, s_{\underline{i}_n^{(t-1)}}, s_m\right)$$

$$\leq \min_{m:s_m \in S_{w_{\underline{n}^{(t+1)}}}, m > \underline{i}^{(t-1)}_{\underline{n}^{(t+1)}}} G\left(w_{\underline{n}^{(t+1)}}, s_{\underline{i}^{(t-1)}_{\underline{n}^{(t+1)}}}, s_m\right)$$

$$= \min_{m:s_m \in S_{w_{\underline{n}^{(t+1)}}}, m > \underline{i}^{(t-1)}_{\underline{n}^{(t+1)}}} G\left(w_{\underline{n}^{(t+1)}}, s_{\underline{i}^{(t)}_{\underline{n}^{(t+1)}}}, s_m\right)$$

$$= \underline{G}^{(t+1)}.$$

Now we consider the case when $\underline{n}^{(t+1)} = \underline{n}^{(t)}$. Since

$$\underline{G}^{(t)} = G\left(w_{\underline{n}^{(t)}}, s_{\underline{i}^{(t-1)}_{\underline{n}^{(t)}}}, s_{\underline{i}^{(t)}_{\underline{n}^{(t)}}}\right) \leq G\left(w_{\underline{n}^{(t)}}, s_{\underline{i}^{(t-1)}_{\underline{n}^{(t)}}}, s_{\underline{i}^{(t+1)}_{\underline{n}^{(t)}}}\right),$$

by Corollary A.1, we have $\underline{G}^{(t)} \leq G\left(w_{\underline{n}^{(t)}}, s_{\underline{i}^{(t)}_{\underline{n}^{(t)}}}, s_{\underline{i}^{(t+1)}_{\underline{n}^{(t)}}}\right) = \underline{G}^{(t+1)}$. Likewise, it can be proven that $\overline{G}^{(t)}$ is a non-increasing function. □

Lemma A.5 *For any* $n = 1, 2, \ldots, N$, $t = 1, 2, \ldots, \underline{T}$ *and* $s_m \in \underline{S}_{w_n}$,

1. $m > \underline{i}_n^{(t)}$ *implies that* $\underline{G}^{(t+1)} \leq G\left(w_n, s_{\underline{i}_n^{(t)}}, s_m\right)$.
2. $m < \underline{i}_n^{(t)}$ *implies that* $\underline{G}^{(t)} \geq G\left(w_n, s_m, s_{\underline{i}_n^{(t)}}\right)$.

Proof Condition (1) holds since

$$\underline{G}^{(t+1)} = \min_{n',m':s_{m'} \in S_{w_{n'}}, m' > \underline{i}^{(t)}_{n'}} G\left(w_{n'}, s_{\underline{i}^{(t)}_{n'}}, s_{m'}\right)$$

$$\leq \min_{m':s_{m'} \in S_{w_n}, m' > \underline{i}^{(t)}_n} G\left(w_n, s_{\underline{i}^{(t)}_n}, s_{m'}\right) \leq G\left(w_n, s_{\underline{i}^{(t)}_n}, s_m\right).$$

Now we prove that (2) holds. By letting $\tilde{t} = \min\left\{t' : n^{(t')} = n, m < \underline{i}_n^{(t')}\right\}$, we have $\underline{i}^{(\tilde{t}-1)}_{\underline{n}^{(t)}} \leq m < \underline{i}_n^{(\tilde{t})} \leq \underline{i}_n^{(t)}$ and $\underline{G}^{(\tilde{t})} = G\left(w_n, s_{\underline{i}_n^{(\tilde{t}-1)}}, s_{\underline{i}_n^{(\tilde{t})}}\right)$. Now we prove that

$$\underline{G}^{(\tilde{t})} \geq G\left(w_n, s_m, s_{\underline{i}_n^{(\tilde{t})}}\right). \tag{A.2}$$

If $m = \underline{i}^{(\tilde{t}-1)}_{\underline{n}^{(t)}}$, (A.2) holds. Otherwise, if $m > \underline{i}^{(\tilde{t}-1)}_{\underline{n}^{(t)}}$, then by Condition (1) we just proved, it holds that

$$G\left(w_n, s_{\underline{i}_n^{(\tilde{t}-1)}}, s_{\underline{i}_n^{(\tilde{t})}}\right) = G^{(\tilde{t})} \leq G\left(w_n, s_{\underline{i}_n^{(\tilde{t}-1)}}, s_m\right).$$

Combining Corollary A.1, (A.2) holds. Therefore, (A.2) holds for all m.

If $t = \tilde{t}$, then (2) holds. Otherwise, if $\tilde{t} < t$, then we define $\mathbb{T} \triangleq \left\{t' : n^{(t')} = n, \tilde{t}+1 \le t' \le t\right\}$. We rewrite $\mathbb{T} = \left\{t_1, t_2, \ldots, t_{|\mathbb{T}|}\right\}$, where $\tilde{t} < t_1 < t_2 < \ldots < t_{|\mathbb{T}|} \le t$. By Lemma A.3, we have

$$\underline{G}^{(\tilde{t})} \le \underline{G}^{(t_1)} \le \underline{G}^{(t_2)} \le \ldots \le \underline{G}^{(t_{|\mathbb{T}|})} \le \underline{G}^{(t)}.$$

That is,

$$\underline{G}^{(\tilde{t})} \le \frac{\varphi\left(w_n, s_{\underline{i}_n^{(t_1)}}\right) - \varphi\left(w_n, s_{\underline{i}_n^{(t_1-1)}}\right)}{\ell\left(s_{\underline{i}_n^{(t_1)}}\right) - \ell\left(s_{\underline{i}_n^{(t_1-1)}}\right)} \le \frac{\varphi\left(w_n, s_{\underline{i}_n^{(t_2)}}\right) - \varphi\left(w_n, s_{\underline{i}_n^{(t_2-1)}}\right)}{\ell\left(s_{\underline{i}_n^{(t_2)}}\right) - \ell\left(s_{\underline{i}_n^{(t_2-1)}}\right)}$$
$$\le \ldots \le \frac{\varphi\left(w_n, s_{\underline{i}_n^{(t_{|\mathbb{T}|})}}\right) - \varphi\left(w_n, s_{\underline{i}_n^{(t_{|\mathbb{T}|}-1)}}\right)}{\ell\left(s_{\underline{i}_n^{(t_{|\mathbb{T}|})}}\right) - \ell\left(s_{\underline{i}_n^{(t_{|\mathbb{T}|}-1)}}\right)} \le \underline{G}^{(t)}.$$

Note that $\underline{i}_n^{(t_1-1)} = \underline{i}_n^{(\tilde{t})}$, $\underline{i}_n^{(t_{|\mathbb{T}|})} = \underline{i}_n^{(t)}$ and $\underline{i}_n^{(t_\tau-1)} = \underline{i}_n^{(t_{\tau-1})}$ for any $2 < \tau < |\mathbb{T}|$. We have

$$\underline{G}^{(\tilde{t})} \le \frac{\varphi\left(w_n, s_{\underline{i}_n^{(t_1)}}\right) - \varphi\left(w_n, s_{\underline{i}_n^{(\tilde{t})}}\right)}{\ell\left(s_{\underline{i}_n^{(t_1)}}\right) - \ell\left(s_{\underline{i}_n^{(\tilde{t})}}\right)} \le \frac{\varphi\left(w_n, s_{\underline{i}_n^{(t_2)}}\right) - \varphi\left(w_n, s_{\underline{i}_n^{(t_1)}}\right)}{\ell\left(s_{\underline{i}_n^{(t_2)}}\right) - \ell\left(s_{\underline{i}_n^{(t_1)}}\right)}$$
$$\le \ldots \le \frac{\varphi\left(w_n, s_{\underline{i}_n^{(t_{|\mathbb{T}|})}}\right) - \varphi\left(w_n, s_{\underline{i}_n^{(t_{|\mathbb{T}|-1})}}\right)}{\ell\left(s_{\underline{i}_n^{(t_{|\mathbb{T}|})}}\right) - \ell\left(s_{\underline{i}_n^{(t_{|\mathbb{T}|-1})}}\right)} = \underline{G}^{(t)}. \tag{A.3}$$

From Lemma A.3 and (A.3), we have

$$\underline{G}^{(\tilde{t})} \le \frac{\varphi\left(w_n, s_{\underline{i}_n^{(t_{|\mathbb{T}|})}}\right) - \varphi\left(w_n, s_{\underline{i}_n^{(\tilde{t})}}\right)}{\ell\left(s_{\underline{i}_n^{(t_{|\mathbb{T}|})}}\right) - \ell\left(s_{\underline{i}_n^{(\tilde{t})}}\right)} \le \underline{G}^{(t)}.$$

Note that

$$\frac{\varphi\left(w_n, s_{\underline{i}_n^{(t_{|\mathbb{T}|})}}\right) - \varphi\left(w_n, s_{\underline{i}_n^{(\tilde{t})}}\right)}{\ell\left(s_{\underline{i}_n^{(t_{|\mathbb{T}|})}}\right) - \ell\left(s_{\underline{i}_n^{(\tilde{t})}}\right)} = G\left(w_n, s_{\underline{i}_n^{(\tilde{t})}}, s_{\underline{i}_n^{(t_{|\mathbb{T}|})}}\right) = G\left(w_n, s_{\underline{i}_n^{(\tilde{t})}}, s_{\underline{i}_n^{(t)}}\right).$$

We have

$$\underline{G}^{(\tilde{t})} \leq G\left(w_n, s_{\underline{i}_n^{(\tilde{t})}}, s_{\underline{i}_n^{(t)}}\right) \leq \underline{G}^{(t)}. \tag{A.4}$$

By (A.2) and (A.4), we have

$$G\left(w_n, s_m, s_{\underline{i}_n^{(\tilde{t})}}\right) \leq G\left(w_n, s_{\underline{i}_n^{(\tilde{t})}}, s_{\underline{i}_n^{(t)}}\right).$$

Combining Corollary A.1, it holds that

$$G\left(w_n, s_m, s_{\underline{i}_n^{(t)}}\right) \leq G\left(w_n, s_{\underline{i}_n^{(\tilde{t})}}, s_{\underline{i}_n^{(t)}}\right). \tag{A.5}$$

By (A.4) and (A.5), we finally have $G\left(w_n, s_m, s_{\underline{i}_n^{(t)}}\right) \leq \underline{G}^{(t)}$. □

Corollary A.2 *For any* $t = 1, 2, \ldots, \overline{T}$ *and any* $s_m \in \overline{S}_{w_{\overline{n}^{(t)}}}$,

1. $m > \overline{m}^{(t)}$ *implies that* $\overline{G}^{(t+1)} \geq G\left(w_{\overline{n}^{(t)}}, s_{\overline{m}^{(t)}}, s_m\right)$.
2. $m < \overline{m}^{(t)}$ *implies that* $\overline{G}^{(t)} \leq G\left(w_{\overline{n}^{(t)}}, s_m, s_{\overline{m}^{(t)}}\right)$.

To ease exposition, we next define a J function. The connections between the G and J functions will be revealed later in Lemma A.6.

Definition A.3 For any $\boldsymbol{U}$ and $\boldsymbol{U}'$, where $L_{\boldsymbol{U}'} \neq L_{\boldsymbol{U}}$, we define

$$J_Q\left(\boldsymbol{U}, \boldsymbol{U}'\right) \triangleq \frac{D_{\boldsymbol{U}',Q} - D_{\boldsymbol{U},Q}}{L_{\boldsymbol{U}'} - L_{\boldsymbol{U}}}.$$

$$\underline{J}^{(t)} \triangleq J_Q\left(\underline{\boldsymbol{U}}^{(t-1)}, \underline{\boldsymbol{U}}^{(t)}\right); \quad \overline{J}^{(t)} \triangleq J_Q\left(\overline{\boldsymbol{U}}^{(t-1)}, \overline{\boldsymbol{U}}^{(t)}\right).$$

Lemma A.6 *For any* $t = 1, 2, \ldots, \underline{T}$, $\underline{J}^{(t)} = \underline{G}^{(t)}$; *for any* $t = 1, 2, \ldots, \overline{T}$, $\overline{J}^{(t)} = \overline{G}^{(t)}$.

Lemma A.7 *For any* $\boldsymbol{U} = \boldsymbol{\Delta}_{i_1, i_2 \ldots, i_N} \in \Psi$, *where* Ψ *is defined in Definition A.1,*

1. $L_{\underline{\boldsymbol{U}}^{(t)}} < L_{\boldsymbol{U}}$, $t = 1, 2, \ldots, \underline{T}$, *implies that* $\underline{J}^{(t+1)} \leq J_Q(\underline{\boldsymbol{U}}^{(t)}, \boldsymbol{U})$.
2. $L_{\overline{\boldsymbol{U}}^{(t)}} < L_{\boldsymbol{U}}$, $t = 1, 2, \ldots, \overline{T}$, *implies that* $\overline{J}^{(t+1)} \leq J_Q(\overline{\boldsymbol{U}}^{(t)}, \boldsymbol{U})$.

Proof We first prove (1). Recall that $\underline{\boldsymbol{U}}^{(t)} = \boldsymbol{\Delta}_{\underline{i}_1^{(t)}, \underline{i}_2^{(t)} \ldots, \underline{i}_N^{(t)}}$ for any t. Then

$$J_Q\left(\underline{\boldsymbol{U}}^{(t)}, \boldsymbol{U}\right) = \frac{\sum_n p\left(w_n\right) \varphi\left(w_n, s_{i_n}\right) - \sum_n p\left(w_n\right) \varphi\left(w_n, s_{\underline{i}_n^{(t)}}\right)}{\sum_n p\left(w_n\right) \ell(s_{i_n}) - \sum_n p\left(w_n\right) \ell(s_{\underline{i}_n^{(t)}})}$$

$$= \frac{\sum_n p(w_n)\left(\varphi\left(w_n, s_{i_n}\right) - \varphi\left(w_n, s_{\underline{i}_n^{(t)}}\right)\right)}{\sum_n p(w_n)\left(\ell\left(s_{i_n}\right) - \ell\left(s_{\underline{i}_n^{(t)}}\right)\right)}.$$

Denote

$$I_0 = \left\{n : i_n = \underline{i}_n^{(t)}\right\}.$$

$$I_+ = \left\{n : i_n > \underline{i}_n^{(t)}\right\}.$$

$$I_- = \left\{n : i_n < \underline{i}_n^{(t)}\right\}.$$

$$a_1 = \sum_n p(w_n)\left(\varphi\left(w_n, s_{i_n}\right) - \varphi\left(w_n, s_{\underline{i}_n^{(t)}}\right)\right).$$

$$b_1 = \sum_n p(w_n)\left(\ell\left(s_{i_n}\right) - \ell\left(s_{\underline{i}_n^{(t)}}\right)\right).$$

$$a_2 = \sum_{n \in I_-} p(w_n)\left(-\varphi\left(w_n, s_{i_n}\right) + \varphi\left(w_n, s_{\underline{i}_n^{(t)}}\right)\right).$$

$$b_2 = \sum_{n \in I_-} p(w_n)\left(-\ell\left(s_{i_n}\right) + \ell\left(s_{\underline{i}_n^{(t)}}\right)\right).$$

Then

$$J_Q\left(\underline{\boldsymbol{U}}^{(t)}, \boldsymbol{U}\right) = \frac{a_1}{b_1}. \tag{A.6}$$

For any $n \in I_0$, we have $i_n = \underline{i}_n^{(t)}$, and then

$$\varphi\left(w_n, s_{i_n}\right) - \varphi\left(w_n, s_{\underline{i}_n^{(t)}}\right) = \ell\left(s_{i_n}\right) - \ell\left(s_{\underline{i}_n^{(t)}}\right) = 0.$$

Note that

$$0 < b_1 + b_2 = \sum_{n \in I_+} p(w_n)\left(-\ell\left(s_{i_n}\right) + \ell\left(s_{\underline{i}_n^{(t)}}\right)\right).$$

It holds that $I_+ \neq \emptyset$. For any $n \in I_+$, we have $\ell\left(s_{i_n}\right) > \ell\left(s_{\underline{i}_n^{(t)}}\right)$. Then, by Lemma A.5,

$$\underline{G}^{(t+1)} \leq G\left(w_n, s_{\underline{i}_n^{(t)}}, s_{i_n}\right)$$

$$= \frac{\varphi\left(w_n, s_{i_n}\right) - \varphi\left(w_n, s_{\underline{i}_n^{(t)}}\right)}{\ell\left(s_{i_n}\right) - \ell\left(s_{\underline{i}_n^{(t)}}\right)}$$

$$= \frac{p\left(w_n\right)\left(\varphi\left(w_n, s_{i_n}\right) - \varphi\left(w_n, s_{\underline{i}_n^{(t)}}\right)\right)}{p\left(w_n\right)\left(\ell\left(s_{i_n}\right) - \ell\left(s_{\underline{i}_n^{(t)}}\right)\right)}.$$

By Lemma A.3, we have

$$\underline{G}^{(t+1)} \leq \frac{\sum_{n\in I_+} p\left(w_n\right)\left(\varphi\left(w_n, s_{i_n}\right) - \varphi\left(w_n, s_{\underline{i}_n^{(t)}}\right)\right)}{\sum_{n\in I_+} p\left(w_n\right)\left(\ell\left(s_{i_n}\right) - \ell\left(s_{\underline{i}_n^{(t)}}\right)\right)} = \frac{a_1 + a_2}{b_1 + b_2}. \tag{A.7}$$

When $I_- = \emptyset$, (1) holds by (A.6) and (A.7). Therefore, we only need to consider the case where $I_- \neq \emptyset$. For any $n \in I_-$, we have $\ell\left(s_{i_n}\right) > \ell\left(s_{\underline{i}_n^{(t)}}\right)$, and by Lemma A.5,

$$\underline{G}^{(t)} \geq G\left(w_n, s_{\underline{i}_n^{(t)}}, s_{i_n}\right)$$

$$= \frac{\varphi\left(w_n, s_{i_n}\right) - \varphi\left(w_n, s_{\underline{i}_n^{(t)}}\right)}{\ell\left(s_{i_n}\right) - \ell\left(s_{\underline{i}_n^{(t)}}\right)}$$

$$= \frac{p\left(w_n\right)\left(-\varphi\left(w_n, s_{i_n}\right) + \varphi\left(w_n, s_{\underline{i}_n^{(t)}}\right)\right)}{p\left(w_n\right)\left(-\ell\left(s_{i_n}\right) + \ell\left(s_{\underline{i}_n^{(t)}}\right)\right)}.$$

By Lemma A.3, we have

$$\underline{G}^{(t)} \geq \frac{a_2}{b_2}. \tag{A.8}$$

By (A.7), (A.8) and Lemma A.4, we have

$$\frac{a_2}{b_2} \leq \underline{G}^{(t)} \leq \underline{G}^{(t+1)} \leq \frac{a_1 + a_2}{b_1 + b_2}.$$

Finally, (1) holds by Lemma A.2.

The second part of Lemma A.7 can be proved in a similar fashion. □

Definition A.4 Theorem 3.1 characterizes a region using the distortion-cost pairs of $\underline{T} + \overline{T} + 2$ semantic encoding schemes $\left\{\underline{\boldsymbol{U}}^{(t)} : t = 0, 1, 2, \ldots, \underline{T}\right\}$ and $\left\{\overline{\boldsymbol{U}}^{(t)} : t = 0, 1, 2, \ldots, \overline{T}\right\}$. We denote the region by $\widetilde{R}_{\text{enc}}$, the piecewise linear connection of $\left(L_{\underline{U}^{(t)}}, D_{\underline{U}^{(t)},Q}\right)$, $t = 0, 1, 2, \ldots, \underline{T}$ by a function $\underline{D}(L)$, and the piecewise linear connection of $\left(L_{\overline{U}^{(t)}}, D_{\overline{U}^{(t)},Q}\right)$, $t = 0, 1, 2, \ldots, \overline{T}$ by a function $\overline{D}(L)$.

Corollary A.3 *For any $L', L'' \in [L_{min}, L_{max}]$, where $L' < L''$, we have*

$$\underline{J}^{(\underline{t}')} \leq \frac{\underline{D}(L'') - \underline{D}(L')}{L'' - L'} \leq \underline{J}^{(\underline{t}'')},$$

$$\overline{J}^{(\overline{t}')} \geq \frac{\overline{D}(L'') - \overline{D}(L')}{L'' - L'} \geq \overline{J}^{(\overline{t}'')},$$

where

$$\underline{t}' \triangleq \min\left\{t : L_{\underline{U}^{(t)}} \geq L'\right\}, \underline{t}'' \triangleq \min\left\{t : L_{\underline{U}^{(t)}} \geq L''\right\}$$
$$\overline{t}' \triangleq \min\left\{t : L_{\overline{U}^{(\overline{t}')}} \geq L'\right\}, \overline{t}'' \triangleq \min\left\{t : L_{\overline{U}^{(t)}} \geq L''\right\}.$$

Proof If $\underline{t}' = \underline{t}''$, it is straightforward that

$$\underline{J}^{(\underline{t}')} = \frac{\underline{D}(L'') - \underline{D}(L')}{L'' - L'} = \underline{J}^{(\underline{t}'')}.$$

If $\underline{t}' < \underline{t}''$, then

$$\begin{aligned}
&\underline{D}(L'') = \underline{D}(L') + \left(\underline{D}_{\underline{U}^{(\underline{t}')}} - \underline{D}(L')\right) \\
&+ \sum_{t=\underline{t}'+1}^{\underline{t}''-1} \left(\underline{D}_{\underline{U}^{(t)}} - \underline{D}_{\underline{U}^{(t-1)}}\right) + \left(\underline{D}(L'') - \underline{D}_{\underline{U}^{(\underline{t}''-1)}}\right) \\
&= \underline{D}(L') + \underline{J}^{(\underline{t}')} \left(L_{\underline{U}^{(\underline{t}')}} - L'\right) \\
&+ \sum_{t=\underline{t}'+1}^{\underline{t}''-1} \underline{J}^{(t)} \left(L_{\underline{U}^{(t)}} - L_{\underline{U}^{(t-1)}}\right) + \underline{J}^{(\underline{t}'')} \left(L'' - L_{\underline{U}^{(\underline{t}''-1)}}\right).
\end{aligned}$$

By Lemmas A.4 and A.6, we have

$$\underline{J}^{(\underline{t}')} \leq \underline{J}^{(\underline{t}'+1)} \leq \ldots \leq \underline{J}^{(\underline{t}'')}.$$

Then,

$$\underline{D}(L') + \underline{J}^{(\underline{t}')} \left(L_{\underline{U}^{(\underline{t}')}} - L'\right)$$

$$
\begin{aligned}
&+\sum_{t=\underline{t}'+1}^{\underline{t}''-1} \underline{J}^{(\underline{t}')}\left(L_{\underline{U}^{(t)}}-L_{\underline{U}^{(t-1)}}\right)+\underline{J}^{(\underline{t}')}\left(L''-L_{\underline{U}^{(\underline{t}''-1)}}\right)\\
&\leq \tilde{D}(L'')\\
&\leq \underline{D}(L')+\underline{J}^{(\underline{t}'')}\left(L_{\underline{U}^{(\underline{t}')}}-L'\right)\\
&+\sum_{t=\underline{t}'+1}^{\underline{t}''-1} \underline{J}^{(\underline{t}'')}\left(L_{\underline{U}^{(t)}}-L_{\underline{U}^{(t-1)}}\right)+\underline{J}^{(\underline{t}'')}\left(L''-L_{\underline{U}^{(\underline{t}''-1)}}\right).
\end{aligned}
$$

That is,

$$
\underline{D}(L')+\underline{J}^{(\underline{t}')}\left(L''-L'\right) \leq \tilde{D}(L'') \leq \underline{D}(L')+\underline{J}^{(\underline{t}'')}\left(L''-L'\right).
$$

Finally,

$$
\underline{J}^{(\underline{t}')}=\frac{\tilde{D}(L'')-\tilde{D}(L')}{L''-L'}=\underline{J}^{(\underline{t}'')}.
$$

The proof of the second part is similar. □

Lemma A.8 *$\underline{D}(L)$ is convex and $\overline{D}(L)$ is concave. The region $\widetilde{R}_{enc}$ is convex.*

Proof We first prove that $\underline{D}(L)$ is convex. It is sufficient to prove that for any L', $L'' \in [L_{\min}, L_{\max}]$ and $\lambda \in (0, 1)$, we have

$$
\underline{D}\left(\hat{L}\right) \leq \lambda \underline{D}(L')+(1-\lambda)\underline{D}(L'')
$$

where $\hat{L} = \lambda L' + (1-\lambda)L''$. Let $t' = \min\left\{t : L_{\underline{U}^{(t)}} \geq L'\right\}$, $t'' = \min\left\{t : L_{\underline{U}^{(t)}} \geq L''\right\}$ and $\hat{t} = \min\left\{t : L_{\underline{U}^{(t)}} \geq \hat{L}\right\}$. If $L' = L''$, then the inequality holds. Otherwise, without loss of generality, assume $L' < L''$. Then, $L' < \hat{L} < L''$. By Corollary 3.11, we have

$$
\frac{\underline{D}\left(\hat{L}\right)-\underline{D}(L')}{\hat{L}-L'} \leq J^{(\hat{t})} \leq \frac{\underline{D}(L'')-\underline{D}\left(\hat{L}\right)}{L''-\hat{L}},
$$

which implies that

$$
\underline{D}\left(\hat{L}\right) \leq \underline{D}(L')+\underline{J}^{(\hat{t})}\left(\hat{L}-L'\right),
$$

$$
\underline{D}\left(\hat{L}\right) \leq \underline{D}(L'')+J^{(\hat{t})}\left(\hat{L}-L''\right).
$$

Thus,

$$\begin{aligned}\underline{D}\left(\hat{L}\right) &= \lambda \underline{D}\left(\hat{L}\right) + (1-\lambda)\,\underline{D}\left(\hat{L}\right) \\ &\le \lambda\left(\underline{D}(L') + \underline{J}^{(\hat{t})}\left(\hat{L} - L'\right)\right) \\ &+ (1-\lambda)\left(\underline{D}(L'') + \underline{J}^{(\hat{t})}\left(\hat{L} - L''\right)\right) \\ &= \lambda \underline{D}(L') + (1-\lambda)\,\underline{D}(L'') \\ &+ \underline{J}^{(\hat{t})}\left(\hat{L} - \left(\lambda L' + (1-\lambda)\,L''\right)\right) \\ &= \lambda \underline{D}(L') + (1-\lambda)\,\underline{D}(L'').\end{aligned}$$

Likewise, it can be proven that $\overline{D}(L)$ is concave. Now we prove that the region $\widetilde{R}_{\text{enc}}$ is convex. It is sufficient to prove that, for any $(L', D'), (L'', D'') \in \widetilde{R}_{\text{enc}}$ and $\lambda \in (0, 1)$, it holds that

$$(\hat{L}, \hat{D}) \in \widetilde{R}_{\text{enc}}.$$

where $\hat{L} = \lambda L' + (1-\lambda)L''$ and $\hat{D} = \lambda D' + (1-\lambda)D''$. It is equivalent to proving that

$$\underline{D}\left(\hat{L}\right) \le \hat{D} \le \overline{D}\left(\hat{L}\right).$$

Since $(L', D'), (L'', D'') \in \widetilde{R}_{\text{enc}}$, we have

$$\underline{D}(L') \le D' \text{ and } \underline{D}(L'') \le D''.$$

On the other hand, $\underline{D}(L)$ is convex, indicating that

$$\begin{aligned}\underline{D}\left(\hat{L}\right) &\le \lambda \underline{D}(L') + (1-\lambda)\,\underline{D}(L'') \\ &\le \lambda D' + (1-\lambda)\,D'' = \hat{D}.\end{aligned}$$

Likewise, we can also prove that $\hat{D} \le \overline{D}\left(\hat{L}\right)$. The lemma is proved. □

We are now ready to prove Theorem 3.1. The achievability can be easily established. According to Algorithm 4, for any $L \in [L_{\min}, L_{\max}]$, $(L, \underline{D}(L))$ and $(L, \overline{D}(L))$ are in R_{enc}. Moreover, Proposition 3.1 shows that R_{enc} is a convex region. Thus, for any $L \in [L_{\min}, L_{\max}]$ and $D \in [\underline{D}(L), \overline{D}(L)]$, we have $(L, D) \in R_{\text{enc}}$. Therefore, $\widetilde{R}_{\text{enc}} \subseteq R_{\text{enc}}$.

Next, we prove the converse, i.e., $R_{\text{enc}} \subseteq \widetilde{R}_{\text{enc}}$. Assuming that there exists an encoding scheme $\boldsymbol{U}^*$ such that $\left(L_{\boldsymbol{U}^*}, D_{\boldsymbol{U}^*,Q}\right) \in R_{\text{enc}}$ and $\left(L_{\boldsymbol{U}^*}, D_{\boldsymbol{U}^*,Q}\right) \notin \widetilde{R}_{\text{enc}}$.

Proposition 3.1 indicates that $(L_{U^*}, D_{U^*,Q})$ can be achieved by the time sharing of a set of deterministic encoding schemes. Denoted by $\boldsymbol{A}$ the set containing all these encoding schemes. Recall from Lemma A.8 that the region $\widetilde{R}_{\text{enc}}$ is convex. Therefore, there exists at least one deterministic encoding scheme $\boldsymbol{U}' \in \boldsymbol{A}$ such that $\left(L_{U'}, D_{U',Q}\right) \in R_{\text{enc}}$ and $\left(L_{U'}, D_{U',Q}\right) \notin \widetilde{R}_{\text{enc}}$. Otherwise, $(L_{U^*}, D_{U^*,Q}) \in \widetilde{R}_{\text{enc}}$, which contradicts the assumption.

By Lemma A.1, there exist four points $\boldsymbol{U}_{\mathsf{l,d}}, \boldsymbol{U}_{\mathsf{r,d}}, \boldsymbol{U}_{\mathsf{l,u}}, \boldsymbol{U}_{\mathsf{r,u}} \in \Psi$ such that

- $D_{U',Q} \geq D_{U_{\mathsf{l,d}},Q}$ and $L_{U'} \geq L_{U_{\mathsf{l,d}}}$;
- $D_{U',Q} \geq D_{U_{\mathsf{r,d}},Q}$ and $L_{U'} \leq L_{U_{\mathsf{r,d}}}$;
- $D_{U',Q} \leq D_{U_{\mathsf{l,u}},Q}$ and $L_{U'} \geq L_{U_{\mathsf{l,u}}}$;
- $D_{U',Q} \leq D_{U_{\mathsf{r,u}},Q}$ and $L_{U'} \leq L_{U_{\mathsf{r,u}}}$.

Hence, there exists at least one encoding scheme

$$\boldsymbol{\Delta} \in \left\{\boldsymbol{U}_{\mathsf{l,d}}, \boldsymbol{U}_{\mathsf{r,d}}, \boldsymbol{U}_{\mathsf{l,u}}, \boldsymbol{U}_{\mathsf{r,u}}\right\} \subseteq \Psi$$

such that one of the following two statements holds.

1. $D_{\boldsymbol{\Delta},Q} < \underline{D}(L_{\boldsymbol{\Delta}})$;
2. $D_{\boldsymbol{\Delta},Q} > \overline{D}(L_{\boldsymbol{\Delta}})$.

Otherwise, $\left(L_{U'}, D_{U',Q}\right) \in \widetilde{R}_{\text{enc}}$.

We first assume that statement (1) holds. Proposition A.3 shows that $\left(L_{\underline{U}^{(0)}}, D_{\underline{U}^{(0)},Q}\right) = \underline{\text{P1}}$. Thus $L_{\min} < L_{\boldsymbol{\Delta}}$. Otherwise, $D_{\boldsymbol{\Delta},Q} \geq \underline{D}(L_{\boldsymbol{\Delta}})$, which contradicts statement (1).

Let $t = \max\{t' : L_{\underline{U}^{(t')}} < L_{\boldsymbol{\Delta}}\}$. Thus, $L_{\underline{U}^{(t)}} < L_{\boldsymbol{\Delta}} \leq L_{\underline{U}^{(t+1)}}$.

Let $\lambda = \frac{L_{\boldsymbol{\Delta}} - L_{\underline{U}^{(t)}}}{L_{\underline{U}^{(t+1)}} - L_{\underline{U}^{(t)}}}$, we have $\lambda \in (0, 1]$ and

$$L_{\boldsymbol{\Delta}} = (1 - \lambda)\, L_{\underline{U}^{(t)}} + \lambda L_{\underline{U}^{(t+1)}}.$$

From Lemma 3.10, it holds that $\underline{J}^{(t+1)} \leq J_Q(\boldsymbol{U}^{(t)}, \boldsymbol{\Delta})$. As a result,

$$D_{\boldsymbol{\Delta},Q} \geq (1 - \lambda)\, D_{\underline{U}^{(t)},Q} + \lambda D_{\underline{U}^{(t+1)},Q} = \underline{D}\,(L_{\boldsymbol{\Delta}})\,.$$

This contradicts statement (1). Therefore, statement (1) does not hold. Similarly, it can be proven that statement (2) does not hold either.

Overall, any encoding scheme $\boldsymbol{U}^*$ with $\left(L_{U^*}, D_{U^*,Q}\right) \in R_{\text{enc}}$ must satisfy $\left(L_{U^*}, D_{U^*,Q}\right) \in \widetilde{R}_{\text{enc}}$. The region characterized by Theorem 3.1 is exactly the distortion-cost region of semantic encoding R_{enc}.

A.4 Proof of Proposition 3.3

Under the Hamming distortion measure, $\psi_q(\hat{w}, \hat{s})$ can be refined as

$$\begin{aligned}\psi_q(\hat{w}, \hat{s}) &= \sum_{w,s} q(w)p(s|w)c(\hat{s}|s)d(w, \hat{w}) \\ &= \sum_{w,s} q(w)p(s|w)c(\hat{s}|s) - q(\hat{w}) \sum_{s} p(s|\hat{w})c(\hat{s}|s) \\ &= q(\hat{s}) - q(\hat{w})p(\hat{s}|\hat{w}),\end{aligned}$$

where $q(\hat{s}) \triangleq \sum_{w,s} q(w)p(s|w)c(\hat{s}|s)$.

Likewise, we have $\psi_p(\hat{w}, \hat{s}) = p(\hat{s}) - p(\hat{w})p(\hat{s}|\hat{w})$, where

$$p(\hat{s}) \triangleq \sum_{w,s} p(w)p(s|w)c(\hat{s}|s).$$

Following Proposition 3.2, semantic decoding gives us

$$\hat{w}_q(\hat{s}) = \arg\min_{\hat{w}} \psi_q(\hat{w}, \hat{s}) = \arg\max_{w} q(w)p(\hat{s}|w), \tag{A.9}$$

$$D_{\boldsymbol{P},\boldsymbol{V}_q^*} = \sum_{\hat{s}} \psi_p\big(\hat{w}_q(\hat{s}), \hat{s}\big) = 1 - \sum_{\hat{s}} p\big(\hat{w}_q(\hat{s})\big)p\big(\hat{s}|\hat{w}_q(\hat{s})\big). \tag{A.10}$$

An intuitive explanation of (A.9) and (A.10) is as follows. Under the prior distribution $q(w)$, (A.9) is simply the maximum a posteriori (MAP) decoding for a given $\hat{s}$. When the decoding scheme is determined, a received message $\hat{s}$ is decoded to $\hat{w}_q(\hat{s})$. As a result, the probability of successful decoding is the summation $\sum_{\hat{s}} p\big(\hat{w}_q(\hat{s})\big)p\big(\hat{s}|\hat{w}_q(\hat{s})\big)$ with each term being the probability of transmitting $\hat{w}_q(\hat{s})$ and receiving $\hat{s}$. The Hamming distortion, i.e., the semantic error, is then given by (A.10).

Likewise, for the true prior p(w), we would have

$$\hat{w}_p(\hat{s}) = \arg\max_{w} p(w)p(\hat{s}|w), \tag{A.11}$$

$$D_{\boldsymbol{P},\boldsymbol{V}_p^*} = 1 - \sum_{\hat{s}} p\big(\hat{w}_p(\hat{s})\big)p\big(\hat{s}|\hat{w}_p(\hat{s})\big). \tag{A.12}$$

Comparing (A.12) and (A.10), $D_{\boldsymbol{P},\boldsymbol{V}_q^*} = D_{\boldsymbol{P},\boldsymbol{V}_p^*}$ if and only if (3.28) holds. Note that $\hat{w}_q(\hat{s})$ and $\hat{w}_p(\hat{s})$ in (A.11) and (A.9) are the solutions to the left- and right-hand sides of (3.28), respectively, but they are not necessarily the only solutions. Therefore, (3.28) does not imply $\hat{w}_q(\hat{s}) = \hat{w}_p(\hat{s})$.

A.5 Proof of Theorem 3.4

The distortion-cost function of CSED $D^*_{\boldsymbol{U},\boldsymbol{V}^*_q}(L)$ can be achieved by the deterministic encoding schemes $\{\underline{\boldsymbol{U}}^{(i)} : i = 0, 1, 2, \ldots, \underline{T}\}$ and the linear combination of two contiguous $\underline{\boldsymbol{U}}^{(i)}$ and $\underline{\boldsymbol{U}}^{(i+1)}$. Thus, (3.42) holds for any point on $D^*_{\boldsymbol{U},\boldsymbol{V}^*_q}(L)$ provided that we can prove (3.42) for all $\underline{\boldsymbol{U}}^{(i)}$.

Under conditions (1) and (2), we have

$$\varphi(w, s) = \frac{\psi_p(\hat{w}, s)}{p(s)}.$$

When operated with CSED, for a meaning $w \in \mathcal{W}$, we define the encoded message s and the decoded message $\hat{w}$ under encoding and decoding schemes $\boldsymbol{U} \in \{\underline{\boldsymbol{U}}^{(i)} : i = 0, 1, 2, \ldots, \underline{T}\}$ and $\boldsymbol{V}^*_q$, respectively, as

$$[s|w] = \arg\min_s \varphi(w, s),$$

$$[\hat{w}|s] = \arg\min_{w'} \psi_q(w', s) = \arg\min_{w'} \varphi(w', s).$$

We further define the subset of messages that $\mathcal{W}$ is mapped to as

$$[\mathcal{S}|\mathcal{W}] = \{[s|w] : w \in \mathcal{W}\}.$$

From conditions (3) and (4), we first prove that the mappings between $\mathcal{W}$ and $[\mathcal{S}|\mathcal{W}]$ and that between $[\mathcal{S}|\mathcal{W}]$ and $\mathcal{W}$ are one-to-one.

We prove this by contradiction. Suppose the mappings between $\mathcal{W}$ and $[\mathcal{S}|\mathcal{W}]$ are not one-to-one, condition (3) gives us $|[\mathcal{S}|\mathcal{W}]| > N$. On the other hand, from (4), we have $N = |\mathcal{W}| \geq [\mathcal{S}|\mathcal{W}]$. Contradiction. The mappings between $\mathcal{W}$ and $[\mathcal{S}|\mathcal{W}]$ are thus one-to-one. Likewise, it can be proven that the mappings between $[\mathcal{S}|\mathcal{W}]$ and $\mathcal{W}$ are also one-to-one.

Next, we prove

$$d\big(w, \big[\hat{w}|[s|w]\big]\big) = \min_{w'} d(w, w'), \ \forall w \in \mathcal{W}. \tag{A.13}$$

First, we sort $\mathcal{W} = \{w_{(n)} : n = 1, 2, \ldots, N\}$ according to $\min_s \varphi(w, s)$ such that $w_{(n_1)} \leq w_{(n_2)}$ implies that $\min_s \varphi(w_{(n_1)}, s) \leq \min_s \varphi(w_{(n_2)}, s)$. Equation (A.13) can be proved by induction.

For $w_{(1)}$, $\boldsymbol{U}$ maps it to

$$[s|w_{(1)}] = \arg\min_s \varphi(w, s),$$

and V_q^* maps $[s|w_{(1)}]$ to

$$w_{(\beta)} = \arg\min_{w} \varphi\left(w, [s|w_{(1)}]\right).$$

where $\beta \geq 1$. If $\beta > 1$,

$$\varphi\left(w_{(\beta)}, [s|w_{(1)}]\right) \leq \varphi\left(w_{(1)}, [s|w_{(1)}]\right).$$

Since $\varphi\left(w_{(\beta)}, [s|w_{(1)}]\right) = \varphi\left(w_{(1)}, [s|w_{(1)}]\right)$ contradicts the condition (4), we have

$$\varphi\left(w_{(\beta)}, [s|w_{(1)}]\right) < \varphi\left(w_{(1)}, [s|w_{(1)}]\right),$$

which means $\beta \leq 1$, and hence, $\beta = 1$. This proves that

$$w_{(1)} = \left[\hat{w}\left|\left[s|w_{(1)}\right]\right.\right].$$

For any $1 < n < N$, we assume $w_{(n)} = \left[\hat{w}\left|\left[s|w_{(n)}\right]\right.\right]$ and prove $w_{(n+1)} = \left[\hat{w}\left|\left[s|w_{(n+1)}\right]\right.\right]$ in the following.

For $w_{(n+1)}$, let

$$w_{(\beta)} = \left[\hat{w}\left|\left[s|w_{(n+1)}\right]\right.\right] = \arg\min_{w} \varphi\left(w, \left[s|w_{(n+1)}\right]\right),$$

we prove below that $\beta = n + 1$.

First, $\beta < n + 1$ contradicts 4). If $\beta > n + 1$,

$$\varphi\left(w_{(\beta)}, [s|w_{(n+1)}]\right) \leq \varphi\left(w_{(n+1)}, [s|w_{(n+1)}]\right),$$

and

$$\min_{s} \varphi\left(w_{(n+1)}, s\right) \leq \min_{s} \varphi\left(w_{(\beta)}, s\right).$$

Since $\min_s \varphi\left(w_{(n+1)}, s\right) = \varphi\left(w_{(n+1)}, \left[s|w_{(n+1)}\right]\right)$, we have

$$\min_{s} \varphi\left(w_{(n+1)}, s\right) = \min_{s} \varphi\left(w_{(\beta)}, s\right) = \\ \varphi\left(w_{(\beta)}, \left[s|w_{(n+1)}\right]\right) = \varphi\left(w_{(n+1)}, \left[s|w_{(n+1)}\right]\right).$$

As a result,

$$\left[s|w_{(n+1)}\right] \in \arg\min_{s} \varphi\left(w_{(n+1)}, s\right) \cap \arg\min_{s} \varphi\left(w_{(\beta)}, s\right).$$

This contradicts (3).

Therefore, $\beta = n + 1$ and $w_{(n+1)} = \left[\hat{w} \,\middle|\, [s|w_{(n+1)}]\right]$.
In conclusion, we have $d\left(w, \left[\hat{w} \,|[s|w]\right]\right) = \min_{w'} d(w, w')$, $\forall w$, and hence,

$$
\begin{aligned}
D_{\boldsymbol{U}, \boldsymbol{V}_q^*}(L) &= \sum_{w,s,\hat{w}} p(w)u(s|w)v(\hat{w}|s)d(w, \hat{w}) \\
&= \sum_{w} p(w)d\left(w, \left[\hat{w} \,|[s|w]\right]\right) = \sum_{w} p(w) \min_{w'} d(w, w') \\
&\le \min_{\boldsymbol{U}', \boldsymbol{V}'} D_{\boldsymbol{U}', \boldsymbol{V}'}(L),
\end{aligned}
$$

for $\underline{\boldsymbol{U}}^{(i)}$, $i = 0, 1, 2, \ldots, \underline{T}$.

Appendix B
Transceiver Design and Subcarrier Allocation for IFDMA

OFDMA, LFDMA, and IFDMA are a class of signal modulation and multiple-access techniques. For these techniques, information of users are multiplexed and carried on subcarriers within a shared spectrum.

Among these schemes, IFDMA has attracted great interest because of its excellent PAPR and its ability to achieve full frequency diversity. Other advantages of IFDMA over OFDMA include smaller sensitivity to frequency offset, and ability to combat frequency nulls.

IFDMA, however, has rigid requirements on subcarrier allocation. Let us refer to a data stream multiplexed onto a set of subcarriers within the shared spectrum using IFDMA as an IFDMA stream. At any one time, the shared spectrum may be occupied by IFDMA streams of multiple users. IFDMA streams are subject to the following two constraints:

- Constraint 1: Subcarrier allocation in IFDMA must be carefully orchestrated so that not only the subcarriers of different IFDMA streams are non-overlapping, but also that the subcarriers of each IFDMA stream are evenly spaced (e.g., subcarriers $\{0, 4, 8, \ldots\}$ are allocated to one IFDMA stream; subcarriers $\{1, 17, 33, \ldots\}$ to another IFDMA stream).
- Constraint 2: The number of subcarriers of an IFDMA stream must be a divisor of the number of the overall available subcarriers (e.g., if there are 64 subcarriers, the number of subcarriers of an IFDMA stream must be one of $\{1, 2, 4, 8, 16, 32, 64\}$; the number of subcarriers of a stream cannot be 3).

These two constraints lead to the impressions that for IFDMA: (1) multiplexing and demultiplexing of data streams by the transceiver are complicated; (2) subcarrier allocation is not only complicated, but also very inflexible.

We that both impressions are not exactly true. In fact, the imposed rigidity belies a nice mathematical structure that can be exploited to simplify both transceiver design and subcarrier allocation. The foundation of our results can be summarized in one sentence [1–3]:

Y. Shao, *AI-Empowered IoT Communications*, Wireless Networks,
https://doi.org/10.1007/978-3-031-98816-5

Multiplexing/demultiplexing and modulation/demodulation of IFDMA data streams can be interpreted as smaller FFTs/IFFTs of different scales embedded in the overall Cooley-Tukey FFT recursive decomposition scheme.

This realization gives rise to the following two IFDMA constructions, one related to resource allocation, and one related to transceiver design:

- A bit-reversal subcarrier allocation scheme whereby, instead of focusing on allocating evenly-spaced subcarriers to an IFDMA stream, the focus can move to allocating contiguous "bins" to an IFDMA stream. After such bin allocation, a bit-reversal mapping of the bin indexes automatically yields the evenly-spaced subcarrier indexes (e.g., if there are 64 subcarriers, an IFDMA stream allocated contiguous bins $\{0, 1, 2, 3\}$ will be mapped to subcarriers $\{0, 32, 16, 48\}$, i.e., evenly-spaced subcarriers $\{0, 16, 32, 48\}$).
- Greatly simplified transceiver designs that multiplex/demultiplex and modulate/demodulate multiple IFDMA streams in parallel within a single FFT/IFFT module.

B.1 Bit-Reversal Subcarrier Allocation

We first introduce the bit-reversal subcarrier allocation scheme for IFDMA.

B.1.1 Subcarrier Allocation in IFDMA

We consider an uplink multiple access system where multiple users transmit signals to a common AP using orthogonal subcarriers. When operated with IFDMA, each user requests a set of subcarriers according to the rate requirement, and constructs an IFDMA stream to transmit the signal.

Let M be the total number of available subcarriers, and N be the number of subcarriers requested by a user. In order for the user to be able to construct an IFDMA stream, the subcarrier allocation is subject to two constraints, as stated earlier:

- Constraint 1: The N subcarriers allocated to a user must be evenly spaced among the M subcarriers. That is, the distance between the indexes of two adjacently allocated subcarriers must be M/N.
- Constraint 2: N must be a divisor of M so that M/N is an integer.

If we index the M subcarriers by $[M] = \{0, 1, \cdots, M-1\}$, then the indexes of the N subcarriers allocated to the user are

$$\left\{d + i\frac{M}{N} : i = [N]\right\}, \tag{B.1}$$

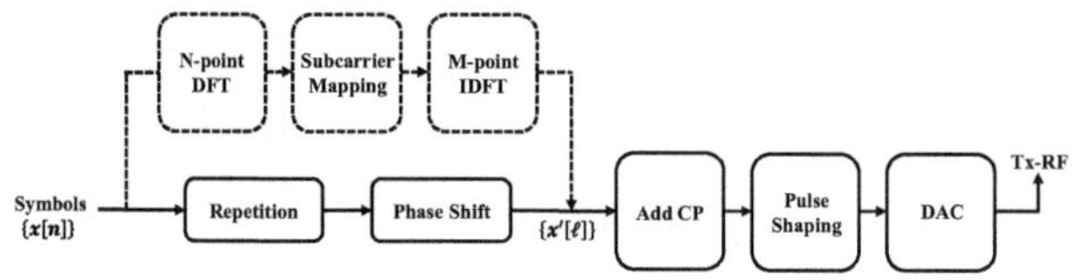

Fig. B.1 Construction of an IFDMA stream of size N at the transmitter

where $d \in [M/N]$ is a constant frequency shift.

Given the N allocated subcarriers, Fig. B.1 shows two possible realizations for the construction of an IFDMA stream at the transmitter. The first is a realization in the frequency domain (the dashed modules), and the second is a realization in the time domain (the solid modules).

Let us first focus on the frequency-domain realization. As shown, the symbols to be transmitted are grouped into blocks of size N, that is, $\{x[n] : n = [N]\}$. Each block of time-domain symbols is transformed to the frequency domain by an N-point DFT, and is then mapped to the N allocated subcarriers according to (B.1), with the $M - N$ unoccupied subcarriers filled with 0. After that, the transmitter performs an M-point IDFT to transform the data on M subcarriers back to the time-domain, and obtains an IFDMA stream $\{x'[\ell] : \ell = [M]\}$. It is easy to verify that

$$x'[\ell] = \frac{N}{M} e^{j\frac{2\pi\ell}{M}d} x[\ell \bmod N], \quad \ell = [M]. \tag{B.2}$$

In particular, if $d = 0$ (i.e., the allocated subcarriers are $\{i\frac{M}{N} : i = [N]\}$), the IFDMA stream is simply $x'[\ell] = \frac{N}{M} x[\ell \bmod N]$, $\ell = [M]$.

Equation (B.2) gives the simpler and more direct time-domain realization of the IFDMA stream. That is, $\{x'[\ell]\}$ can be directly constructed from $\{x[n]\}$ by repetition and frequency upshifting, hence bypassing the N-point DFT, subcarrier mapping/multiplexing and the M-point IDFT.

Most of the nice attributes of IFDMA can be understood from its simple time-domain representation in (B.2), e.g., the simple transmitter design in uplink and the superior PAPR property compared with OFDMA.

Despite the advantages, a common view on IFDMA is that the subcarrier allocation, as well as the multiplexing/demultiplexing of IFDMA streams, are quite inflexible due to the above two constraints. Specifically,

- Constraint 1 requires the subcarriers allocated to each user to be evenly spaced among the M subcarriers. This is even more challenging in asynchronous systems (defined below) where the requests from different users can come and go in an asynchronous manner.
- Constraint 2 prevents the users from requesting an arbitrary number of subcarriers, which is possible in OFDMA. IFDMA thus has coarse granularity in resource allocation.

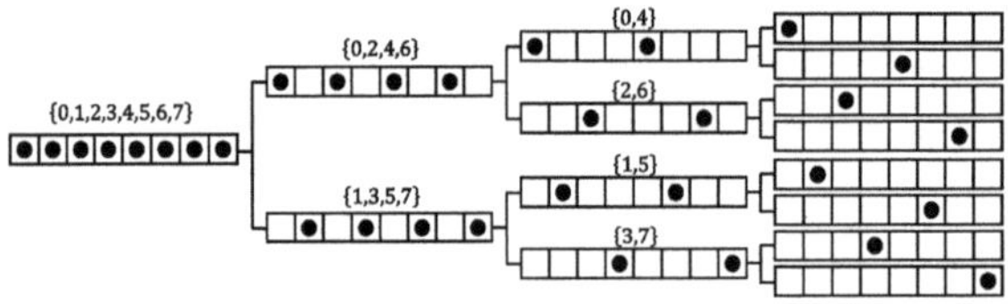

Fig. B.2 A tree structure for synchronous subcarrier allocation wherein $M = 8$. Each vertex of the tree is associated with a set of subcarriers (represented by black points) that can be assigned to users

Definition B.1 (Synchronous and Asynchronous Subcarrier Allocation) In synchronous subcarrier allocation, all the users' requests arrive at and depart from the AP simultaneously. A new batch of requests will not arrive until all the earlier requests have left the system. In asynchronous subcarrier allocation, the users' requests arrive at the AP asynchronously, and the service times of different requests are non-uniform. Thus, a newly arriving request may see some subcarriers already occupied by existing requests.

Existing solutions on subcarrier allocation address only Constraint 1 in synchronous systems. The general idea is to construct a tree structure, and associate the vertices of the tree with eligible sets of subcarriers that can be allocated to users.

In [4], the author assumed M is a power of 2, and constructed a tree similar to the one used for generating Orthogonal Variable Spreading Factor (OVSF) codes. An example of the tree structure is given in Fig. B.2.

As can be seen, each vertex of the tree is associated with a set of subcarriers (represented by black points) that can be assigned to users. Given a number of synchronous requests, the AP performs the following procedure to allocate subcarriers

1. Sort the requests according to their sizes in descending order.
2. For the request with the largest size, randomly choose a vertex on the tree whose associated set has the same size with the request. Assign the subcarriers in the associated set to the request.
3. Eliminate the vertex chosen in step 2 and all its descendent vertex.
4. Repeat steps 2 and 3 until all requests are satisfied.

In [5], the authors studied the synchronous subcarrier allocation problem for more general composite M. Unlike the case of power-of-2 M where the binary tree is well-determined, the tree structure for general composite M has many variations. For example, Fig. B.3 presents the two variations when $M = 6$.

To summarize,

- Existing solutions address only the synchronous subcarrier allocation. Their idea is to build a tree structure so that the vertices of the tree can be mapped to different sets of evenly-spaced subcarriers of various sizes.

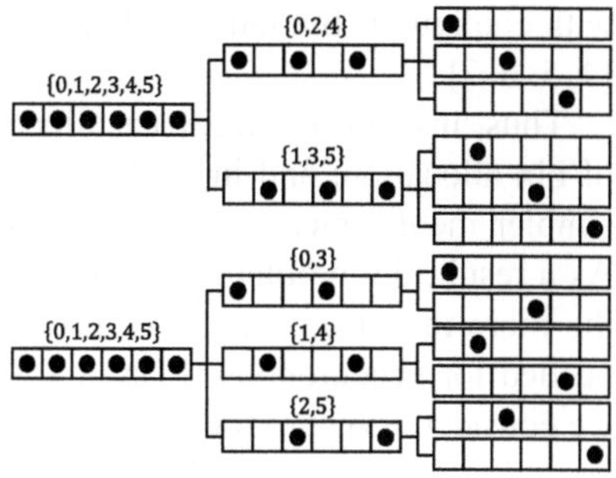

Fig. B.3 The two tree structures for subcarrier allocation when $M = 6$

- Existing solutions cannot be generalized to asynchronous subcarrier allocation because sorting the requests is no longer possible when request arrivals are asynchronous. Without sorting, the random vertex-selection policy leads to a large blocking probability of incoming requests, as will be shown in this appendix.
- Existing solutions still suffer from Constraint 2. N must be a divisor of M, but not any integers, limiting the granularity of the allocated resources.

B.1.2 Overview of Our Key Results

This section makes advances on all the three problems faced by existing solutions, as elaborated below.

B.1.2.1 Synchronous Subcarrier Allocation

We put forth a bit-reversal subcarrier allocation scheme, where no tree construction is required, for synchronous subcarrier allocation (bit reversal is for power-of-2 M; for general composite M, instead of bit-reversal, we can use digit-reversal). Given this scheme, we prove that resource allocation of IFDMA is on an equal footing with OFDMA in that "full loading" is possible. That is, provided constraints 1 and 2 are fulfilled by all user requests, as long as the total number of subcarriers requested by the users is no more than M, all the requests can be fulfilled.

An important finding which form the basis for our bit-reversal subcarrier allocation scheme is that, the indexes of the evenly-spaced subcarriers is contiguous in a bit-reversal permutation [6]. For example, the bit-reversal permutation for $M = 8$ is $(0, 4, 2, 6, 1, 5, 3, 7)$. For IFDMA, a user requesting two subcarriers can be allocated subcarriers indexed by $\{0, 4\}$, $\{1, 5\}$, $\{2, 6\}$, or $\{3, 7\}$; a user requesting four subcarriers can be allocated subcarriers indexed by $\{0, 2, 4, 6\}$ or $\{1, 3, 5, 7\}$. Suppose that each subcarrier is put into a bin with the bin index being the bit reversal of the subcarrier index. Then the subcarriers allocated to a user will occupy bins with contiguous bin indexes. For example, subcarriers $\{1, 3, 5, 7\}$ will occupy bins $\{4, 5, 6, 7\}$.

Thus, instead of subcarrier allocation, we can consider an equivalent problem of bin allocation; and instead of evenly-spaced subcarriers, we can now have the simpler mental picture of contiguous bins when performing resource allocation. As a result, the problem of allocating evenly-spaced subcarriers is transformed to the more intuitive problem of filling contiguous bins, and no tree construction is needed for synchronous IFDMA systems. This model also comes in handy when we consider asynchronous subcarrier allocation.

B.1.2.2 Asynchronous Subcarrier Allocation

We put forth a new bin-filling policy, the "min-small-change policy" (to be elaborated in Sect. B.1.3), for asynchronous subcarrier allocation to avoid fragmentation of contiguous bins, hence reducing blocking probability. Compared with the random bin-filling policy, our min-small-change policy reduces the blocking probability of IFDMA systems by 17.21% under moderate offered load in a system where the arrivals are a Poisson process and the holding times of accepted requests are exponentially distributed.

Under random arrivals, an incoming request can be rejected due to instantaneous system overload (i.e., not enough subcarriers left to fulfil the number of subcarriers requested). An OFDMA system can suffer from blocking due to instantaneous overload only, as illustrated in Fig. B.4. For an IFDMA system, besides blocking due to overload, an incoming request can also be rejected due to fragmentation of bins (i.e., the number of vacant bins is larger than the size of the request, but a set of contiguous bins is not available to fulfill the request).

An important metric to measure the quality of a bin-filling policy in asynchronous IFDMA systems is the ability to reduce the amount of rejected load due to fragmentation. Our min-small-change bin-filling policy aims to minimize the amount of such rejections.

However, even with our min-small-change policy, the asynchronous IFDMA systems still have higher blocking probability than asynchronous OFDMA systems because the amount of rejections caused by fragmentation cannot be reduced to

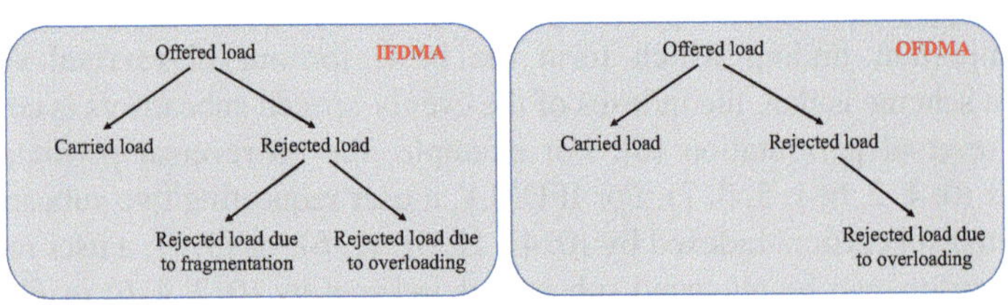

Fig. B.4 The partition of the offered load in asynchronous IFDMA and OFDMA systems

0. In this context, this section further proposes a multi-stream IFDMA scheme, as described below.

B.1.2.3 Multi-Stream IFDMA

We put forth a new multi-stream IFDMA scheme to serve two purposes. First, it enables us to remove Constraint 2. The request size N can be any positive integer smaller than or equal to M in multi-stream IFDMA systems. Second, the multi-stream IFDMA can achieve the same level of flexibility with OFDMA even in asynchronous systems (i.e., a request can be rejected only due to system overload, but not due to bin fragmentation).

Definition B.2 (Multi-Stream IFDMA) Let M be the total number of subcarriers. Suppose a user request for N subcarriers, where N can be any positive integer smaller than or equal to M. A multi-stream IFDMA system allows the user to construct multiple IFDMA streams, as opposed to only one stream in traditional IFDMA systems. Specifically, the AP will partition the request to Q sub-requests, and the size of each sub-request is a divisor of M. That is, $N = \sum_{i=1}^{Q} N_i$ and each N_i is a divisor of M. The AP then allocates subcarriers to each sub-request following the evenly-spaced-subcarrier rule. The user will then construct Q IFDMA streams and transmit the sum of the Q streams.

For example, suppose that $M = 16$ and $N = 6$. In a conventional IFDMA system, the user has to request for 8 instead of 6 subcarriers to fulfill its rate requirement because 6 is not a divisor of 16, causing bandwidth wastage. However, in multi-stream IFDMA systems, the user can partition 6 into $4 + 2$, and construct two IFDMA streams, one of size 4 and the other of size 2. The transmitted signal of this user is then the sum of the two IFDMA streams. Note that the partition is not unique. In the extreme case, the user can construct 6 streams, each of size 1.

With the multi-stream IFDMA, the amount of rejections caused by fragmentation can be minimized to 0 in asynchronous systems. This can be easily understood from the definition because we can always partition the request to smaller sub-requests. In the extreme case, multi-stream IFDMA becomes OFDMA if we partition any incoming request to multiple sub-requests of size 1.

Despite the advantages in subcarrier allocation, multi-stream IFDMA also raises two new issues:

1. The multiplexing/demultiplexing of multiple IFDMA streams at both transmitter and receiver may result in more complex transceiver design.
2. The perfect PAPR property of conventional IFDMA is no longer retained, because the sum of IFDMA streams causes fluctuations in the amplitude of the transmitted signal.

These two concerns will be addressed in the next section on IFDMA transceiver design.

B.1.3 Synchronous Subcarrier Allocation

This section focuses on the subcarrier allocation problem in synchronous IFDMA systems. We show that IFDMA is as flexible as OFDMA in terms of synchronous subcarrier allocation. To ease exposition, this section assumes (a) the total number of available subcarriers, M, is a power of 2; (b) the number of subcarriers requested by each user, N, is a divisor of M (hence is also a power of 2). The results obtained in this section can be well-extended to general composite M, and general N which may not be a divisor of M with our multi-stream IFDMA scheme.

B.1.3.1 Bit-Reversal Subcarrier Allocation Scheme

To begin with, we put forth a new bit-reversal subcarrier allocation scheme. Instead of allocating evenly-spaced subcarriers to users, we consider the problem of filling contiguous bins with users' requests. Below, we present a basic version of the bit-reversal scheme with a sort-first bin-filling policy.

The Bit-Reversal Subcarrier Allocation Scheme

Suppose that we have $M = 2^m$ bins (instead of subcarriers) labeled with $k = 0, 1, \cdots, 2^m - 1$.

- Sorting. We sort the users according to the number of requested subcarriers in descending order.
- Bin Allocation. We allocate bins to the sorted users in the order of bin 0 to bin $2^m - 1$. That is, each user who requests 2^i subcarriers is allocated the next 2^i unallocated bins. Users with larger request size is considered first.
- Bit-reversal mapping. We perform a bit-reversal mapping from bins to subcarriers. That is, if a bin with binary index $b_{m-1}b_{m-2}\cdots b_1 b_0$ is allocated to a user, then the subcarrier with binary index $b_0 b_1 \cdots b_{m-2} b_{m-1}$ will be allocated to this user.

Table B.1 gives an example of the bit-reversal subcarrier allocation scheme. In this example, we assume $M = 8$ ($m = 3$), and three users A, B and C requests for 1, 4 and 2 subcarriers, respectively. To use the bit-reversal subcarrier allocation scheme, we first sort the users according to their requests. The sorted users are B (4 subcarriers), C (2 subcarriers) and A (1 subcarrier). Following the second step, we allocate bins $\{0, 1, 2, 3\}$ to user B, bins $\{4, 5\}$ to user C, and then bins $\{6\}$ to user A. After bit-reversal mapping, users A, B and C finally get subcarriers indexed by $\{0, 2, 4, 6\}$, $\{1, 5\}$ and $\{3\}$, respectively.

Essentially, our bit-reversal subcarrier allocation scheme reveals that the underlying relationship between evenly-spaced subcarriers and contiguous bins is a bit-reversal mapping. In this context, we can focus on filling the contiguous bins

Table B.1 An example of the bit-reversal subcarrier allocation

Bins	Binary index of bins	Bit-reversal	Subcarriers
0	000	000	0
1	001	100	4
2	010	010	2
3	011	110	6
4	100	001	1
5	101	101	5
6	110	011	3
7	111	111	7

instead of evenly-spaces subcarriers, and no tree construction is needed as in [4, 5]. In fact, if we go back to tree in Fig. B.2, it can be found that the subcarrier locations in the leaf vertices form a bit-reversal permutation. The tree structure and the bit-reversal mapping are closely related.

Given the bit-reversal scheme, we prove in Theorem B.1 below that IFDMA is as flexible as OFDMA in terms of synchronous subcarrier allocation. In particular, provided that the total number of subcarriers requested by all the users does not exceed M, all the requests can be fulfilled, and full loading is possible for IFDMA systems. The validity of bit-reversal mapping is also proved within the proof of Theorem B.1 (i.e., bit reversing the indexes of allocated contiguous bins give rise to subcarrier indexes that are evenly spaced).

Theorem B.1 (Full Loading Is Possible in IFDMA) *Let a_n be the number of users requesting for 2^n subcarriers, and $M = 2^m$. If the inequality in* (B.3) *below is satisfied, we can use the bit-reversal subcarrier allocation scheme to assign subcarriers to users such that the spacing between the subcarriers assign to a user with request size 2^n is 2^{m-n} (i.e., the evenly-spaced-subcarrier constraint is satisfied).*

$$\sum_{n=0}^{m} a_n 2^n \leq 2^m. \tag{B.3}$$

Proof We prove Theorem B.1 by induction. First, the theorem can be easily verified to be valid for $m = 1$ or 2. We prove below that if the theorem is valid for any $m \geq 2$, it is also valid for $m + 1$.

A bit-reversal permutation is a permutation of a sequence of $M = 2^m$ items. It is defined by indexing the elements of the sequence by the numbers from 0 to $M - 1$ and then reversing the binary representations of each of these numbers [6]. Each item is then mapped to the new position given by this reversed value. For example, a length-4 permutation is $(0, 2, 1, 3)$, and a length-8 permutation is $(0, 4, 2, 6, 1, 5, 3, 7)$.

An observation is that, the length-$(m + 1)$ bit-reversal permutation can be generated from the length-m bit-reversal permutation by doubling the numbers in the length-m permutation and concatenate the resulting sequence with the same sequence with each value increased by one. For example, doubling the length-4 permutation $(0, 2, 1, 3)$ gives $(0, 4, 2, 6)$; adding one to each value in $(0, 4, 2, 6)$ gives $(1, 5, 3, 7)$; and concatenating these two sequences gives the length-8 permutation $(0, 4, 2, 6, 1, 5, 3, 7)$.

Consider bin allocation associated with the second step of our bit-reversal allocation algorithm. We have two possible cases as far as the set $\{a_n : n = 1, 2, \cdots, m + 1\}$ is concerned. Case I: $a_{m+1} = 1$; Case II: $a_{m+1} = 0$. For case I, there is a single user wanting all the 2^{m+1} subcarriers. The theorem is trivially true for this case.

For case II, we divide the bins into two subsets: subset i consists of bins $\{0, 1, \cdots, 2^m - 1\}$ and subset ii consists of bins $\{2^m, 2^m + 1, \cdots, 2^{m+1} - 1\}$. Since $a_{m+1} = 0$, and that all the users want only a multiple of 2^n $(n = 0, 1, 2, \cdots, m)$ bins, the bins allocated to a user must all belong to one of the two subsets according to our bin allocation algorithm in the second step (i.e., we cannot have some bins of a user belonging to subset i while other bins of the same user belonging to subset ii).

Let us first consider the users being allocated bins in subset i. If we were to consider the length-m bit reversal permutation for this group of users, the subcarriers allocated to them would satisfy the length-m problem by the assumption of our inductive argument. Specifically, if a user wants 2^n subcarriers, $0 \leq n \leq m$, then the spacing between subcarriers allocated would be 2^{m-n} by assumption. When we migrate to the length-$(m + 1)$ problem, the indexes of the subcarriers allocated to these users would be doubled. Specifically, the new spacing would be $2 \times 2^{m-n} = 2^{m+1-n}$, thus satisfying the new IFDMA constraint.

We next consider the users being allocated bins in subset ii. If we were to index these bins by $\{0, 1, \cdots, 2^m - 1\}$ rather than $\{2^m, 2^m + 1, \cdots, 2^{m+1} - 1\}$, we see that, by the same reasoning as in the previous paragraph, these users would also satisfy the new IFDMA constraint for the length-$(m + 1)$ problem. The rest is adding one to the indexes of the allocated subcarriers so that these subcarriers are distinct from the subcarriers allocated to users of the first group in the previous paragraph. Adding one to all subcarrier indexes preserves the spacing between subcarriers. Thus, the IFDMA constraint for the length-$(m + 1)$ problem remains satisfied. □

B.1.3.2 Bin-Filling Policies

With the bit-reversal scheme, the problem of allocating evenly-spaced subcarriers is transformed to a more intuitive problem of filling contiguous bins. The bit-reversal subcarrier allocation scheme in Sect. B.1.3.1 is a bin-filling policy that satisfies the subcarrier allocation constraints. There could be other bin-filling policies that also satisfy the subcarrier allocation constraints. In this section, we give another policy that does so without the need for the sorting step. By getting rid of the need for sorting, this policy also suggests what we should do for asynchronous subcarrier allocation where user requests do not arrive together (in which case sorting the requests before allocating subcarriers to them is not possible). This part considers synchronous subcarrier allocation with this new policy, and asynchronous subcarrier allocation will be considered in Sect. B.1.4.

To satisfy the IFDMA constraint, an eligible bin-filling policy must allocate an "unoccupied fillable bin subset" to each user. An unoccupied fillable bin subset is basically a subset of as-yet unallocated bins with contiguous indexes that can be used to fulfill a new request, defined below:

Definition B.3 (Fillable Bin Subset) A fillable bin subset of size N, denoted by $[iN, iN+N-1]$, is a subset of contiguous bins with labels $\{iN, iN+1, \cdots, iN+N-1\}$ such that a request of size N can be allocated the bins in the subset without violating the IFDMA constraint. In particular, given a request of size N, there are M/N "potential fillable bin subsets" that can be used to fulfill it, i.e., $i = 0, 1, 2, \cdots, M/N - 1$. A potential fillable bin subset can be occupied or unoccupied. For the "occupied fillable bin subset", at least one bin in the subset is already occupied by another request. For the "unoccupied fillable bin subset", all the N bins in the subset are as yet unallocated and therefore can be used to fulfill a request of size N.

The Min-Small-Change Bin-Filling Policy The idea of the min-small-change policy is to partition the unallocated bins into as few fillable bin subsets as possible to prevent fragmentation. With this notion, initially, without any request, we have a single unoccupied fillable bin subset, $[0, M-1]$. Suppose that the first request presented to us is a request for two bins. We can give it bins $\{0, 1\}$. Now, the remaining unoccupied fillable bin subsets are $\{2, 3\}$, $\{4, 5, 6, 7\}$, $\{8, 9, 10, 11, 12, 13, 14, 15\}$, $\cdots$. If the next request presented to us is also for two bins, we should give it the subset $\{2, 3\}$ rather than fragmenting other subsets for the allocation. If the request is for four bins, we should give it $\{4, 5, 6, 7\}$. The idea is analogous to keeping as little small change as possible when we spend our money. Large fillable bin subsets correspond to bills of large denominations, and small fillable bin subsets correspond to bills of small denominations. That is where the name "min-small-change" is coined from.

Theorem B.2 (The Min-Small-Change Policy Enables Fully Loaded IFDMA Systems) *Let $M = 2^m$ and $N = 2^n$ where $n \leq m$. Suppose that a set of requests are presented to us in a random order in synchronous IFDMA systems. When using the min-small-change policy to fill the bins for the bit-reversal subcarrier allocation scheme, a new request for N subcarriers is guaranteed to be granted provided that* (B.3) *is satisfied (i.e., the total number of requested subcarriers in the system does not exceed M).*

Proof We prove Theorem B.2 by contradiction. Suppose at some point during the allocation process, we are faced with a situation where the next request is for N bins but we cannot find an unoccupied fillable bin subset with $N = 2^n$ or more bins. If (B.3) is satisfied, there must be multiple smaller unoccupied fillable bin subsets such that together they have N bins. These smaller bin subsets must each have 2^i

bins for some $i < n$. In other words, we have

$$2^n = \sum_{i=0}^{n-1} \psi_i 2^i, \tag{B.4}$$

where ψ_i is the number of unoccupied fillable bin subsets of the size 2^i in the system.

On the other hand, a key observation from the bin allocation process of the min-small-change policy is that at any one time as we allocate bins to successive requests, we cannot have two unoccupied fillable bin subsets of the same size thanks to the min-small-change policy. That is, for each i, there can be at most one such bin subset. This means ψ_i can only be either 1 or 0. However, in this case, we must have

$$2^n > \sum_{i=0}^{n-1} \psi_i 2^i, \tag{B.5}$$

because the LHS is larger than the RHS even if $\psi_i = 1$ for all $i < n$.

Equation (B.4) contradicts (B.5). Thus, the next request for N bins can always be granted as long as (B.3) is satisfied. □

Theorem B.2 shows that fully loaded IFDMA systems can be achieved even without sorting the synchronous requests. In fact, the benefits of the min-small-change policy go far beyond this. Unlike the sort-first bin-filling policy, the min-small-change policy can be further applied in (1) asynchronous IFDMA systems where the requests come and go, as will be expounded in Sect. B.1.4; and (2) systems wherein a subset of subcarriers are not available for use, e.g., the direct-conversion receivers in Sect. B.1.3.3 below.

B.1.3.3 Subcarrier Allocation for Direct-Conversion Receiver

For many communication systems, the RF signals are directly down-converted to the zero frequency (baseband) instead of the intermediate frequency (IF) for the simple design of direct-conversion receivers (DCR). However, such systems typically incur a strong interfering signal at the DC because of the leakage of the local oscillator (LO) or the nonzero DC offset in the receiver's analog-to-digital converter (ADC). This is sidestepped in many systems, e.g., WiFi and LTE, by just not transmitting a subcarrier at the center of the band. Thus, it is desirable to avoid the use of signals containing DC components.

For the sort-first bin-filling policy, the null subcarrier at DC will be a problem. Considering the example in Table B.1. When using DCR, the fourth subcarrier (the first bin) is unavailable, hence the request of user B cannot be satisfied. On the other hand, for the min-small-change policy, avoiding the DC subcarrier is not a problem.

Corollary B.1 below is a rephrase of the Theorem B.1 when the DC subcarrier is unavailable in DCR and with the use of the min-small-change policy.

Corollary B.1 (Subcarrier Allocation for DCR) *Let a_n be the number of users requesting for 2^n subcarriers, and $M = 2^m$. Consider direct-conversion receivers where the DC subcarrier is unavailable. Provided that the inequality in* (B.6) *below is satisfied, then we can use the bit-reversal subcarrier allocation scheme with the min-small-change bin-filling policy to assign subcarriers to users such that their IFDMA constraints are satisfied.*

$$\sum_{n=0}^{m} a_n 2^n \leq 2^m - 1. \tag{B.6}$$

The RHS of (B.6) *is due to the fact that we are giving up one subcarrier.*

Proof We can treat the unavailable subcarrier as one that has already been allocated to a fictitious request of size 1 that was presented to us first in the allocation process. □

To illustrate Corollary B.1, consider the example of $M = 8$ ($m = 3$). At the beginning, instead of having a complete unoccupied fillable bin subset $[0, M - 1]$, we have multiple smaller unoccupied fillable bin subsets. Suppose that the DC subcarrier is subcarrier $\{4\}$. After bit reversal, the corresponding bin to be avoided is bin $\{1\}$. Thus, the initial fillable bin subsets are $\{0\}$, $\{2, 3\}$, $\{4, 5, 6, 7\}$. That is, it is as if we have already granted bin 1 to a fictitious request that asks for one subcarrier before we begin the real bin allocation.

For the original IFDMA system with the DC subcarrier, we have 2^{m-n} potential fillable bin subsets for a request that asks for 2^n subcarriers. For an IFDMA system that avoids the DC subcarriers, the number of possible allocations is just one less than the above. That is, we have $2^{m-n} - 1$ potential allocations for requests that ask for 2^n subcarriers. It is not possible to grant a request for 2^m subcarriers now; but this is the case even for an OFDM system that avoids the DC subcarrier. For requests that ask for 2^{m-1} subcarriers, we have one potential fillable bin subset in IFDMA that avoids the DC subcarrier; but even for the OFDMA system that avoids the DC subcarrier, one can only grant one (but not two) request that asks for 2^{m-1} subcarriers. Compared with non-DC OFDMA system, non-DC IFDMA has the same "capacity performance" in that the fully loaded capacity is decreased by 1 only.

To conclude this section, we reiterate that, other than the limit that the request sizes has to be a divisor of M (Constraint 2), IFDMA has no limitations compared with OFDMA in terms of synchronous resource allocation. Even Constraint 2 can be removed with our multi-stream IFDMA scheme. IFDMA/multi-stream IFDMA, on the other hand, have better PAPR performance than OFDMA.

B.1.4 Asynchronous Subcarrier Allocation

In asynchronous IFDMA systems, the users' requests arrive at the AP asynchronously. At any time, a request is granted only if there is an unoccupied fillable bin subset in the system whose size is equal to or larger than the size of the request (if the size is larger, the unoccupied fillable bin subset can be broken to multiple unoccupied smaller fillable bin subsets, one of which can be used to fulfill the request). Otherwise, the request will be rejected. Each accepted request has a holding time and it will leave the system upon the expiry of the holding time.

This section studies the subcarrier allocation problem in asynchronous IFDMA systems. Two issues of interest are as follows: (1) What is required in order for the IFDMA system to be *strictly nonblocking* (to be defined later)? (2) What is the performance gap between asynchronous IFDMA and OFDMA systems as far as the blocking probability is concerned?

To ease exposition, we denote the state of an IFDMA systems with M subcarriers by the occupancies of M bins $h_{M-1}h_{M-2}\cdots h_1 h_0$. In particular, each binary symbol $h_i = \{0, 1\}$ indicates whether the i-th bin is occupied: $h_i = 0$ means it is vacant, and $h_i = 1$ means it is occupied. Further, underlines below $h_i = 1$ indicate grouping of the bins for a request. For example, consider the case of $M = 4$. State $0\,0\,\underline{1\,1}$ means the last two bins are allocated to a single request of size two, and state $0\,0\,\underline{1}\,\underline{1}$ means the last two bins are allocated to two different requests of size one.

B.1.4.1 Blocking Properties of Asynchronous IFDMA Systems

With the sort-first or the min-small-change policy, no request will be rejected in synchronous IFDMA systems as long as the total size of all requests does not exceed M. The reason is that the two bin-filling policies minimize fragmentation of unoccupied fillable bin subsets. For example, when $M = 4$, $\underline{1}\,\underline{1}\,0\,0$ is a legal state, but $\underline{1}\,0\,0\,\underline{1}$ is illegal with the min-small-change policy when applied to synchronous IFDMA systems: in the first case, the two 0's represent an unoccupied fillable bin subset of size 2, while in the second case, the two 0's represent two unoccupied fillable bin subsets of size 1, which does not conform to the min-small-change policy.

On the other hand, even with the min-small-change bin-filling policy, $\underline{1}\,0\,0\,\underline{1}$ is legal in asynchronous IFDMA systems because of departures of ongoing requests. For example, state $\underline{1}\,0\,\underline{1}\,\underline{1}$ is legal because there could be three ongoing requests of size 1. In an asynchronous IFDMA system, state $\underline{1}\,0\,\underline{1}\,\underline{1}$ can evolve to state $\underline{1}\,0\,0\,\underline{1}$ upon the departure of the request holding the third bin. If a new request of size 2 then arrives, the new request cannot be granted even though there are two vacant bins in $\underline{1}\,0\,0\,\underline{1}$ because the two vacant bins cannot form a fillable bin subset of size 2.

For asynchronous IFDMA, if we allow the rearrangement of the occupied bins, there will also be no rejections in asynchronous IFDMA systems provided (B.3) is satisfied. For the same example in the above, a new request of size two can still be granted even when the system is in state $\underline{1}\;0\;0\;\underline{1}$, because we can reassign bins to the existing requests so that the state becomes $\underline{1}\;\underline{1}\;0\;0$ or $0\;0\;\underline{1}\;\underline{1}$, creating an unoccupied fillable bin subset of two.

For concreteness, let us state the formal definitions of "Rearrangeably Nonblocking" and "Strictly Nonblocking".

Definition B.4 (Rearrangeably Nonblocking) A system is said to be rearrangeably nonblocking if a new request can always be granted, after rearranging the assigned bins to the existing requests if necessary, provided that (B.3) is satisfied (i.e., the total size of the requests, including the new request, does not exceed M).

Corollary B.2 *Asynchronous IFDMA systems are rearrangeably nonblocking.*

Proof Since we are allowed to reassign bins for the existing requests, we can rerun the synchronous resource allocation algorithm for all the requests (the existing requests plus the new request) as if they arrive synchronously. □

Definition B.5 (Strictly Nonblocking) A system is said to be strictly nonblocking if a new request can always be granted without rearranging the assigned bins to the existing requests, provided that (B.3) is satisfied (i.e., the total size of the requests, including the new request, does not exceed M).

Corollary B.3 *OFDMA systems are strictly nonblocking.*

Proof This is obvious because a new request can be granted in OFDMA systems as long as there are enough vacant subcarriers, i.e., (B.3) is satisfied. □

Asynchronous IFDMA poses a stringent requirement on the offered load if the system is to be strictly nonblocking. The RHS of (B.3) must be relaxed to guarantee that there is no rejection/blocking of new requests.

Definition B.6 (Strictly Nonblocking under Offered Load Limit) A system is said to be strictly nonblocking with number of occupied bins limited to K if a new request can always be granted, provided the total size of all requests (existing plus the new request) is no more than K.

Theorem B.3 (Strictly Nonblocking Property of Asynchronous IFDMA Systems Under Offered Load Limit) *Suppose that $M = 2^m$ is the total number of subcarriers, and there is a new request arriving at an asynchronous IFDMA system. Denote by a_n the number of requests (including the new request) requiring 2^n subcarriers. The asynchronous IFDMA system is strictly nonblocking if and only if the following inequality is satisfied:*

$$\sum_{n=0}^{m} a_n 2^n < \begin{cases} 2^{\frac{m}{2}+1} & \text{if } m \text{ is even}, \\ 3 \times 2^{\frac{m-1}{2}} & \text{if } m \text{ is odd}. \end{cases} \tag{B.7}$$

Proof Subcarrier occupancy can translate from bin occupancy through bit-reversal mapping. Thus, we can think in terms of bin occupancy. We ask the following question. Suppose that there is a new request for 2^n subcarriers. What is the minimum number of bins occupied by existing users that can block the new request? For this new request, there are $M/2^n = 2^{m-n}$ potential fillable bin subsets that can be used to fulfill it. However, if some bin in each and every of these potential fillable bin subsets is already occupied by an existing request, then none of them can be used to fulfill the new request. The minimum number of occupied bins to cause this situation is 2^{m-n}, wherein one bin in each of the potential fillable bin subsets is occupied by an existing request of size 1. If, on the other hand, the number of occupied bins is less than 2^{m-n}, then at least one of the potential fillable bin subsets is empty, and blocking cannot occur for the new request. Thus, to guarantee nonblocking property, the number of occupied bins plus the number of newly requested bins must be less than $2^{m-n} + 2^n$. The RHS of (B.7) is found by $\min_{0 \le n \le m}(2^{m-n} + 2^n)$ (i.e., finding the worst-case n). □

Theorem B.3 indicates that, for asynchronous IFDMA systems, the price to pay for having the strictly nonblocking property is pretty steep and may not be worthwhile. Compared with (B.3), the total number of requested subcarriers allowed at any one time has to be severely limited in order to have the strictly nonblocking property as in (B.7).

Multi-Stream IFDMA A multi-stream IFDMA scheme can enable strictly nonblocking IFDMA systems with full loading (i.e., a multi-stream IFDMA system is strictly nonblocking in the sense of Definition B.5).

As stated in Definition B.2, a multi-stream IFDMA scheme allows the user to construct multiple IFDMA streams, as opposed to only one stream in traditional IFDMA systems. Given a new request of size N, the AP can partition the request to multiple sub-requests, and the size of each sub-request is a divisor of M. In this context, the problem of finding an unoccupied fillable bin subset of size N is transformed to a problem of finding multiple unoccupied fillable bin subsets of smaller size. In the extreme case, the AP can partition the request to N sub-requests, each of size one.

Corollary B.4 *Multi-stream IFDMA systems are strictly nonblocking.*

Proof If the number of total requested subcarriers does not exceed M, a new request can always be partitioned to multiple sub-requests so that multiple empty fillable bin subsets can be used to fulfill the request—in the extreme case, we partition the incoming request to multiple sub-requests of size one. □

In addition to the strictly nonblocking property, multi-stream IFDMA also enables fine-grained asynchronous subcarrier allocation. The size of requests is restricted to a divisor of M in Theorem B.3. This restriction can be removed with the multi-stream scheme. Each user can request an arbitrary number of subcarriers, and the multi-stream IFDMA systems are strictly nonblocking as long as the total number of requested subcarriers is no more than M.

B.1.4.2 Blocking Probability of Asynchronous IFDMA Systems

Without the multi-stream scheme, the strictly nonblocking property is expensive to achieve for asynchronous IFDMA systems. Let $M = 1024$ ($m = 10$), for example. To have a strictly nonblocking IFDMA system, the load needs to be limited to $2^{\frac{m}{2}+1} = 64$ subcarriers according to Theorem B.3. On the other hand, OFDMA systems are strictly nonblocking even with full loading of 1024 subcarriers.

However, given a certain amount of blocking tolerance (i.e., we give up making the system strictly nonblocking by allowing blocking as long as the blocking probability is not excessive), how does an IFDMA system perform when benchmarked against an OFDMA system in the asynchronous scenario? This subsection answers this question.

In asynchronous IFDMA systems, the bin-filling policy is very important in that it determines the probability that a new request gets blocked. We will consider two bin-filling polices for asynchronous IFDMA: the random policy considered in [4, 5], and our min-small-change policy proposed in Sect. B.1.3.

The Markov Chain Analytically, the state transitions in asynchronous IFDMA systems can be modeled as a Markov Chain (MC) for a given request arrival model (e.g., Poisson arrival), a given holding-time model (e.g., exponential holding time), and a given bin-filling policy.

The complete state transitions of the MC when $M = 4$ is shown in Fig. B.5, where the min-small-change policy is used to fill the bins (we omit the state-transition probabilities to avoid cluttering the picture). As can be seen, there are 26 states. A bidirectional edge between two states means the transition can be from left to right (upon a request arrival), and also from right to left (upon a request departure). A unidirectional links means the state can only transit in one direction,

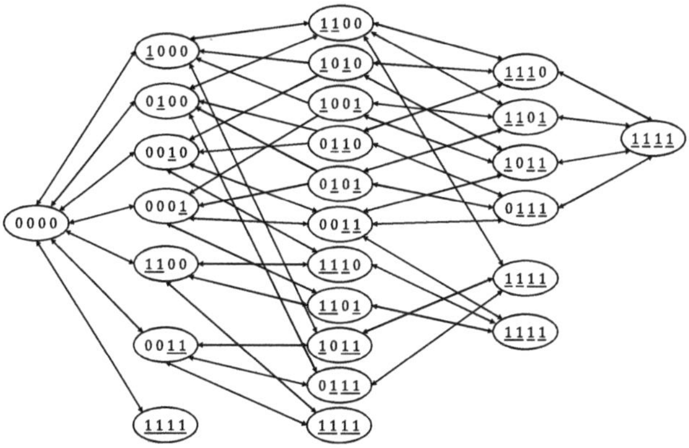

Fig. B.5 An example of the MC associated with the state transitions of asynchronous IFDMA systems. The total number of subcarriers $M = 4$, and we use the min-small-change policy to fill the bins. If the random bin-filling policy is used, all the edges in the MC are bidirectional

in particular, from right to left. For instance, suppose the system is in state $\underline{1}\,0\,0\,0$. Upon an arrival of request of size one, the state can become $\underline{1}\,\underline{1}\,0\,0$, but not $\underline{1}\,0\,\underline{1}\,0$ or $\underline{1}\,0\,0\,\underline{1}$ according to the min-small-change policy. On the other hand, all three states $\underline{1}\,\underline{1}\,0\,0$, $\underline{1}\,0\,\underline{1}\,0$ and $\underline{1}\,0\,0\,\underline{1}$ can potentially evolve to state $\underline{1}\,0\,0\,0$ upon a departure of a request of size one. Thus, the edge between states $\underline{1}\,0\,0\,0$ and $\underline{1}\,\underline{1}\,0\,0$ is bidirectional, while the edge between $\underline{1}\,0\,0\,0$ and $\underline{1}\,0\,\underline{1}\,0$ (or $\underline{1}\,0\,0\,\underline{1}$) is unidirectional.

To derive the blocking probability of the IFDMA system, we have to compute the steady-state distribution of the associated MC. However, the number of the states in the MC grows at an extremely fast rate (even faster than exponential) with the increase of M. As indicated by Corollary B.5, the MC becomes analytically intractable very quickly.

Corollary B.5 (Number of States in the MC) *Let $M = 2^m$ be the total number of subcarriers, and $f(m)$ be the number of states in the MC associated with the asynchronous IFDMA systems. We have*

$$f(m) = f^2(m-1) + 1 \geq 2^{2^m}. \tag{B.8}$$

A possible way to reduce the number of states in the MC is to merge the states that have the same blocking probabilities into a super state. For example, the states $\underline{1}\,0\,0\,0$, $0\,\underline{1}\,0\,0$, $0\,0\,\underline{1}\,0$, and $0\,0\,0\,\underline{1}$ can be merged together as a super state because they are equivalent in terms of blocking probability. The MC then becomes a consolidated MC, in which the number of super states is much less than the number of fine states in the original MC. As an example, the fine-state MC in Fig. B.5 can be transformed to the consolidated MC in Fig. B.6, where the number of super states is 11, while the number of fine states is 26.

Corollary B.6 (Number of Super States in the Consolidated MC) *Let $M = 2^m$, and $g(m)$ be the number of super states in the consolidated MC associated with the*

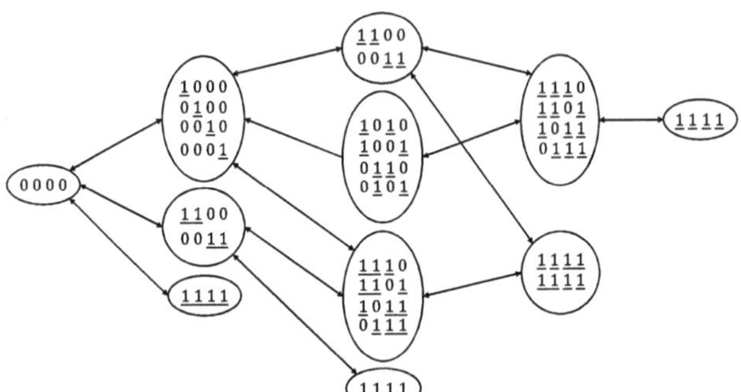

Fig. B.6 An example of the consolidated MC associated with the state transitions of asynchronous IFDMA systems. The total number of subcarriers $M = 4$, and we use the min-small-change policy to fill the bins *subcarrier allocation in asynchronous IFDMA systems. We have*

$$g(m) = \frac{1}{2}g^2(m-1) + \frac{1}{2}g(m-1) + 1 \geq 2^{2^{m-1}+1}. \tag{B.9}$$

Corollary B.6 shows that the number of super states is still too huge to handle. In view of the analytical intractability of the problem, we resort to simulation analysis to investigate the blocking probability of asynchronous IFDMA systems.

Monte-Carlo Simulations We simulate a multiple access system with one AP and multiple users. The total number of subcarriers is $M = 2^m = 1024$, and the users can request $N = 2^n, n = 0, 1, 2, \cdots, m$ subcarriers. The system is either IFDMA or OFDMA.

For requests of different sizes, we assume their arrivals follow independent Poisson process. In particular, the arrival rate (the number of arrivals per unit time) of a request is inversely proportional to its size, i.e., for a request of size 2^n, the arrival rate $\lambda_n = \frac{1}{2^n}\lambda$ where λ is a constant. Further, we assume the holding time of a request (i.e., the duration that the user stays in the system) follows exponential distribution with mean μ. Without loss of generality, we assume $\mu = 1$ for all requests.

For this system, the offered load (normalized by the total number of subcarriers) is given by

$$G = \frac{1}{2^m}\sum_{n=0}^{m}\frac{2^n\lambda_n}{\mu} = (m+1)2^{-m}\lambda. \tag{B.10}$$

The blocking probability can be computed by

$$P_B = \frac{\sum_{n=0}^{m} 2^n r_B(n)}{\sum_{n=0}^{m} 2^n r(n)} \tag{B.11}$$

where $r(n)$ is the number of arrived requests of size 2^n, and $r_B(n)$ is the number of requests of size 2^n that gets blocked. The normalized throughput of the system $S = 1 - P_B$.

Under random arrivals, a request is sure to be blocked in the case of instantaneous overload, wherein there are not enough unoccupied subcarriers left to fulfill the number of subcarriers requested. We note that even for a strictly nonblocking system as defined in Definition B.5, blocking due to instantaneous overload can still occur under the random request arrival model.

As summarized in Fig. B.4, instantaneous overload is the only cause for blocking in OFDMA. On the other hand, in IFDMA, a request can be blocked due to either instantaneous overload or fragmentation of bins, in which case the number of vacant

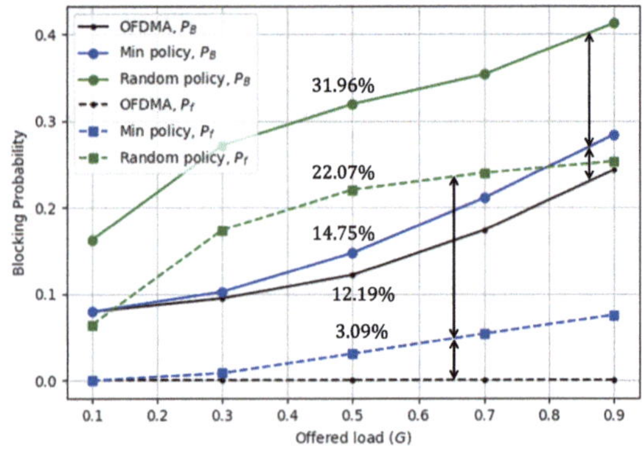

Fig. B.7 The blocking probabilities versus the offered load of OFDMA and IFDMA systems

bins is larger than the size of the request, but there is no unoccupied fillable bin subset available to fulfill the request.

In this context, we further define the "the probability of blocking due to fragmentation" to characterize the proportion of the "rejected load due to fragmentation" in the offered load:

$$P_f = \frac{\sum_{n=0}^{m} 2^n r_f(n)}{\sum_{n=0}^{m} 2^n r(n)} \tag{B.12}$$

where $r_f(n)$ is the number of requests of size 2^n that gets blocked when the system backlog (i.e., the number of occupied subcarriers) is less than or equal to $2^m - 2^n$ (i.e., blocking only due to the fragmentation of bins). Note that $P_f = 0$ in OFDMA, because a request of size 2^n can be granted as long as the system backlog is less than or equal to $2^m - 2^n$.

Figure B.7 presents the blocking probabilities (P_B and P_f) versus the offered load to the system. Three schemes are simulated: OFDMA, IFDMA with random bin-filling policy, and IFDMA with the min-small-change bin-filling policy. They are abbreviated as "OFDMA", "Random policy", and "Min policy" in the figure, respectively.

Let us first compare the two bin-filling polices in IFDMA systems. As shown, the min-small-change policy performs significantly better than the random policy. At a moderate offered load 0.5, the min-small-change policy reduces the blocking probability P_B by 17.21%.

The performance of P_f is even more revealing to showcase the importance of a good bin-filling policy. For the min-small-change policy, 3.09% of the offered load are rejected due to the fragmentation of bins. On the other hand, for the random policy, 22.07% of the offered load are rejected due to the fragmentation of bins. The min-small-change policy reduces the blocking probability due to the fragmentation of bins by 18.98%.

Then, we benchmark the blocking probabilities of IFDMA systems (with the min-small-change policy) against the OFDM systems. As can be seen, as expected, IFDMA systems has higher blocking probability than the OFDMA systems, but the

Fig. B.8 The blocking probabilities of OFDMA and IFDMA versus the offered load to the system. The solid curves are copied from Fig. B.7, and the dashed curves are the results of limiting the stream size

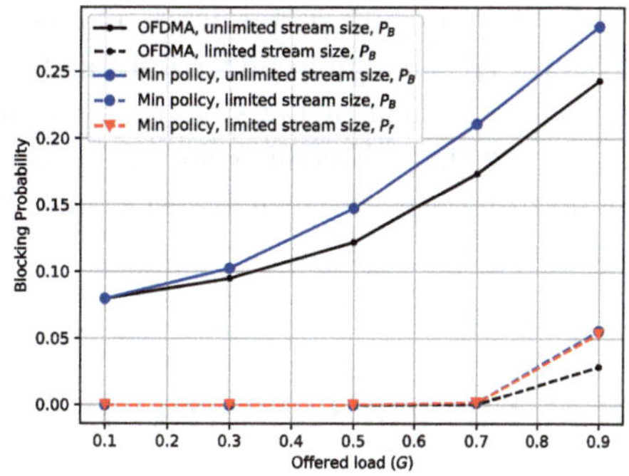

gap is not large. When the offered load is 0.5, the performance gap is only 2.56% in terms of P_B, and 3.09% in terms of P_f.

In the simulation of Fig. B.7, requests of different sizes (from size 1 to size M) are mixed together and fed into the systems. Additional simulations are performed by us to investigate the blocking probabilities of requests of different sizes. We found that, unsurprisingly, the requests with larger size have higher blocking probabilities, in particular, the requests of size M are mostly blocked (again this is obvious because even there is one occupied bin only, a request of size M will be blocked). This is the case for both IFDMA and OFDMA.

Whether for IFDMA or OFDMA systems, it is not a good idea to support a mix of large and small requests at the same time. This is because large requests will be blocked most of the time, effectively eliminating the use of the system to fulfill large requests.

In this light, we advocate not mixing requests of small and large sizes into the same system. To verify the benefits of this, we repeat the simulations in Fig. B.7, but change the composition of the traffic fed into the system. Specifically, given the same offered load to the system, the sizes of the request in Fig. B.7 can be 2^n where $n = 0, 1, \cdots, m$; however in the new simulation, we limit the size of the requests to 2^n where $n = 0, 1, \cdots, m/2$.[1]

The simulation results are presented in Fig. B.8. As shown, for the same offered load, the blocking probabilities P_B of both OFDMA and IFDMA systems are reduced by much under the new traffic model where the mixing of small and large streams is prevented. In particular, for IFDMA, almost all the blocking are caused by the fragmentation of bins at a heavy load 0.9. This matches with our traffic design that the worst case of blocking due to fragmentation is retained, only the blocking due to instantaneous overload is largely reduced.

[1] The proof of Theorem B.3 indicates that the condition for strictly nonblocking is the most stringent when $n = m/2$. That is, $\arg\min_{0 \geq n \geq m}(2^{m-n} + 2^n)$ is in the ballpark of $m/2$. Thus, we limit the size of the request to be no more than $2^{m/2}$ so that we are still considering the "worst-case" of blocking due to fragmentation. Essentially, we are reducing blocking due to instantaneous overload while retaining the worst cases of blocking due to fragmentation.

Furthermore, the performance gaps between asynchronous OFDMA and IFDMA systems are negligible under light and moderate offered load. This is reasonable because the span of the stream size is limited. We note in passing also that, to the extent that all requests are of the same size (i.e., 2^n for a fixed n), then there is no difference between IFDMA and OFDMA as far as blocking probability is concerned.

B.2 Efficient Transceiver Design

In this section, we will present an efficient transceiver design for IFDMA. To start with, we shall analyze the complexities of conventional IFDMA transceiver designs.

We consider an IFDMA system where multiple end nodes exchange messages with a common AP. In the uplink, each end node constructs an IFDMA data stream carried over a number of subcarriers and transmits the stream to the AP. The streams of different end nodes are carried on different subcarriers and they arrive at the AP at the same time. The receiver then extracts the information streams from different end nodes according to their subcarrier allocations. In the downlink, the AP maps different end node's IFDMA data streams to their respective subcarriers and transmits the aggregated IFDMA streams of all end nodes to the end nodes. The receiver at each end node then extracts its own IFDMA stream.

Most research works present a model of IFDMA as shown in Fig. B.9, which gives the schematic of the transceiver design for a single IFDMA stream.

IFDMA: Transmitter

The transmitter designs for uplink and downlink in IFDMA systems differ only in the number of streams to be processed. In the uplink, the transmitter (end node) has to construct only one IFDMA stream, while in the downlink, the transmitter (AP) has to construct multiple IFDMA streams targeted for different end nodes.

Let us first focus on the transmitter design of a single IFDMA stream. As discussed in the previous section, Fig. B.9 also presents the frequency-domain real-

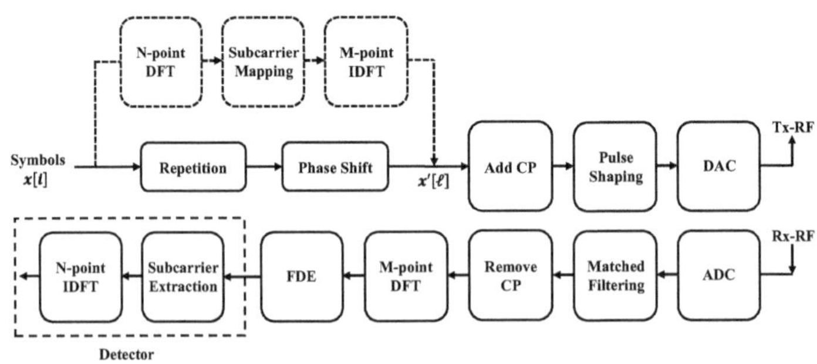

Fig. B.9 The conventional transceiver of a single IFDMA stream ization (the dashed modules) and the time-domain realization (the solid modules). Again, let M be the number of subcarriers available in the overall IFDMA system. To construct an IFDMA stream with $N \le M$ subcarriers, the transmitter (i.e., the end node in uplink or the AP in downlink) groups the symbols to be transmitted into blocks of size N. Denote a block of N time-domain samples by $\{x[n] : n = [N]\}$, where $[N] = 0, 1, 2, \ldots, N - 1$. For the frequency-domain realization, the time-domain samples are transformed to frequency-domain samples by an N-point DFT, and then mapped to N evenly spaced subcarriers among the M subcarriers. In practice, M can be large and is typically chosen to be a power of 2 for simple and efficient implementation of radix-2 FFT/IFFT. We assume M is a power of 2 throughout this appendix. For an IFDMA stream, the distance between the indexes of two adjacent subcarriers must be M/N. That is, N must be a divisor of M so that M/N is an integer. Specifically, the indexes of the N selected subcarriers are

$$\left\{d + i\frac{M}{N} : i = [N]\right\}, \tag{B.13}$$

where $d \in [M/N]$ is a constant frequency shift.

In this way, the N subcarriers are evenly spaced among the M subcarriers, and the unoccupied subcarriers are filled with 0. The transmitter then performs an M-point IDFT to generate the time-domain IFDMA stream $x'[\ell] : \ell = [M]$. We have

$$x'[\ell] = \frac{N}{M} e^{j\frac{2\pi\ell}{M}d} x[\ell \bmod N], \quad \ell = [M]. \tag{B.14}$$

When $d = 0$, the IFDMA time-domain signal is simply $x'[\ell] = \frac{N}{M} x[\ell \bmod N]$, and the subcarriers occupied are $\{iM/N : i = [N]\}$. In other words, $x'[\ell]$ is simply the block of the original time-domain samples $x[n]$ repeated M/N times normalized by N/M. When $d \neq 0$, the extra phase term $e^{j\frac{2\pi\ell}{M}d}$ in (B.14) corresponds to frequency upshifting of the selected subcarriers. Equation (B.14) also indicates that a simpler and more direct time-domain transmitter design is possible. This design bypasses the N-point DFT, subcarrier mapping and M-point IDFT.

Complexity Analysis Figure B.9 only shows the generation of one IFDMA stream at the transmitter. In general, if the transmitter has to construct K IFDMA streams (such as that in downlink), the blocks between $x[n]$ and $x'[\ell]$ in Fig. B.9 have to be repeated K times. Table B.2 compares the complexities of the time-domain transmitter and frequency-domain transmitter in terms of the number of complex multipliers required.

For uplink, we assume that an end node transmits only one IFDMA stream. When implemented in hardware (e.g., FPGA), a time-domain transmitter following (B.14) has to deploy M complex multipliers. A frequency-domain transmitter, on the other hand, needs to deploy many N-point DFTs of various sizes, i.e., $\{N = 2^n : n = 1, 2, \ldots, \log M\}$, so that the end nodes can be allocated different numbers of subcarriers at different times according to their needs. For OFDM modulation,

Table B.2 Complexities of conventional transceivers for IFDMA in terms of number of complex multipliers

	Transmitter (time domain)	Transmitter (frequency domain)	Receiver
Uplink	M	$\frac{3}{2}M\log_2 M$	$\frac{1}{4}M\log_2^2 M + \frac{1}{2}M\log_2 M$
Downlink	M^2	$\frac{1}{4}M\log_2^2 M + \frac{1}{2}M\log_2 M$	$\frac{3}{2}M\log_2 M$

an M-point IDFT is required. Let $N = 2^n$ and $M = 2^m$ where n and m are positive integers. The overall complexity of the frequency-domain implementation is given by

$$\sum_{n=1}^{m}\left(\frac{N}{2}\log_2 N\right) + \frac{M}{2}\log_2 M \approx \frac{3}{2}M\log_2 M. \tag{B.15}$$

Equation (B.15) follows from that an N-point IDFT requires $\frac{N}{2}\log_2 N$ complex multipliers using Cooley-Tukey FFT.

For downlink, the signal transmitted by the AP is an aggregate of multiple IFDMA streams to the end nodes. The number of IFDMA streams equals the number of end nodes. In the extreme case, there are M end nodes, each of which occupies one subcarrier. A time-domain transmitter has to deploy M^2 complex multipliers, because the transmitter has to construct M IFDMA streams in the extreme case, each requiring M complex multipliers. The complexity of a frequency-domain transmitter, on the other hand, is

$$\sum_{n=1}^{m}\frac{M}{N}\left(\frac{N}{2}\log_2 N\right) + \frac{M}{2}\log_2 M \approx \frac{1}{4}M\log_2^2 M + \frac{M}{2}\log_2 M, \tag{B.16}$$

because (1) the M-point IDFT is common to all IFDMA streams; (2) we need N-point IFFTs of various sizes (from $n = 1$ to $n = m$) to accommodate IFDMA streams of different sizes; (3) for a given N, up to M/N end nodes can simultaneously request for N subcarriers. Thus, we need M/N N-point IDFTs for each possible N.

IFDMA: Receiver

For a single IFDMA stream, the receiver can be designed as an exact inverse mapping of the transmitter design in the frequency domain, as shown in Fig. B.9. Specifically, for both uplink and downlink, the received signal is an aggregate of multiple IFDMA streams, and the M-point DFT and frequency-domain equalization (FDE) module are common to all received streams. To extract and decode a single IFDMA stream with N subcarriers, a detector containing a subcarrier extraction module and an N-point IDFT is used.

Complexity Analysis For downlink, there is only one desired IFDMA stream at the receiver (end node). The receiver, however, needs detectors of all possible values of $\{N = 2^n : n = 1, 2, \ldots, m\}$ to decode since the receiver may be allocated IFDMA streams of different sizes at different times. The complexity of the downlink receiver is the same as (B.15).

For uplink, the receiver (AP) has to decode all incoming IFDMA streams transmitted by different end nodes. If there are K end nodes, we need K detectors after the FDE, one for each IFDMA stream/end node. When implemented in hardware (e.g., FPGA), the IFDMA receiver not only needs to deploy N-point IFFTs of various sizes, but also for each N, M/N N-point IDFTs are required because up to M/N end nodes can simultaneously request for N subcarriers. The overall complexity of the uplink receiver is then the same as (B.16).

Definition of Complexity

This appendix measures the complexities of different transceiver designs by the number of complex multipliers, not the number of complex multiplications. In other words, we are evaluating the hardware implementation complexity as opposed to the computational complexity. To understand why the FFTs/IFFTs of different size should be considered when calculating the implementation complexity, it is important to look at the transceiver design from a hardware perspective instead of a software perspective. When implemented in hardware (e.g., FPGA or ASIC), if we want to compute a 2^n-point IFFT, there must be an IFFT hardware module of size 2^n; and if we want to compute a 2^m-point IFFT, there must be an IFFT hardware module of size 2^m. In other words, hardware implementation is static. All signal processing modules must be deployed on the chip beforehand, and be ready for incoming signal. Thus, FFTs/IFFTs of all possible sizes must be deployed to meet all possible rate requirements. That is at least the case with the conventional IFDMA transceiver design. The thesis of this appendix is that, by exploiting the coincidence between the structure of IFDMA and the structure of Cooley-Tukey FFT/IFFT decomposition, the FFTs/IFFTs of different sizes that are required of the IFDMA streams of different sizes can be computed using one single FFT/IFFT of a large size. This leads to a much simplified transceiver design.

B.2.1 Complexities of Conventional Transceivers for Multi-IFDMA

A key problem in IFDMA systems is how to allocate subcarriers to end nodes. Compared with OFDMA/LFDMA, IFDMA have stricter requirements on the subcarrier mapping: (1) the subcarriers of an IFDMA stream must be evenly spaced; (2) each IFDMA stream can only have a power-of-2 of subcarriers when M is a power of 2. These two restrictions lead to a misconception that IFDMA is less flexible and efficient than LFDMA.

In the previous section, we have shown that when numbers of subcarriers requested by all end nodes are powers of 2, subcarrier allocation in IFDMA is no less efficient than in LFDMA and that full loading is possible.

Bit-Reversal Subcarrier Allocation

- Instead of allocating subcarriers to nodes, we first focus on allocating $M = 2^m$ bins to nodes. Sort all requests in the descending order of their n, and allocate bins to them in the order of bin 0 to bin M. That is, the nodes requesting for more subcarriers will be considered first.
- Perform bit-reversal mapping from bins to subcarriers. If a bin with binary index $b_{m-1}b_{m-2}\ldots b_1b_0$ is allocated to a node, then the subcarrier with binary index $b_0b_1b_{m-2}b_{m-1}$ will be allocated to this node.

When the request sizes are non-power-of-2, the above scheme does not work since M/N is not an integer. In this case, we can use the multiple IFDMA scheme. For example, consider a node that requires 6 subcarriers. Obviously, we cannot construct a single IFDMA stream because 6 is not a power of 2. However, we can allocate two IFDMA streams to the node, one with 4 subcarriers and one with 2 subcarriers. The signal of the node is the sum of the two IFDMA streams.

With the new Multi-IFDMA scheme, the resource allocation of IFDMA system is as flexible as LFDMA/OFDMA. We will refer to the conventional IFDMA system that allocates only one IFDMA stream to each end node as the *Single-IFDMA*. Although Multi-IFDMA is more flexible than Single-IFDMA, the implementation of Multi-IFDMA is also more complex than the Single-IFDMA if we use the framework of the conventional transceiver in Fig. B.9.

Table B.3 summarizes the complexities of the multi-IFDMA systems if we use the conventional transceiver designs. Multi-IFDMA is more complex than Single-IFDMA because in Multi-IFDMA, both the transmitter and receiver must be able to process multiple IFDMA streams, whether in uplink or downlink. In the following, we shall present IFDMA transceivers with significantly lower complexity than the conventional designs. Importantly, our transceiver designs are unified designs compatible with both Single-IFDMA and Multi-IFDMA systems.

Our transceivers are frequency-domain transceivers. Unlike the conventional frequency-domain design in Fig. B.9, we do not require multiple N-point DFTs/IDFTs of various N (the main source of the complexity there). It is possible

Table B.3 Complexities of conventional transceivers for Multi-IFDMA in terms of number of complex multipliers

	Transmitter (time domain)	Transmitter (frequency domain)	Receiver
Uplink	M^2	$\frac{1}{4}M\log_2^2 M + \frac{1}{2}M\log_2 M$	$\frac{1}{4}M\log_2^2 M + \frac{1}{2}M\log_2 M$
Downlink			

Table B.4 Comparison of the complexities of different IFDMA receivers

		Conventional receiver	Our design (with FDE)	Our design (without FDE)
Single IFDMA	Uplink	$\frac{1}{4}M\log_2^2 M + \frac{1}{2}M\log_2 M$	$\frac{1}{2}M\log_2^2 M + \frac{1}{2}M\log_2 M$	$\frac{1}{2}M\log_2 M$
	Downlink	$M\log_2 M + \frac{1}{2}M\log_2 M$		
Multi IFDMA	Uplink	$\frac{1}{4}M\log_2 M + \frac{1}{2}M\log_2 M$		
	Downlink			

Table B.5 Comparison of the complexities of different IFDMA transmitters

		Transmitter (time domain)	Transmitter (frequency domain)	Our design
Single-IFDMA	Uplink	M	$\frac{3}{2}M\log_2 M$	$\frac{1}{2}M\log_2 M$
	Downlink	M^2	$\frac{1}{4}M\log_2^2 M + \frac{1}{2}M\log_2 M$	
Multi-IFDMA	Uplink	M^2	$\frac{1}{4}M\log_2^2 M + \frac{1}{2}M\log_2 M$	
	Downlink			

to use a single M-point DFT/IDFT to multiplex (at the transmitter) or demultiplex (at the receiver) the desired IFDMA streams of different sizes, regardless of the number of IFDMA streams and the number of end nodes. The detailed receiver and transmitter designs are presented in Sects. B.2.2 and B.2.3, respectively.

Tables B.4 and B.5 summarize the complexities of our transceiver designs and the conventional IFDMA transceiver in terms of the number of complex multipliers required. As shown, our designs require the minimum resources among all designs except for the uplink Single-IFDMA case, where an end node only needs to construct a single IFDMA stream. In that case, the time-domain implementation in (B.14) is the simplest way to generate an IFDMA stream. Combining Tables B.4 and B.5, the complexities of our new transceivers are simpler than the conventional transceiver by at least an order of $\log_2 M$ in all communication scenarios.

B.2.2 Receiver Design

In the conventional design, the receiver for a single IFDMA stream is an exact inverse mapping of the transmitter. An M-point DFT followed by a detector are required. When there are K IFDMA streams, the M-point DFT is common to all streams, but the detector has to be repeated K times. As given in Table B.4, to accommodate requests of all sizes, the complexity of the receiver is $M\log_2 M + \frac{1}{2}M\log_2 M$ in the downlink of the Single-IFDMA systems, and $\frac{1}{4}M\log_2^2 M + \frac{1}{2}M\log_2 M$ otherwise.

In this section, we first dissect the inner structure of IDFT to gain insight into its relationship with IFDMA streams. Based on the insight, we devise a simpler unified

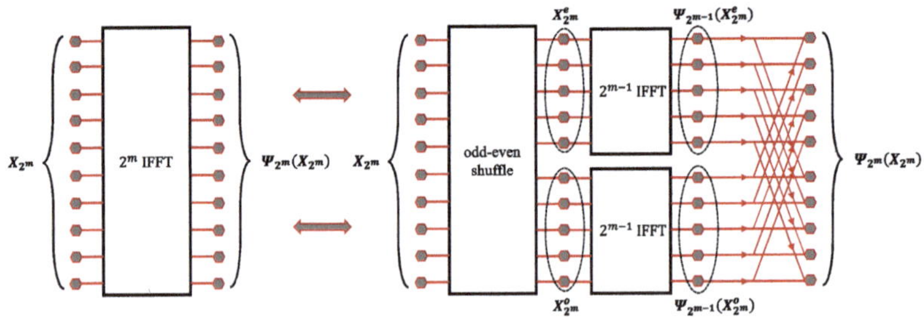

Fig. B.10 Decomposition of 2^m-point IDFT into two 2^{m-1}-point IDFTs. The weights in the butterfly networks are omitted

detector which uses a single M-point IDFT to demultiplex and extract the streams from all transmitters. This unified receiver is compatible with both Single-IFDMA and Multi-IFDMA systems.

B.2.2.1 IDFT Decomposition

A 2^m-point DFT/IDFT can be decomposed into two 2^{m-1}-point DFT/IDFTs, one for odd-indexed samples, and the other for even-indexed samples. Specifically, denote by Ψ_{2^m} the 2^m-point IDFT operation, and X_{2^m} the input vector to the IDFT. We have

$$\Psi_{2^m}\left(X_{2^m}\right)[\ell] = \frac{1}{2}\Psi_{2^{m-1}}\left(X^e_{2^m}\right)[\ell \bmod 2^{m-1}] + \frac{1}{2}W^{\ell}_{2^m}\Psi_{2^{m-1}}\left(X^o_{2^m}\right)[\ell \bmod 2^{m-1}], \tag{B.17}$$

where $\ell = [2^m]$, and vectors $X^e_{2^m}$ and $X^o_{2^m}$ contain even-indexed elements and odd-indexed elements of X_{2^m}, and $W_{2^m} = e^{j\frac{2\pi}{2^m}}$.

The decomposition in (B.17) is illustrated in Fig. B.10. As shown, the 2^m-point IDFT operation on input vector X_{2^m} is equivalent to performing (1) an odd-even shuffle to split X_{2^m} into two vectors $X^e_{2^m}$ and $X^o_{2^m}$; (2) two parallel 2^{m-1}-point IDFT to get $\Psi_{2^{m-1}}\left(X^e_{2^m}\right)$ and $\Psi_{2^{m-1}}\left(X^o_{2^m}\right)$; and (3) multiple butterfly combinations following (B.17) to finally get $\Psi_{2^m}\left(X_{2^m}\right)$.

In particular, this decomposition process is recursive. Each 2^{m-1}-point IDFT embedded in the 2^m-point IDFT can be further decomposed into two 2^{m-2}-point IDFTs. The overall idea of divide and conquer is the foundation of Cooley-Tukey FFT/IFFT algorithms.

B.2.2.2 IFDMA Detector Design

Theorem B.4 below is a corollary of the above IDFT decomposition. It forms the basis of our IFDMA transceiver designs.

Theorem B.4 *Denote by* $\{X[k] : k = [2^m]\}$ *the input to a* 2^m*-point IDFT. After* t *recursive decompositions inside the IDFT, there are* 2^t *parallel* 2^{m-t}*-point IDFTs. The inputs to the* d'*-th* 2^{m-t}*-point IDFT,* $d' = [2^t]$*, are* $\{X[d + j2^t] : j = [2^{m-t}]\}$*, where* d *is the* t*-bit-reversal mapping of* d'*.*

Proof Let $b_{m-1}b_{m-2}\ldots b_1b_0$ be the binary representation of positions occupied by subcarriers at the input of the overall 2^m-point IDFT. With respect to Fig. B.10, the positions are arranged in an ascending order from top to bottom (i.e., position 00…00 is at the top and position 11…11 is at the bottom). Thus, $b_{m-1}b_{m-2}\ldots b_1b_0$ is also the subcarrier index k.

With respect to Fig. B.10, after the first-stage odd-even shuffle of size 2^m, a subcarrier at original position $b_{m-1}b_{m-2}\ldots b_1b_0$ will end up at position $b_0b_{m-1}b_{m-2}\ldots b_1$ (i.e., the odd-even shuffle executes a cyclic right-shift of $b_{m-1}b_{m-2}\ldots b_1b_0$ so that the new position is $b_0b_{m-1}b_{m-2}\ldots b_1$). After the two second-stage odd-even shuffles of size 2^{m-1} within the 2^{m-1}-point IDFTs in Fig. B.10, a subcarrier at original position $b_{m-1}b_{m-2}\ldots b_1b_0$ will end up at position $b_0b_1b_{m-1}b_{m-2}\ldots b_2$ (i.e., the two smaller odd-even shuffles execute a cyclic right shift of $b_0b_{m-1}b_{m-2}\ldots b_1$, excluding b_0, so that the new position is $b_0b_1b_{m-1}b_{m-2}\ldots b_2$). In general, after the t-th odd-even shuffles of size 2^{m-t+1}, a subcarrier at original position $b_{m-1}b_{m-2}\ldots b_1b_0$ will end up at position $b_0\ldots b_{t-1}b_{m-1}b_{m-2}\ldots b_t$. That is, the position $b_0\ldots b_{t-1}b_{m-1}b_{m-2}\ldots b_t$ is the position occupied by subcarrier $b_{m-1}b_{m-2}\ldots b_1b_0$ when it reaches the input to a 2^{m-t}-point IDFT.

Consider the d'-th 2^{m-t}-point IDFT. The positions of the inputs to this IDFT are $\{2^{m-t}d' + 0, 2^{m-t}d' + 1, \ldots, 2^{m-t}d' + 2^{m-t} - 1\}$. Let the binary representation of the input positions be $b_0\ldots b_{t-1}b_{m-1}b_{m-2}\ldots b_t$, where $b_0\ldots b_{t-1}$ are fixed to be the binary representation of integer d', and $b_{m-1}b_{m-2}\ldots b_t$ ranges from 00…00 (corresponding to integer 0) to 11…11 (corresponding to integer $2^{m-t} - 1$). This binary representation is consistent with the decimal representation.

According to the subcarrier-to-position mapping, the subcarriers occupying the above positions are subcarriers $b_{m-1}b_{m-2}\ldots b_tb_{t-1}\ldots b_0$ where $b_{m-1}b_{m-2}\ldots b_t$ ranges from 0 to $2^{m-t} - 1$, and $b_{t-1}\ldots b_0$ is the t-bit-reversal mapping of $b_0\ldots b_{t-1}$. Let d be the integer corresponding to $b_{t-1}\ldots b_0$. Thus, the subcarriers mapped to the inputs of the d'-th 2^{m-t}-point IDFT are indexed by $\{d + j2^t : j = [2^{m-t}]\}$. □

Consider an IFDMA system with a total of $M = 2^m$ subcarriers. Suppose a node A requests for N subcarriers, where N is a power of 2 (i.e., $N = 2^n$, $n = 0, 1, \ldots, m$). For non-power-of-2 N, we can use the Multi-IFDMA scheme to divide the N requests to several sub-requests, each of which is a power of 2.

Following the principle of IFDMA, the subcarriers allocated to node A are indexed by $\{d + j2^{m-n} : j = [2^n]\}$, where constant d can be any integer in $[2^{m-n}]$. To decode this stream at the AP, the detector has to perform (1) demultiplexing, to extract the subcarriers $\{d + j2^{m-n} : j = [2^n]\}$ assigned to node A, and (2) 2^n-point IDFT, to transform the frequency domain equalized data to the time domain.

Let $m - n = t$ in Theorem B.4. An immediate result is that after $m - n$ recursive decompositions inside the 2^m-point IDFT, the inputs to the d'-th 2^n-point IDFT are

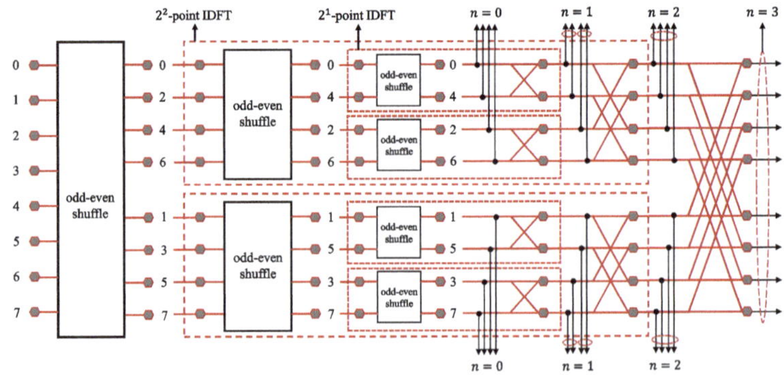

Fig. B.11 Illustration of the IFDMA detector for $M = 2^3$. A single M-point IDFT can accommodate IFDMA streams with all possible request size $N = 2^n$. Switching elements are inserted within the IDFT network to tap the target streams

the data on subcarriers $\{d + j2^{m-n} : j = [2^n]\}$, $d = [2^{m-n}]$, and d' is a $(m - n)$-bit-reversal mapping of d. Notice that these subcarriers are exactly the subcarriers allocated to node A.

The above result implies that the two processes required to extract the data stream from node A, demultiplexing and 2^n-point IDFT, can be inherently performed within the 2^m-point IDFT. In other words, a single 2^m-point IDFT at the receiver can be used to simultaneously demultiplex and decode IFDMA streams of all sizes. This is in contrast to the conventional design implied by Fig. B.9, which uses multiple detectors of different sizes to do so.

Specifically, an IFDMA stream with 2^n subcarriers can be extracted after a 2^n-point IDFT within the overall 2^m-point IDFT. We can simply insert 2^n 1×2 switching elements at the outputs of the 2^n-point IDFTs to extract the target data on the 2^n subcarriers.

Figure B.11[2] shows an example of the unified detector where $m = 3$ ($M = 8$). The detector works in the following way:

[2] We note that the schematic of Fig. B.11 also gives an interpretation as to why the bit-reversal subcarrier allocation algorithm works. In the algorithm, a user is first allocated a set of "bins" with contiguous indexes. The bin indexes then undergo a bit-reversal to obtain the corresponding indexes of the interspersed subcarriers allocated to the node. In the schematic of Fig. B.11, the subcarriers allocated are at the input on the left-hand side. The m stages of odd-even shuffles as a whole are equivalent to a bit-reversal mapping. Hence, it maps the subcarriers back to contiguous "bins". Throughout the different stages of the subsequent butterfly networks, horizontal line 000 is bin 000, horizontal line 001 is bin 001, horizontal line 010 is bin 010, ..., and horizontal line 111 is bin 111. The butterfly networks do not cause the bin positions to change; just that the content of one bin may be modified by contents of other bins through linear combinations within the butterfly networks. The net effect of the butterfly networks is to convert the frequency-domain signals in the bins to the original time-domain signals.

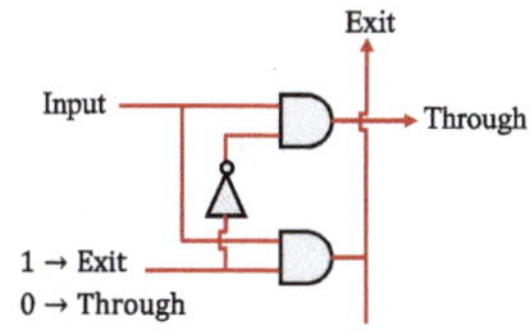

Fig. B.12 A design for the 1×2 switching element using AND and NOT gates. When the control port is set to '0', the "Input" will be connected to "Through"; when the control port is set to '1', the "Input" will be connected to 'Exit'

1. The data after FDE is indexed from 0 to 7 on the LHS of Fig. B.11. The overall structure is a 2^3-point IDFT, hence an IFDMA stream that occupies 2^3 subcarriers ($n = 3$) can be extracted at the rear end (if there is such an IFDMA stream allocated to an end node).
2. In the first decomposition, the 2^3-point IDFT is split into two 2^2-point IDFTs. In accordance with Theorem B.4, the input to the d'-th 2^2-point IDFT is indexed by $\{d + j2^1 : j = [2^2]\}$, where d and d' can be $d = 0$ ($d' = 0$) or $d = 1$ ($d' = 1$). Thus, IFDMA streams that occupy 2^2 subcarriers ($n = 2$) can be extracted at the outputs of the 2^2-point IDFTs (if there are such allocated IFDMA streams).
3. In the second decomposition, each 2^2-point IDFT is again split into two 2^1-point IDFTs. In accordance with Theorem B.4, the input to the d'-th 2^1-point IDFT is indexed by $\{d + j2^2 : j = [2^1]\}$, where d and d' can be $d = 0$ ($d' = 0$), $d = 1$ ($d' = 2$), $d = 2$ ($d' = 1$), or $d = 3$ ($d' = 3$). Thus, IFDMA streams that occupy 2^1 subcarriers ($n = 1$) can be extracted at the outputs of the 2^1-point IDFTs (if there are such allocated IFDMA streams).
4. Lastly IFDMA streams that occupy only 1 subcarrier ($n = 0$) can be extracted right after the odd-even shuffles of size 2.

Figure B.12 shows a possible design for the 1×2 switching element. As can be seen, the switch element has one input port "Input", one control port, and two output ports "Exit" and "Through". The "Input" port is connected to the previous stage within the butterfly networks (see Fig. B.11), and the "Through" port is connected to the next stage within the butterfly networks. When the control port is set to '0', the switch is in "Through" state, and the input data will flow through the switch to the next stage in the butterfly networks. On the other hand, when the control port is set to '1', the switch is in Exit state, and the input data will be diverted to the "Exit" port and be extracted.

Overall, the detector consumes $\frac{1}{2}M \log_2 M$ complex multipliers. Together with the M-point DFT, the receiver consumes $M \log_2 M$ complex multipliers (see Table B.4 for additional reference). The number of 1×2 switching elements needed to tap the desired streams is $Mlog_2M$.

Selection of Taps

Another problem after tapping is selection of taps that contain the desired data. As illustrated in Fig. B.11, there are $M(log_2 M + 1)$ taps in total. Only M of the taps may contain the desired data of the allocated IFDMA streams. Thus, a subsequent switch is required to extract the M target symbols from the $M(log_2 M + 1)$ taps. We can consider the switching problem separately for each subcarrier. Specifically, if a particular subcarrier belongs to an IFDMA stream with only one subcarrier, it will be extracted after the m-stage odd-even shuffle (before the first butterfly combination); if it belongs to an IFDMA stream with two subcarriers, the particular subcarrier will be extracted after a 2-point IDFT, etc. Thus, the switching problem for each particular subcarrier is essentially a $(\log_2 M + 1)$ choose 1 problem. We can then use a $(\log_2 M + 1) \times 1$ switch to export the designed output. For all M subcarriers, a total of M $(\log_2 M + 1) \times 1$ switches are needed. Rather than considering the design of these switches as a separate problem, we will next present a design in which the tapping and selection of taps are simultaneously performed within the IDFT structure.

B.2.2.3 Design Refinement

Our first refinement is straightforward. In Fig. B.11, the net effect of the $\log_2 M$ stages of odd-even shuffles of different sizes is essentially a bit-reversal shuffle. Thus, we can replace all the odd-even shuffles with a single bit-reversal shuffle at the front-end. This is implied by the IDFT decomposition anyway.

The next refinement uses a tapping bus to combine the functions of tapping and selection of taps for each subcarrier. The tapping bus makes use of 2×2 switching elements rather than 1×2 switching elements. A 2×2 switching element connects two input ports "Input 1" and "Input 2" to two possible output ports "Through" and "Exit". It is ordinarily set to the "Through" state wherein "Input 1" is connected to "Through", and "Input 2" is connected to "Exit". In the "Exit" state, "Input 1" is connected to "Exit", and "Input 2" is connected to "Through". The "Exit" state allows a sample to be tapped.

An example is given in Fig. B.13, where $M = 8$. We assume subcarriers $\{1, 3, 5, 7\}$ are assigned to a node A, subcarriers $\{0, 4\}$ are assigned to a node B, subcarrier $\{6\}$ is assigned to a node C. As shown, after the bit-reversal shuffle, we equip each subcarrier with a tapping bus, into which $(\log_2 M + 1)$ 2×2 switching elements are inserted (right after the 2^n-point IDFT for $n = 0, 1, 2, 3$).

For node A, the data should be extracted at the exit of the 4-point IDFT. Thus, on the tapping buses of subcarriers $\{1, 3, 5, 7\}$, the 2×2 switching elements right after the 4-point IDFT are set to the Exit states, while the other switching elements will stay at the Through states. In this way, node A's data can be tapped off at the exit of the tapping bus. Note that the function of tap selection is integrated into the tapping-bus design. The Exit state at a particular stage and the Through state at all

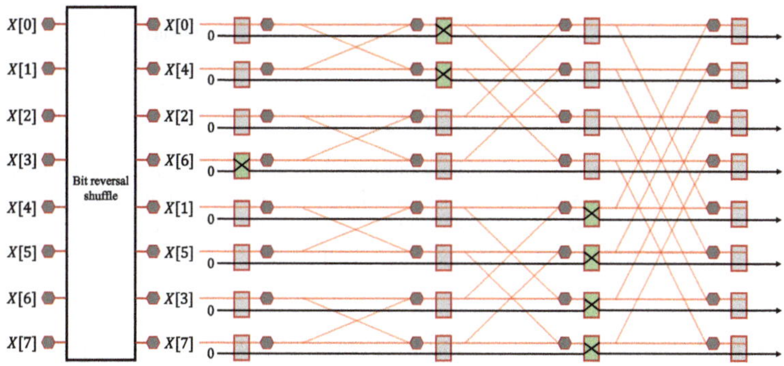

Fig. B.13 An example of the refined M-point IDFT detector for $M = 8$. The odd-even shuffles are combined into a single bit-reversal shuffle. The tapping and switching problems are solved together by a tapping-bus design

other stages basically selects the tap with the Exit state and eliminates the taps with the Through state.

Similarly, for the tapping buses of subcarriers {0, 4}, the 2×2 switching elements right after the 2-point IDFT are set to the Exit states, while the other switching elements stay in the Through states. Node B's data will be tapped out at the exit of the tapping bus. For the tapping bus of subcarriers {6}, the first 2×2 switching element is set to the Exit state, the switching elements after stay at the Through states. Note that subcarrier {2} is unassigned, hence all the switching elements stay at the Through states.

Simple Design of 2×2 **Switching Element** The implementation of a 2×2 switching element is in general more complex than that of a 1×2 switching element in Figs. B.11 and B.12. However, by exploiting an IFDMA property, we can simplify the design of the 2×2 switching elements tailored for our purpose. This custom design comes from an observation of Fig. B.13 that if an IFDMA stream being tapped off at a stage is also passed through to the subsequent butterfly combinations at later stages, other IFDMA streams will not be contaminated. This is because the IFDMA streams that could have been affected through the subsequent butterfly combinations must have already been tapped off at the same stage or at a prior stage. For example, consider subcarrier $X[6]$ in Fig. B.13. Suppose that $X[6]$ gets passed to both outputs Exit and Through. Via the Through output, $X[6]$ will affect the arithmetic at other subcarriers (all $X[i]$ in this example) through the subsequent butterfly combinations. However, the butterfly combinations affected by it are not useful anymore because the associated subcarriers have already been tapped off before they are affected. The reader may ask "How about $X[2]$?" If $X[2]$ were associated with an IFDMA stream, given that $X[6]$ is already allocated to another IFDMA stream with one subcarrier, $X[2]$ cannot be associated with an IFDMA stream with more than one subcarrier and must be an IFDMA stream with one subcarrier, in which case it will be tapped also at the same stage as $X[6]$.

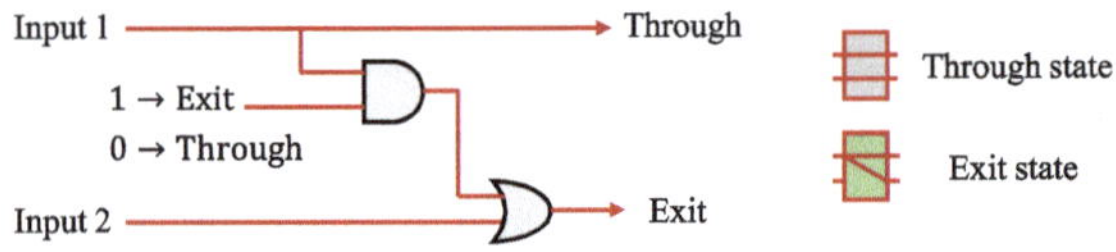

Fig. B.14 A simple design of the 2×2 switching element tailored for the IFDMA detector, using AND and OR gates. When the control port is set to '0', the "Input 1" will be connected to "Through", and the "input 2" will be connected to "Exit"; when the control port is set to '1', the "Input 1" will be connected to both "Through" and 'Exit' ports

This observation implies that in the Exit state, we can connect "Input 1" to both the "Exit" and "Through" output ports simultaneously. The design of the associated switching element is rather simple, as shown in Fig. B.14. When the control port is set to '0', the switch is in the Through state, "Input 1" will be connected to the "Through" port, and "Input 2" will be connected to the "Exit" port. On the other hand, when the control port is set to '1', the switch is in the Exit state, the "Input 1" will be connected to both "Through" and "Exit" ports. We note that in this case, all other 2×2 switching elements of the tapping bus must be set to Through state, so that Input 2 of this 2×2 must contain 0, and the output of the OR gate will simply follow Input 1.

We can thus replace the 2×2 switching elements in Fig. B.13 by our tailored 2×2 switching elements for our final design. Overall, this IFDMA detector is simply an M-point IFFT with tapping buses. It consumes $\frac{1}{2}M \log_2 M$ complex multipliers, and $M(\log_2 M + 1)$ 2×2 switch elements.

B.2.2.4 Extension: Receiver Without FDE

In the original IFDMA receiver design, the input will first go through an M-point DFT, followed by a FDE module, and an M-point IDFT. Extractions of IFDMA streams are performed within the M-point IDFT.

The objective of an equalizer is to compensate for the linear distortion (Inter-Symbol Interference) of the channel. However, after FDE and transformation back to the time domain, the noise on different time samples will be correlated. If we feed such samples to a channel decoder that treats the noise as uncorrelated, we will have suboptimal performance. A more advanced system will use a believe propagation (BP) algorithm or a Viterbi algorithm to perform channel decoding directly on the samples with ISI. For such systems, FDE is not required and our detector can be further simplified (another example where FDE is not required is the air-to-ground radar for surveillance or tracking.).

We first note that FFT can be designed as a reflection of IFFT, in which case the input data will first go through stages of butterfly networks before a bit reversal shuffle. Specifically, an IFFT can be written in a matrix form as

$$\mathbf{\Psi}_M(\mathbf{X}) = \mathbf{S}_m \mathbf{S}_{m-1} \ldots \mathbf{S}_2 \mathbf{S}_1 \mathbf{P}_0 \mathbf{X}, \tag{B.18}$$

where X is the input frequency-domain data to the M-point IDFT, P_0 is a bit-reversal permutation matrix, and $\{S_n : n = 1, 2, \ldots, m\}$ is the n-th stage butterfly network inside the IFFT.

Conversely, an FFT can be written in a matrix form as

$$X = P_0^{-1} S_1^{-1} S_2^{-1} \ldots S_{m-1}^{-1} S_m^{-1} \Psi_M(X), \tag{B.19}$$

where $S'_n = S^{-1}_{m+1-n}$, $n = 1, 2, \ldots, m$, is the n-th stage butterfly network inside the FFT.

In our previous design, the receiver is composed of an M-point DFT, an FDE module, and an M-point IDFT with tapping buses. If we bypass the FDE, the receiver is an M-point DFT followed by an M-point IDFT with tapping buses.

Given (B.18) and (B.19), we notice that the rear-end bit-reversal shuffle of the FFT cancels with the front-end bit reversal of IFFT. Operations of some of the stages of their butterfly networks also get canceled. Let us consider an IFDMA stream requesting 2^n subcarriers. In our previous design, this stream will be extracted after the n-th stage butterfly network inside the M-point IDFT. However, it is equivalent to extracting this stream right after the $(m - n)$-th stage butterfly network inside the M-point DFT, because all the butterfly combinations of the DFT and IDFT in between cancel out. Essentially, the M-point IDFT will not be needed anymore and can be eliminated.

As an example, Fig. B.15 shows the design corresponding to that of Fig. B.13 for which the FDE and M-point IDFT are removed. The IFDMA streams are directly extracted within the M-point DFT shown in Fig. B.15.

Note that in Fig. B.13, we extract an IFDMA stream with 2^n subcarriers after the n-th stage butterfly network of the IFFT. In Fig. B.15, we extract the same IFDMA stream at the $(m-n)$-th stage butterfly network of the FFT. Thus, an IFDMA stream with only one subcarrier will have to go through all stages of the FFT, while an IFDMA stream with all M subcarriers does not need to go through the FFT at all.

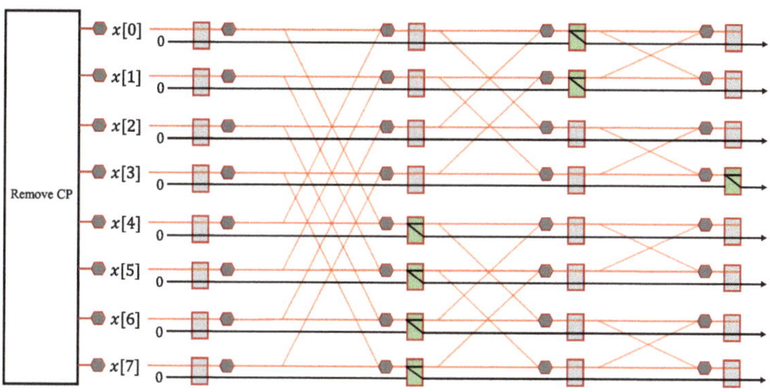

Fig. B.15 Receiver design for systems without frequency domain equalization. Only an M-point DFT is required to extract the desired IFDMA streams

To conclude this section, let us go back to Table B.4 and summarize the advantages of our new receiver design over the conventional receiver design.

1. Our new receiver design is a unified receiver design. It is compatible with both Single-IFDMA and Multi-IFDMA, and is independent of the number of IFDMA streams and the number of end nodes.
2. Our new receiver design is a much simpler receiver design with a significantly reduced number of complex multipliers. The tapping and switching problems are also solved by the clean tapping-bus design. Only $\frac{1}{2}M(\log_2 M + 1)$ 2×2 switching elements are needed.

We emphasize that the conventional design also faces a switching problem after the IDFTs of various sizes. That is, the desired streams still need to be selected from among the outputs of the IDFTs of various sizes. The associated switching problem is much more complex than that in our design because of the much higher output dimensionality of the multiple IDFTs.

B.2.3 Transmitter Design

The key insight that led to our simple IFDMA receiver design in the preceding section is that the inner structure of IDFT is coincident with the subcarrier mapping rule of IFDMA. It is not surprising that the same insight can also help simplify IFDMA transmitters. In this section, we first present a simple IFDMA transmitter design. After that, we examine the PAPR characteristics of multi-IFDMA transmitters.

B.2.3.1 Transmitter Design

Figure B.9 shows two transmitter designs for single-IFDMA, one in the time domain, and one in the frequency domain. If the transmitter has to construct multiple IFDMA streams (e.g., the downlink transmitter or the uplink transmitter in Multi-FDMA systems), there are also two ways to generate the aggregated IFDMA signal in the conventional designs. Specifically, the transmitter can

- generate each data stream separately in the time domain, then sum up the multiple streams thus generated;
- transform the data of each stream to the frequency domain using an N-point DFT (thus, multiple N-point DFT of different N will be needed, one for each stream), map the multiple frequency-domain streams to interleaved subcarriers, then transform the aggregated streams back to the time domain using a common M-point IDFT.

The insight we gained from Sect. B.2.2 can be used to simplify the frequency-domain transmitter. In particular, the multiple small N-point DFTs before subcarrier mapping can be replaced by a single M-point DFT. We have shown in Fig. B.11 how the N-point IDFTs of different sizes at the receiver can be replaced by an M-point IFFT. The same idea applies here. The only difference is that, instead of tapping off the decoded streams, the transmitter has to insert streams into the M-point DFT at different stages of the butterfly networks. We could use a DFT design that corresponds to a reflection of the IDFT design in Fig. B.11. With this design, an IFDMA stream with 2^n subcarriers should be inserted after the $(m - n)$-th stage butterfly network inside the DFT.

After replacing multiple N-point DFTs by one M-point DFT, the transmitter is now composed of an M-point DFT followed by an M-point IDFT. Following (B.18) and (B.19), the mirror-image designs of DFT and IDFT enable us to further simplify the transmitter architecture because the two bit-reversal shuffles (at the rear-end of DFT and the front-end of IDFT) cancel each other and they both can be removed. Also, the operations of some butterfly networks of DFT and IDFT also cancel each other. An IFDMA stream with 2^n subcarriers, originally be inserted after the $(m-n)$-th stage butterfly network inside the M-point DFT, can be inserted right after the n-th stage butterfly network inside the M-point IDFT. All the butterfly combinations in between cancel out. As a result, the M-point DFT can be removed altogether, and we only need an M-point IDFT at the transmitter.

As an example, Fig. B.16 presents a transmitter design with $M = 8$ $(m = 3)$. As shown, the transmitter design requires only a single M-point IDFT with tapping buses. The reader may observe that this design is a mirror image of the DFT design of the receiver in Fig. B.15. Over there, the IDFT is redundant and has been removed, here the DFT is redundant and has been removed. In Fig. B.16, the three IFDMA streams are the same as those in Fig. B.15. The first stream occupies 4 subcarriers indexed by {1, 3, 5, 7}, the second stream occupies 2 subcarriers indexed by {0, 4}, and the third stream occupies 1 subcarrier indexed by {6}. In general, an IFDMA stream with 2^n subcarriers is inserted right after the n-th stage butterfly network inside the M-point IDFT.

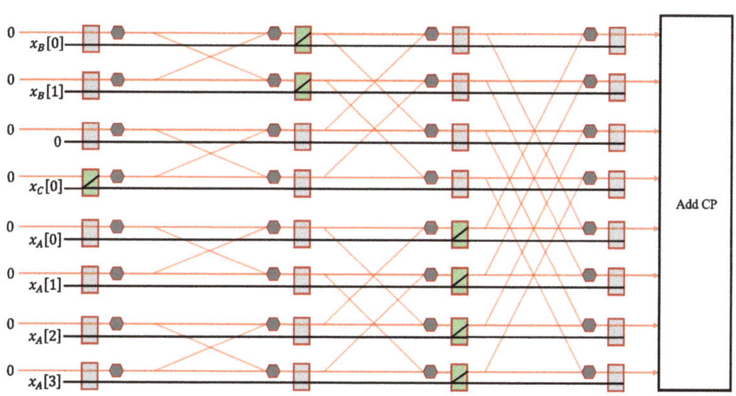

Fig. B.16 An example of the new transmitter design which requires an M-point IDFT only

Finally, let us go back to Table B.5 and compare the complexities of different transmitter designs. As can be seen, our design requires only $\frac{1}{2}M \log_2 M$ complex multipliers at the transmitter. This is the minimal resources required among all other designs except for the uplink Single-IFDMA case. In that case, each end node only needs to transmit a single IFDMA stream, and (B.14) gives the simplest implementation.

Figures B.15 and B.16 present a final receiver-transmitter pair for IFDMA. An interesting observation is that the input and output to the receiver and transmitter are both time-domain signals. The IFFT/FFT are actually not transforming signals between time domain and frequency domain. For the transmitter, it dispatches each data symbol to a number of multiple time-domain samples, each of which is the original data symbol multiplying by a phase term. For the receiver, it collects the multiple time-domain samples back into the data symbol. If we combine Tables B.4 and B.5, the complexity of our new transceivers is simpler than the conventional transceiver by at least an order of $\log M$ in all communication scenarios.

B.2.3.2 PAPR Performance of Multi-IFDMA

Multi-IFDMA systems achieve the same level of flexibility as OFDMA/LFDMA systems in terms of subcarrier allocation by allowing each end node to be allocated multiple IFDMA streams. The complexity of the Multi-IFDMA transceiver is much lower than that of LFDMA, and is equal to that OFDMA if the IFDMA receiver does not perform FDE.

However, an issue left open is the PAPR property of the Multi-IFDMA systems. In particular, does the Multi-IFDMA transmitter inherit the low PAPR property of the Single-IFDMA transmitter? This subsection addresses this question.

The signal of a single IFDMA stream is a single-carrier modulated signal. With pulse shaping, single-carrier modulated signals in general do not have a Gaussian distribution. This makes it difficult to analytically derive the statistical properties of a single IFDMA stream. In our case, the transmitted signal in the Multi-IFDMA systems is the sum of multiple (but finite) IFDMA streams, which is also difficult to analyze. Therefore, we resort to Monte Carlo simulations for numerical analysis.

In the simulations, we fix $M = 128$, and set different values of $N \leq M$. Our objective is to study the PAPR performance of the Multi-IFDMA system benchmarked against the LFDMA and OFDMA systems. The simulation setups are as follows:

1. Modulation: QPSK.
2. Subcarrier mapping rule: For OFDMA, the N subcarriers are randomly chosen. For LFDMA, the N contiguous subcarriers are randomly chosen. For Multi-IFDMA, we first partition N to the sum of several integers, each of which is a power of 2. In particular, the partition is a "minimal partition", that is, the partition results in the minimum number of power-of-2 integers (e.g., for $N = 7$, the partition is $4 + 2 + 1$). We construct an IFDMA stream for each power-of-

Fig. B.17 The CCDF performance of IFDMA/Multi-IFDMA, LFDMA, OFDMA when $N = 64$, 65, and 127

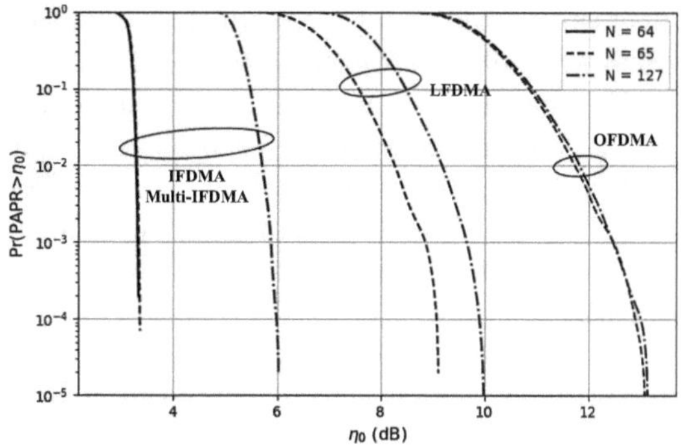

2 integer, and sum up the signals of the IFDMA streams in the time domain to construct the Multi-IFDMA stream.

3. Pulse shaping: A root-raised-cosine (RRC) pulse with roll-off factor $\beta = 0.5$ is used at the transmitter to shape the transmitted signal. In particular, the RRC pulse is truncated so that it spreads over 20 symbols only, and it is oversampled 10 times.
4. PAPR metric: Each packet contains 10 OFDM symbols, with each OFDM symbol containing 20 time symbols including CPs. Thus, with the aforementioned 10-time oversampling, there are 2000 samples per packet over which we compute the packet's PAPR. We compute the PAPR of 10,000 packets. The performance is measured by the complementary cumulative distribution function (CCDF) of PAPR, defined as the probability that the PAPR is higher than a certain threshold η_0 (in dB), i.e., $\Pr(\text{PAPR} > \eta_0)$.

Figure B.17 shows the CCDF performance[3] of IFDMA/Multi-IFDMA, LFDMA, OFDMA when $N = 64$, 65, and 127. When $N = 64$, the Multi-IFDMA scheme is essentially the single-IFDMA stream, thus the PAPR performance is the best among all systems. When $N = 65$, the Multi-IFDMA stream consists of two IFDMA streams, one occupying 64 subcarriers and one occupying 1 subcarrier. Considering the 99.9-percentile PAPR $\left(\Pr(\text{PAPR} > \eta_0) = 10^{-3}\right)$, the CCDF performance is very close to the $N = 64$ case, meaning that the sum of the two streams only fluctuates the amplitude slightly. On the other hand, the CCDF gain of Multi-IFDMA over LFDMA is about 5.5 dB, and that over OFDMA is about 9.2 dB. When $N = 127$, the Multi-IFDMA stream consists of seven IFDMA streams, the CCDF performance is 2.7 dB worse than the $N = 64$ case. This matches with our intuition that the more IFDMA streams are mixed together, the higher amplitude

[3] For Multi-IFDMA, the subcarriers are allocated to each stream using our bit-reversal subcarrier allocation scheme. For LFDMA and OFDMA, the CCDF performance is averaged over all possible subcarrier allocation schemes. In the other words, the CCDF performances for LFDMA and OFDMA are the mean performance rather than the worst performance.

fluctuations there will be. Nevertheless, compared with LFDMA and OFDMA, the 99.9-percentile PAPR gains of Multi-IFDMA are 3.7 dB and 6.5 dB, respectively.

To summarize, for Multi-IFDMA, a single IFDMA stream has the most favorable PAPR performance when N is a power of 2. As more and more IFDMA streams are added, the CCDF performance gets worse. The worst PAPR happens when $N = 2^m - 1$ where N has the most number of partitions. However, even in the worst case, the PAPR performance of Multi-IFDMA is still significantly better than LFDMA or IFDMA.

Our overall conclusions are as follows. Multi-IFDMA has better PAPR than LFDMA and significantly better PAPR than OFDMA. In terms of transceiver complexity, IFDMA and OFDMA are comparable. Note that for OFDMA, only an IDFT is needed at the transmitter and a DFT is needed at the receiver, and the complexity in terms of number of complex multipliers is therefore $\frac{1}{2}M \log_2 M$. For LFDMA, the complexities of both transmitter and receiver are in the order of $M^2 \log_2 M$, if we use the conventional SC-FDMA design (similar to that shown in Fig. B.9).

B.3 Summary

IFDMA is a multiplexing and signal modulation technique for multi-user systems. It has attracted great interest because of its excellent PAPR compared with OFDMA and LFDMA. IFDMA, however, has strict constraints on subcarrier mapping:

Constraint 1: Subcarriers assigned to an IFDMA data stream must be evenly spaced over the spectrum used, and subcarriers assigned to different IFDMA data streams must be orthogonal.

Constraint 2: The number of subcarriers assigned to an IFDMA data stream must be a divisor of the total number of available subcarriers.

These constraints led to the impressions that (1) transceiver designs for multiplexing and demultiplexing IFDMA streams must be complex; (2) resource allocation in IFDMA systems must be inflexible.

The results of this appendix indicate otherwise. To debunk (1), we devise new IFDMA transceivers that can reduce complexity by an order of $\log M$ or more. Our designs are inspired by a key finding that multiplexing/demultiplexing and modulation/demodulation of IFDMA data streams are coincident with FFTs/IFFTs of different scales embedded within the overall Cooley-Tukey FFT recursive decomposition. To debunk (2), we put forth a new Multi-IFDMA scheme that allocates multiple IFDMA streams to each user. In conjunction with a bit-reversal subcarrier allocation algorithm, resource allocation in Multi-IFDMA can be as flexible as those of OFDMA and LFDMA. Importantly, our new transceiver designs are unified designs compatible with both conventional IFDMA and Multi-IFDMA.

Overall, our results yield the following new understandings:

- IFDMA is amenable to simple transceiver design. In particular, IFDMA transceivers are much simpler than LFDMA transceivers, and comparable to OFDMA transceivers, in complexity.
- Flexible resource allocation is possible with Multi-IFDMA.
- Multi-IFDMA, like IFDMA, has better PAPR characteristics than OFDMA and LFDMA.

Appendix C
Misaligned Over-the-Air Computation

This appendix provides supplementary materials for Chap. 4.

C.1 The Generalized System Model for AirComp

In this appendix, we generalize the system model in Sect. 4.4.2.1 by considering fast fading channels $\tilde{h}_m(t)$ and residual CFO in the received signal $r(t)$.

At each transmitter, to compensate the fast channel fading and CFO between the transmitter and the receiver, α_m in (4.42) is set as $\alpha_m(t) = e^{-j\bar{\varepsilon}_m t}/\bar{h}_m(t)$, where $\bar{\varepsilon}_m$ is the estimated CFO at the m-th device.

The received signal $r(t)$ is then given by

$$r(t) = \sum_{m=1}^{M} \tilde{h}_m(t) e^{j\tilde{\varepsilon}_m t} x_m(t - \tau_m) + z(t). \tag{C.1}$$

Unlike (4.43), each $\tilde{h}_m(t)$ is now a fast fading channel and $\tilde{\varepsilon}_m$ is the CFO between the m-th device and the PS.

Substituting (4.42) into (C.1) gives us

$$\begin{aligned} r(t) &= \sum_{m=1}^{M} \tilde{h}_m(t) e^{j\tilde{\varepsilon}_m t} \alpha_m(t) \sum_{\ell=1}^{L} s_m[\ell] p(t - \tau_m - \ell T) + z(t) \\ &= \sum_{\ell=1}^{L} \sum_{m=1}^{M} h'_m(t) e^{j\varepsilon'_m t} s_m[\ell] p(t - \tau_m - \ell T) + z(t), \end{aligned} \tag{C.2}$$

Y. Shao, *AI-Empowered IoT Communications*, Wireless Networks,
https://doi.org/10.1007/978-3-031-98816-5

where $h'_m(t) = \tilde{h}_m(t)/\bar{h}_m(t)$ is the residual channel-fading coefficient and $\varepsilon'_m = \tilde{\varepsilon}_m - \bar{\varepsilon}_m$ is the residual CFO between the m-th device and the PS. As can be seen, CFO introduces additional channel-gain misalignments among devices.

With the revised signal $r(t)$, the discrete samples can be written in the same form as (4.47) with different coefficients $c_{m,k}[i]$ and $c'_{m,k}[i]$, giving,

$$c_{m,k}[i] = \frac{1}{T}\int_{(i-1)T+\tau_k}^{(i-\mathbb{1}_{m>k})T+\tau_m} h'_m(t)e^{j\varepsilon'_m\zeta}\,d\zeta = \frac{1}{j\varepsilon'_m T}\times \tag{C.3}$$

$$\left(e^{j\varepsilon'_m[(i-\mathbb{1}_{m>k})T+\tau_m]} - e^{j\varepsilon'_m[(i-1)T+\tau_k]}\right)\int_{(i-1)T+\tau_k}^{(i-\mathbb{1}_{m>k})T+\tau_m} h'_m(t)\,dt,$$

$$c'_{m,k}[i] = \frac{1}{T}\int_{(i-\mathbb{1}_{m>k})T+\tau_m}^{iT+\tau_k} h'_m(t)e^{j\varepsilon'_m\zeta}\,d\zeta = \frac{1}{j\varepsilon'_m T}\times \tag{C.4}$$

$$\left(e^{j\varepsilon'_m[iT+\tau_k]} - e^{j\varepsilon'_m(i-\mathbb{1}_{m>k})T+\tau_m}\right)\int_{(i-\mathbb{1}_{m>k})T+\tau_m}^{iT+\tau_k} h'_m(t)\,dt.$$

On the other hand, for our WMFS scheme, the $y_k[i]$ samples in (4.54) can be written as

$$y_k[i] = \sum_{m=1}^{M} g_{m,k}[i]s_m[i-\mathbb{1}_{m>k}] + \tilde{z}_k[i], \tag{C.5}$$

where

$$g_{m,k}[i] = h'_m\frac{1}{d_k}\int_{(i-1)T+\tau_k}^{(i-1)T+\tau_{k+1}} e^{j\varepsilon'_m\zeta}\,d\zeta = \frac{h'_m}{d_k}\frac{e^{j\varepsilon'_m d_k}-1}{j\varepsilon'_m}e^{j\varepsilon'_m[(i-1)T+\tau_k]}.$$

As a result, if the channel is fast fading and there is residual CFO in the received signal, we can simply modify the coefficient matrix $\boldsymbol{D}$ as

$$\boldsymbol{D} = \tag{C.6}$$

$$\begin{bmatrix}
g_{1,1}[1] & & & & & & & & \\
g_{1,2}[1] & g_{2,1}[1] & & & & & & & \\
\cdots & g_{2,2}[1] & \cdots & & & & & & \\
g_{1,M}[1] & \cdots & \cdots & g_{M,1}[1] & & & & & \\
 & g_{2,M}[1] & \cdots & g_{M,2}[1] & g_{1,1}[2] & & & & \\
 & & \cdots & \cdots & g_{1,2}[2] & g_{2,1}[2] & & & \\
 & & & g_{M,M}[1] & \cdots & g_{2,2}[2] & \cdots & & \\
 & & & & g_{1,M}[2] & \cdots & \cdots & g_{M,1}[2] & \\
 & & & & & g_{2,M}[2] & \cdots & g_{M,2}[2] & \cdots \\
 & & & & & & \cdots & \cdots & \cdots \\
 & & & & & & & g_{M,M}[2] & \cdots \\
 & & & & & & & & \cdots
\end{bmatrix}.$$

C.2 Proof of Theorem 4.2

This appendix proves that the marginal likelihood function $f(\boldsymbol{y}|\boldsymbol{s}[i])$ is a multivariate Gaussian and derive its moment parameters by a sum-product process.

To begin with, we point out that a multivariate Gaussian distribution can be parameterized by two sets of parameters [7, 8]: the moment parameter $(\boldsymbol{\mu}, \boldsymbol{\Sigma})$ and the canonical parameter $(\boldsymbol{\eta}, \boldsymbol{\Lambda})$. The two sets of parameters can be transformed into one another and they are useful in different circumstances, as detailed below.

Definition C.1 (The Moment Parameters $(\boldsymbol{\mu}, \boldsymbol{\Sigma})$) For a multivariate real Gaussian random variable $\boldsymbol{w}$ of dimension $2M$, its moment parameters are defined as

$$\boldsymbol{\mu} = \mathbb{E}[\boldsymbol{w}], \quad \boldsymbol{\Sigma} = \mathbb{E}\left[(\boldsymbol{w} - \boldsymbol{\mu})(\boldsymbol{w} - \boldsymbol{\mu})^\top\right].$$

The moment form of the Gaussian distribution is given by

$$\mathcal{N}(\boldsymbol{w}; \boldsymbol{\mu}, \boldsymbol{\Sigma}) = \frac{1}{(2\pi)^M |\boldsymbol{\Sigma}|^{\frac{1}{2}}} \exp\left\{-\frac{1}{2}(\boldsymbol{w} - \boldsymbol{\mu})^\top \boldsymbol{\Sigma}^\dagger (\boldsymbol{w} - \boldsymbol{\mu})\right\}.$$

Moment parameters are useful when we marginalize a multivariate Gaussian [8]. Let us partition $\boldsymbol{w} = [\boldsymbol{v}_1, \boldsymbol{v}_2]^\top$ where $\boldsymbol{w}_1$, $\boldsymbol{w}_2$ are multivariate Gaussians of dimensions κ and $2M - \kappa$, respectively. The moment parameters can be partitioned accordingly as

$$\boldsymbol{\mu} = \begin{bmatrix} \boldsymbol{\mu}_1 \\ \boldsymbol{\mu}_2 \end{bmatrix}, \quad \boldsymbol{\Sigma} = \begin{bmatrix} \boldsymbol{\Sigma}_{11} & \boldsymbol{\Sigma}_{12} \\ \boldsymbol{\Sigma}_{21} & \boldsymbol{\Sigma}_{22} \end{bmatrix}.$$

If we marginalize out $\boldsymbol{w}_2$ from $\boldsymbol{w}$, the marginal $f(\boldsymbol{w}_1)$ is still a multivariate Gaussian distribution:

$$\begin{aligned} f(\boldsymbol{w}_1) &= \int_{\boldsymbol{w}_2} \mathcal{N}\left(\begin{bmatrix} \boldsymbol{w}_1 \\ \boldsymbol{w}_2 \end{bmatrix}; \begin{bmatrix} \boldsymbol{\mu}_1 \\ \boldsymbol{\mu}_2 \end{bmatrix}, \begin{bmatrix} \boldsymbol{\Sigma}_{11} & \boldsymbol{\Sigma}_{12} \\ \boldsymbol{\Sigma}_{21} & \boldsymbol{\Sigma}_{22} \end{bmatrix}\right) d\boldsymbol{w}_2 \\ &\propto \mathcal{N}(\boldsymbol{w}_1; \boldsymbol{\mu}_1, \boldsymbol{\Sigma}_{11}). \end{aligned} \tag{C.7}$$

Definition C.2 (The Canonical Parameters $(\boldsymbol{\eta}, \boldsymbol{\Lambda})$) For a Gaussian random variable $\boldsymbol{w}$ of dimension $2M$, its canonical parameters are defined as

$$\boldsymbol{\eta} = \Sigma^\dagger \boldsymbol{\mu}, \quad \boldsymbol{\Lambda} = \Sigma^\dagger.$$

The canonical form of the Gaussian distribution is given by

$$\mathcal{N}(\boldsymbol{w}; \boldsymbol{\eta}, \boldsymbol{\Lambda}) = \exp\left\{-\frac{1}{2}\boldsymbol{w}^\top \boldsymbol{\Lambda} \boldsymbol{w} + \boldsymbol{w}^\top \boldsymbol{\eta} + \rho\right\},$$

where ρ is a constant

$$\rho = -\frac{1}{2}\left(2M \ln 2\pi - \ln |\boldsymbol{\Lambda}| + \boldsymbol{\eta}^\top \boldsymbol{\Lambda}^\dagger \boldsymbol{\eta}\right).$$

The canonical parameters are useful when we multiply two or more Gaussians [9]. Let $\{\boldsymbol{w}_k : k = 1, 2, \ldots, K\}$, $\boldsymbol{w}_k \sim \mathcal{N}(\boldsymbol{w}; \boldsymbol{\eta}_k, \boldsymbol{\Lambda}_k)$ be a set of multivariate real Gaussian random variables of dimension $2M$. Then, their product is still a Gaussian whose canonical parameters are the sum of the canonical parameters of the original K Gaussians:

$$\prod_{k=1}^{K} \mathcal{N}(\boldsymbol{w}_k; \boldsymbol{\eta}_k, \boldsymbol{\Lambda}_k) \propto \mathcal{N}\left(\boldsymbol{w}; \sum_{k=1}^{K} \boldsymbol{\eta}_k, \sum_{k=1}^{K} \boldsymbol{\Lambda}_k\right) \tag{C.8}$$

$$= \exp\left\{-\frac{1}{2}\boldsymbol{w}^\top \sum_{k=1}^{K} \boldsymbol{\Lambda}_k \boldsymbol{w} + \boldsymbol{w}^\top \sum_{k=1}^{K} \boldsymbol{\eta}_k + \sum_{k=1}^{K} \rho_k\right\}.$$

Let us set out to show that the marginal likelihood function $f(\boldsymbol{y}|\boldsymbol{s}[i])$ is an M-dimensional complex Gaussian distribution.

Now that the factor graph in Fig. 4.21 has a tree structure, we only need to pass the messages from left to right (forward message passing) and then from right to left (backward message passing). Each message needs to be computed only once, after which the exact marginal posterior distribution converges.

C.2.1 Forward Message Passing

We first investigate how the messages are passed from left to right in Fig. 4.21. Without loss of generality, we shall focus on message passing from one variable $\boldsymbol{W}_{k,i}$ to another variable $\boldsymbol{W}_{k+1,i}$ on the right.

Notice that $\boldsymbol{W}_{k,i} = \mathcal{V}(y_k[i]) = \{s_1[i], \ldots, s_k[i], s_{k+1}[i-1], s_{k+2}[i-1], \ldots, s_M[i-1]\}$ and $\boldsymbol{W}_{k+1,i} = \mathcal{V}(y_{k+1}[i]) = \{s_1[i], \ldots, s_k[i], s_{k+1}[i], s_{k+2}[i-1], \ldots, s_M[i-1]\}$. Thus, the only difference between $\boldsymbol{W}_{k,i}$ and $\boldsymbol{W}_{k+1,i}$ is the $(k+1)$-th symbol. We consider each complex random variable $s_k[i]$ as a pair of real random vector: the real part and the imaginary part. Then, each $\boldsymbol{W}_{k,i}$ can be viewed as a $2M$-dimensional real random variable. To simplify the notation, we denote the $2M$-dimensional real variates corresponding to $\boldsymbol{W}_{k,i}$ and $\boldsymbol{W}_{k+1,i}$, respectively, by

$$\boldsymbol{w}_{k,i} = \left(b_1^{\mathfrak{r}}, \ldots, b_k^{\mathfrak{r}}, b_{k+1}^{\mathfrak{r}}, b_{k+2}^{\mathfrak{r}}, \ldots, b_M^{\mathfrak{r}}, b_1^{\mathfrak{i}}, \ldots, b_k^{\mathfrak{i}}, b_{k+1}^{\mathfrak{i}}, b_{k+2}^{\mathfrak{i}}, \ldots, b_M^{\mathfrak{i}}\right),$$

$$\boldsymbol{w}_{k+1,i} = \left(b_1^{\mathfrak{r}}, \ldots, b_k^{\mathfrak{r}}, c_{k+1}^{\mathfrak{r}}, b_{k+2}^{\mathfrak{r}}, \ldots, b_M^{\mathfrak{r}}, b_1^{\mathfrak{i}}, \ldots, b_k^{\mathfrak{i}}, c_{k+1}^{\mathfrak{i}}, b_{k+2}^{\mathfrak{i}}, \ldots, b_M^{\mathfrak{i}}\right),$$

as shown in Fig. C.1.

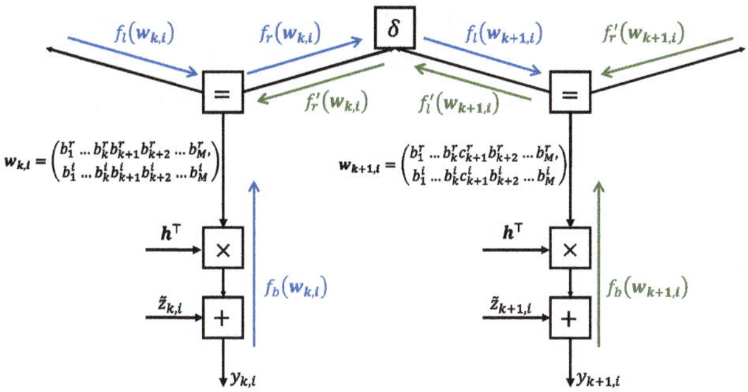

Fig. C.1 The forward message passing from $W_{k,i}$ to $W_{k+1,i}$ and the backward message passing from $W_{k+1,i}$ to $W_{k,i}$

On the left half of Fig. C.1, there are three edges centered around the equality function " =" (marked in blue). Due to the equality constraint, these four edges are associated with the same high-dimensional variable $\boldsymbol{w}_{k,i}$. In the forward message passing, there are three messages to be computed.

C.2.1.1 The Message Passed from the Bottom

The message is the message passed from the bottom, denoted by $f_b(\boldsymbol{w}_{k,i})$. This message carries the information about $\boldsymbol{w}_{k,i}$ contained in the sample $y_{k,i}$. As per (4.54),

$$\begin{aligned} y_{k,i} &= \sum_{m=1}^{M} (h_m^{\mathfrak{r}} + jh_m^{\mathfrak{i}})(b_m^{\mathfrak{r}} + jb_m^{\mathfrak{i}}) + (\tilde{z}_{k,i}^{\mathfrak{r}} + j\tilde{z}_{k,i}^{\mathfrak{i}}) \\ &= \sum_{m=1}^{M} (h_m^{\mathfrak{r}} b_m^{\mathfrak{r}} - h_m^{\mathfrak{i}} b_m^{\mathfrak{i}}) + \tilde{z}_{k,i}^{\mathfrak{r}} + j\sum_{m=1}^{M} (h_m^{\mathfrak{r}} b_m^{\mathfrak{i}} + h_m^{\mathfrak{i}} b_m^{\mathfrak{r}}) + j\tilde{z}_{k,i}^{\mathfrak{i}}, \end{aligned}$$

where $\tilde{z}_{k,i}^{\mathfrak{r}}, \tilde{z}_{k,i}^{\mathfrak{i}} \sim \mathcal{N}(0, \frac{N_0}{2d_k})$. Thus, the likelihood function $f(y_{k,i}|\boldsymbol{w}_{k,i})$ is Gaussian, giving

$$\begin{aligned} f(y_{k,i}|\boldsymbol{w}_{k,i}) \propto \ &\exp\left\{-\frac{d_k}{N_0}\left[y_{k,i}^{\mathfrak{r}} - \sum_{m}(h_m^{\mathfrak{r}} b_m^{\mathfrak{r}} - h_m^{\mathfrak{i}} b_m^{\mathfrak{i}})\right]^2\right\} \\ &\times \exp\left\{-\frac{d_k}{N_0}\left[y_{k,i}^{\mathfrak{i}} - \sum_{m}(h_m^{\mathfrak{r}} b_m^{\mathfrak{i}} + h_m^{\mathfrak{i}} b_m^{\mathfrak{r}})\right]^2\right\}. \end{aligned}$$

When we pass the information bottom up, $y_{k,i}$ is our observation (hence a constant) and $\boldsymbol{w}_{\boldsymbol{k},\boldsymbol{i}}$ is the variable. Therefore, $f_b(\boldsymbol{w}_{\boldsymbol{k},\boldsymbol{i}}) = f(y_{k,i}|\boldsymbol{w}_{\boldsymbol{k},\boldsymbol{i}})$. After some manipulations, we can write $f_b(\boldsymbol{w}_{\boldsymbol{k},\boldsymbol{i}})$ as a $2M$-dimensional Gaussian distribution:

$$f_b(\boldsymbol{w}_{\boldsymbol{k},\boldsymbol{i}}) \propto \mathcal{N}(\boldsymbol{w}_{\boldsymbol{k},\boldsymbol{i}}, \boldsymbol{\eta}_b, \boldsymbol{\Sigma}_b), \tag{C.9}$$

where $\boldsymbol{\eta}_b$ and $\boldsymbol{\Sigma}_b$ are defined as

$$\boldsymbol{\eta}_b = \frac{2d_k}{N_0}\begin{bmatrix}\boldsymbol{\beta}_1\\ \boldsymbol{\beta}_2\end{bmatrix}\begin{bmatrix}y_{k,i}^{\mathfrak{r}}\\ y_{k,i}^{\mathfrak{i}}\end{bmatrix}, \quad \boldsymbol{\Sigma}_b = \frac{2d_k}{N_0}\begin{bmatrix}\boldsymbol{\beta}_1\boldsymbol{\beta}_1^\top & \boldsymbol{\beta}_1\boldsymbol{\beta}_2^\top\\ \boldsymbol{\beta}_2\boldsymbol{\beta}_1^\top & \boldsymbol{\beta}_1\boldsymbol{\beta}_1^\top\end{bmatrix}, \tag{C.10}$$

and the matrices $\boldsymbol{\beta}_1$ and $\boldsymbol{\beta}_1$ are composed of channel coefficients as follows:

$$\boldsymbol{\beta}_1 = \begin{bmatrix} h_1^{\mathfrak{r}} & h_1^{\mathfrak{i}}\\ h_2^{\mathfrak{r}} & h_2^{\mathfrak{i}}\\ \vdots & \vdots\\ h_M^{\mathfrak{r}} & h_M^{\mathfrak{i}}\end{bmatrix}, \quad \boldsymbol{\beta}_2 = \begin{bmatrix} -h_1^{\mathfrak{i}} & h_1^{\mathfrak{r}}\\ -h_2^{\mathfrak{i}} & h_2^{\mathfrak{r}}\\ \vdots & \vdots\\ -h_M^{\mathfrak{i}} & h_M^{\mathfrak{r}}\end{bmatrix}.$$

In (C.10), we have assumed that the dimensionality of $\boldsymbol{w}_{\boldsymbol{k},\boldsymbol{i}}$ is $2M$, that is, $y_{k,i} = y_k[i]$ is related to M complex variables. However, this is only valid when the number of neighbor symbols of $y_k[i]$ is M. As shown in (4.60), $|\mathcal{V}(y_k[i])| = M$ only when $1 < i \leq L$. Thus, we can compute $\boldsymbol{\eta}_b$ and $\boldsymbol{\Sigma}_b$ by (C.10) only when $1 < i \leq L$.

For the boundary samples ($i = 1, L+1$) whose neighbor symbols are less than M, we further multiply the parameters $\boldsymbol{\eta}_b$ and $\boldsymbol{\Sigma}_b$ in (C.10) by an indicator vector $\boldsymbol{\gamma}$ and an indicator matrix $\boldsymbol{\Gamma}$, respectively, to ensure that $f_b(\boldsymbol{w}_{\boldsymbol{k},\boldsymbol{i}})$ does not contain information about the symbols that do not belong to $\mathcal{V}(y_k[i])$. The general form of $\boldsymbol{\eta}_b$ and $\boldsymbol{\Sigma}_b$ are

$$\boldsymbol{\eta}_b = \frac{2d_k}{N_0}\begin{bmatrix}\boldsymbol{\beta}_1\\ \boldsymbol{\beta}_2\end{bmatrix}\begin{bmatrix}y_{k,i}^{\mathfrak{r}}\\ y_{k,i}^{\mathfrak{i}}\end{bmatrix} \circ \boldsymbol{\gamma}_{\boldsymbol{k}}, \tag{C.11}$$

$$\boldsymbol{\Sigma}_b = \frac{2d_k}{N_0}\begin{bmatrix}\boldsymbol{\beta}_1\boldsymbol{\beta}_1^\top & \boldsymbol{\beta}_1\boldsymbol{\beta}_2^\top\\ \boldsymbol{\beta}_2\boldsymbol{\beta}_1^\top & \boldsymbol{\beta}_1\boldsymbol{\beta}_1^\top\end{bmatrix} \circ \boldsymbol{\Gamma}_{\boldsymbol{k}}, \tag{C.12}$$

where $\circ$ is an elementwise multiplication. The indicator vector $\boldsymbol{\gamma}$ and the indicator matrix $\boldsymbol{\Gamma}$ are defined as follows.

First, for the first M samples (i.e., $i = 1$), we have $|\mathcal{V}(y_k[i])| = k$ from (4.60). Thus, we define

$$\boldsymbol{\gamma}_{\boldsymbol{k}} = \begin{bmatrix}\mathbf{1}_{k\times 1}\\ \mathbf{0}_{(M-k)\times 1}\\ \mathbf{1}_{k\times 1}\\ \mathbf{0}_{(M-k)\times 1}\end{bmatrix},$$

$$\boldsymbol{\Gamma}_k = \begin{bmatrix} \mathbf{1}_{k\times k} & \mathbf{0}_{k\times(M-k)} & \mathbf{1}_{k\times k} & \mathbf{0}_{k\times(M-k)} \\ \mathbf{0}_{(M-k)\times k} & \mathbf{0}_{(M-k)\times(M-k)} & \mathbf{0}_{(M-k)\times k} & \mathbf{0}_{(M-k)\times(M-k)} \\ \mathbf{1}_{k\times k} & \mathbf{0}_{k\times(M-k)} & \mathbf{1}_{k\times k} & \mathbf{0}_{k\times(M-k)} \\ \mathbf{0}_{(M-k)\times k} & \mathbf{0}_{(M-k)\times(M-k)} & \mathbf{0}_{(M-k)\times k} & \mathbf{0}_{(M-k)\times(M-k)} \end{bmatrix},$$

where **1** and **0** are all-ones and all-zero matrices with subscripts denoting their dimensions.

Second, for the last M samples (i.e., $i = L + 1$), we have $|\mathcal{V}(y_k[i])| = M - k$ from (4.60). Thus, we define

$$\gamma_k = \begin{bmatrix} \mathbf{0}_{k\times 1} \\ \mathbf{1}_{(M-k)\times 1} \\ \mathbf{0}_{k\times 1} \\ \mathbf{1}_{(M-k)\times 1} \end{bmatrix},$$

$$\boldsymbol{\Gamma}_k = \begin{bmatrix} \mathbf{0}_{k\times k} & \mathbf{0}_{k\times(M-k)} & \mathbf{0}_{k\times k} & \mathbf{0}_{k\times(M-k)} \\ \mathbf{0}_{(M-k)\times k} & \mathbf{1}_{(M-k)\times(M-k)} & \mathbf{0}_{(M-k)\times k} & \mathbf{1}_{(M-k)\times(M-k)} \\ \mathbf{0}_{k\times k} & \mathbf{0}_{k\times(M-k)} & \mathbf{0}_{k\times k} & \mathbf{0}_{k\times(M-k)} \\ \mathbf{0}_{(M-k)\times k} & \mathbf{1}_{(M-k)\times(M-k)} & \mathbf{0}_{(M-k)\times k} & \mathbf{1}_{(M-k)\times(M-k)} \end{bmatrix}.$$

Finally, for all other samples ($1 < i \le L$), we simply set

$$\gamma_k = \mathbf{1}_{2M\times 1}, \quad \boldsymbol{\Gamma}_k = \mathbf{1}_{2M\times 2M}.$$

This is consistent with (C.10).

C.2.1.2 The Message Passed from the Left

The second message, denoted by $f_\ell(\boldsymbol{w}_{k,i})$ in Fig. C.1, is the message passed from $\boldsymbol{w}_{k-1,i}$ on the left. This message is obtained in the same way as $f_\ell(\boldsymbol{w}_{k+1,i})$ and we will analyze it at the end of forward message passing. For now, let us assume it is Gaussian and denote it by

$$f_\ell(\boldsymbol{w}_{k,i}) \propto \mathcal{N}\left(\boldsymbol{w}_{k,i}; \boldsymbol{\eta}_\ell, \boldsymbol{\Lambda}_\ell\right). \tag{C.13}$$

C.2.1.3 The Message Out of a Local Function

As per the sum-product rule, the message out of a local function along an edge is the product of all incoming messages to this local function along all other edges. Thus, the message out of the equality function "=", denoted by $f_r(\boldsymbol{w}_{k,i})$ in Fig. C.1, can be obtained by

$$f_r(\boldsymbol{w}_{k,i}) = f_b(\boldsymbol{w}_{k,i}) f_\ell(\boldsymbol{w}_{k,i}). \tag{C.14}$$

This is the "product" step of the sum-product algorithm. From (C.8), we know $f_r(\boldsymbol{w}_{k,i})$ is a Gaussian distribution, and

$$f_r(\boldsymbol{w}_{k,i}) \propto \mathcal{N}\left(\boldsymbol{w}_{k,i}; \boldsymbol{\eta}_r, \boldsymbol{\Lambda}_r\right), \tag{C.15}$$

where $\boldsymbol{\eta}_r = \boldsymbol{\eta}_b + \boldsymbol{\eta}_\ell$, $\boldsymbol{\Lambda}_r = \boldsymbol{\Lambda}_b + \boldsymbol{\Lambda}_\ell$. We emphasize that this message is an aggregation of all the known information about $\boldsymbol{w}_{k,i}$ from the left side of the graph.

The next step is to pass the message $f_r(\boldsymbol{w}_{k,i})$ through the compatibility function δ. Notice that the compatibility function connects two different variables: on the LHS, the variable associated with the edge is $\boldsymbol{w}_{k,i}$; on the RHS, the variable associated with the edge is $\boldsymbol{w}_{k+1,i}$. Therefore, we have to integrate $f_r(\boldsymbol{w}_{k,i})$ over all variates that are in $\boldsymbol{w}_{k,i}$ but not in $\boldsymbol{w}_{k+1,i}$. This is the "sum" step of the sum-product algorithm.

Notice that the common variates of $\boldsymbol{w}_{k,i}$ and $\boldsymbol{w}_{k+1,i}$ are

$$\boldsymbol{w}_\cap = \left(b_1^{\mathfrak{r}}, \ldots, b_k^{\mathfrak{r}}, b_{k+2}^{\mathfrak{r}}, \ldots, b_M^{\mathfrak{r}}, b_1^{\mathfrak{i}}, \ldots, b_k^{\mathfrak{i}}, b_{k+2}^{\mathfrak{i}}, \ldots, b_M^{\mathfrak{i}}\right),$$

and the two different variates are $b_{k+1}^{\mathfrak{r}}$ and $b_{k+1}^{\mathfrak{i}}$—in $\boldsymbol{w}_{k+1,i}$, these two variates are $c_{k+1}^{\mathfrak{r}}$ and $c_{k+1}^{\mathfrak{i}}$.

Let us integrate $f_r(\boldsymbol{w}_{k,i})$ over $b_{k+1}^{\mathfrak{r}}$ and $b_{k+1}^{\mathfrak{i}}$, giving,

$$f(\boldsymbol{w}_\cap) = \int_{b_{k+1}^{\mathfrak{r}}} \int_{b_{k+1}^{\mathfrak{i}}} f_r(\boldsymbol{w}_{k,i}) db_{k+1}^{\mathfrak{r}} db_{k+1}^{\mathfrak{i}}.$$

As per (C.7), $f(\boldsymbol{w}_\cap)$ is also Gaussian. In particular, if we write $f_r(\boldsymbol{w}_{k,i})$ and $f(\boldsymbol{w}_\cap)$ in moment form as

$$f_r(\boldsymbol{w}_{k,i}) \propto \mathcal{N}\left(\boldsymbol{w}_{k,i}; \boldsymbol{\mu}_r, \boldsymbol{\Sigma}_r\right),$$
$$f_r(\boldsymbol{w}_\cap) \propto \mathcal{N}\left(\boldsymbol{w}_\cap; \boldsymbol{\mu}_\cap, \boldsymbol{\Sigma}_\cap\right),$$

then $\boldsymbol{\mu}_\cap$ can be obtained by deleting the $(k+1)$-th and $(k+1+M)$-th rows of $\boldsymbol{\mu}_r$; $\boldsymbol{\Sigma}_\cap$ can be obtained by deleting the $(k+1)$-th and $(k+1+M)$-th rows and columns of $\boldsymbol{\Sigma}_r$.

However, $\boldsymbol{w}_\cap$ is a $(2M-2)$ dimensional variable. To obtain $f_\ell(\boldsymbol{w}_{k+1,i})$, we have to expand the dimensionality of $\boldsymbol{w}_\cap$ by adding $c_{k+1}^{\mathfrak{r}}$ and $c_{k+1}^{\mathfrak{i}}$ in the $(k+1)$-th and $(k+1+M)$-th positions. After expansion, $f_\ell(\boldsymbol{w}_{k+1,i})$ is still multivariate Gaussian:

$$f_\ell(\boldsymbol{w}_{k+1,i}) \propto \mathcal{N}\left(\boldsymbol{w}_{k+1,i}; \boldsymbol{\mu}_\ell, \boldsymbol{\Sigma}_\ell\right), \tag{C.16}$$

where $\boldsymbol{\mu}_\ell$ can be obtained by adding two zeros to $\boldsymbol{\mu}_\cap$; and $\boldsymbol{\Sigma}_\ell$ can be obtained by adding two all-zero rows and two all-zero columns to $\boldsymbol{\Sigma}_\cap$.

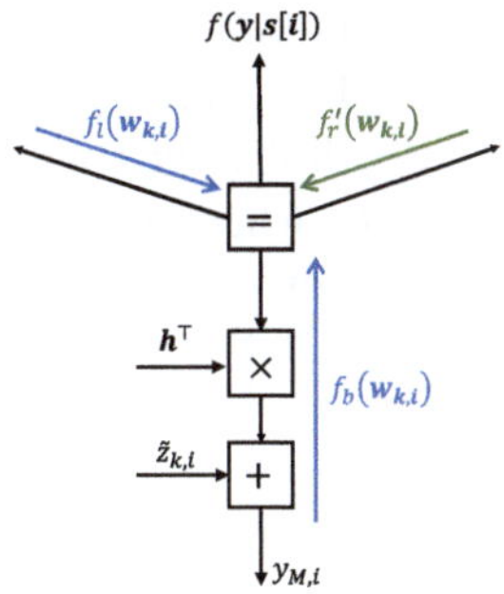

Fig. C.2 The marginalization process in the sum-product algorithm

To summarize, we have shown that all the messages involved in the forward message passing are $2M$-dimensional multivariate Gaussians and can be parameterized by (C.9), (C.13), (C.15), (C.16), respectively.

C.2.2 Backward Message Passing

Our tree structure is symmetric. Thus, backward message passing is symmetric to forward message passing. As shown in Fig. C.1, to pass the messages from $\boldsymbol{w}_{k+1,i}$ and $\boldsymbol{w}_{k,i}$, we first compute two incoming messages $f_b(\boldsymbol{w}_{k+1,i})$ and $f_r'(\boldsymbol{w}_{k+1,i})$, where $f_b(\boldsymbol{w}_{k+1,i})$ is the same as that in the forward message passing and $f_r'(\boldsymbol{w}_{k+1,i})$ is the message passed from $\boldsymbol{w}_{k+2,i}$ on the right.

Then, $f_\ell'(\boldsymbol{w}_{k+1,i})$ and $f_r'(\boldsymbol{w}_{k,i})$ are computed from "product" and "sum", respectively, by

$$f_\ell'(\boldsymbol{w}_{k+1,i}) = f_b(\boldsymbol{w}_{k+1,i}) f_r'(\boldsymbol{w}_{k+1,i}), \tag{C.17}$$

$$f_r'(\boldsymbol{w}_{k,i}) = \int_{c_{k+1}^{\mathfrak{r}}} \int_{c_{k+1}^{\mathfrak{i}}} f_\ell'(\boldsymbol{w}_{k,i}) dc_{k+1}^{\mathfrak{r}} dc_{k+1}^{\mathfrak{i}}. \tag{C.18}$$

C.2.3 Marginalization

After one forward message passing from left to right and one backward message passing from right to left, the marginal likelihood function of each variable $\boldsymbol{w}_{k,i}$ converges and can be computed by

$$f(\boldsymbol{w}_{k,i} | y_1, y_2, \ldots, y_M) = f_b(\boldsymbol{w}_{k,i}) f_\ell(\boldsymbol{w}_{k,i}) f_r'(\boldsymbol{w}_{k,i}), \tag{C.19}$$

as illustrated in Fig. C.2. Therefore, $f(\boldsymbol{w}_{k,i} | y_1, y_2, \ldots, y_M)$ is a $2M$-dimensional real Gaussian distribution. In particular, if we write the four messages on the RHS of (C.19) in the canonical form, then the canonical parameters of $f(\boldsymbol{w}_{k,i}|\boldsymbol{y}_1, \boldsymbol{y}_2, \ldots, \boldsymbol{y}_M)$ is the sum of them.

Recall that $\boldsymbol{w}_{k,i} = (b_1^{\mathrm{r}}, \ldots, b_M^{\mathrm{r}}, b_1^{\mathrm{i}}, \ldots, b_M^{\mathrm{i}})$ is a $2M$ dimensional real random variable, where b_m^{r} and b_m^{i} are the real and imaginary parts of the m-th complex element of $\boldsymbol{W}_{k,i} = \mathcal{V}(y_k[i]) = (s_1[i], s_2[i], \ldots, s_k[i], s_{k+1}[i-1], s_{k+2}[i-1], \ldots, s_M[i-1])$, thus, $f(\boldsymbol{y}|\boldsymbol{W}_{k,i})$ is an M-dimensional complex Gaussian distribution.

Let $k = M$, we have $\boldsymbol{W}_{M,i} = (s_1[i], s_2[i], \ldots, s_M[i]) = \boldsymbol{s}[\boldsymbol{i}]$. This means

$$f(\boldsymbol{y}|\boldsymbol{s}[\boldsymbol{i}]) = f(\boldsymbol{y}|\boldsymbol{W}_{k,i}) \tag{C.20}$$

is an M-dimensional complex Gaussian, the mean and covariance of which can be computed from (C.19).

Appendix D
The One-Step Look-Ahead Policy

This appendix investigates the behaviors of the one-step look-ahead policy for significant sampling in Sect. 5.2.

D.1 Analysis of the One-Step Look-Ahead Policy

Suppose $X(t)$ is a zero-mean OU process parameterized by $\{\mu = 0, \theta, \sigma\}$. Without loss of generality, we sample $X(t)$ at time $t = 0$, and obtain $X(0) = \varepsilon$. Then, following the one-step look-ahead policy ϕ' in (5.3), the optimal sampling interval T^* can be obtained from

$$T^* = \arg\min_{T>0} \frac{\eta + \mathbb{E}[C_{[0,T]}|X(0) = \varepsilon]}{T} \triangleq \arg\min_{T>0} h(T),$$

where $C_{[0,T]}$ is the accumulated cost of error in $[0, T]$. Set $\frac{d\,h(T)}{dT} = 0$, we have

$$\begin{aligned} \frac{d\,h(T)}{d\,T} &= \frac{d\,\mathbb{E}[C_{[0,T]}|X(0) = \varepsilon]}{d\,T} T - (\eta + \mathbb{E}[C_{[0,T]}|X(0) = \varepsilon]) = 0 \\ \eta &= \frac{d\,\mathbb{E}[C_{[0,T]}|X(0) = \varepsilon]}{d\,T} T - \mathbb{E}[C_{[0,T]}|X(0) = \varepsilon]. \end{aligned} \tag{D.1}$$

To compute $\mathbb{E}[C_{[0,T]}|X(0) = \varepsilon]$, we can write $C_{[0,T]}$ as $C_{[0,T]} = \int_0^T \varphi(t)dt$, in which $\varphi(t)$ is the instantaneous cost of error at epoch $t \in [0, T]$, and is given by $\varphi(t) = -X(t)\mathbb{1}_{(-\infty,0)}(X(t))$ (that is, when $X(t) \le 0$, $\varphi(t) = -X(t)$; else, $\varphi(t) = 0$).
Thus,

$$\mathbb{E}[C_{[0,T]}|X(0) = \varepsilon] = \int_0^T \mathbb{E}[\varphi(t)|X(0) = \varepsilon]dt \triangleq \int_0^T g(t)\,dt,$$

Y. Shao, *AI-Empowered IoT Communications*, Wireless Networks,
https://doi.org/10.1007/978-3-031-98816-5

and (D.1) can be refined as

$$\eta = \frac{d \int_0^T g(t)\,dt}{dT} T - \int_0^T g(t)\,dt = T g(T) - \int_0^T g(t)\,dt. \tag{D.2}$$

We then compute $g(t)$. Given $X(0) = \varepsilon$, we have

$$X(t) = \varepsilon e^{-\theta t} + \frac{\sigma}{\sqrt{2\theta}} e^{-\theta t} W(e^{2\theta t} - 1), \quad t > 0, \tag{D.3}$$

where $W(e^{2\theta t} - 1)$ is a time-scaled Winner process, $W(e^{2\theta t} - 1) \sim \mathcal{N}(0,\ e^{2\theta t} - 1)$. This means $X(t) \sim \mathcal{N}(\mu_t, \sigma_t^2)$ at time $t > 0$, where $\mu_t = \varepsilon e^{-\theta t}$ and $\sigma_t^2 = \frac{\sigma^2}{2\theta}(1 - e^{-2\theta t})$.

Function $g(t)$ can then be computed as

$$\begin{aligned} g(t) &= \int_{-\infty}^{0} -x f_{X(t)}(X(t) = x) dx \qquad (D.4) \\ &= \frac{\sigma_t}{\sqrt{2\pi}} e^{-\frac{\mu_t^2}{2\sigma_t^2}} - \frac{\mu_t}{2} \operatorname{erfc}\left(\frac{\mu_t}{\sqrt{2\sigma_t^2}}\right) \\ &= \sqrt{\frac{\sigma^2(1 - e^{-2\theta t})}{4\pi\theta}} e^{-\frac{\theta\varepsilon^2}{\sigma^2(e^{2\theta t} - 1)}} - \frac{\varepsilon e^{-\theta t}}{2} \operatorname{erfc}\left(\sqrt{\frac{\theta\varepsilon^2}{\sigma^2(e^{2\theta t} - 1)}}\right) \end{aligned}$$

In fact, $g(t)$ is the expected instantaneous cost of error at t given $X(0) = \varepsilon$. It describes the transient behavior of $X(t)$ after sampling. Two key observations from (D.4) are

- $g(t)$ is an increasing function of t, and a decreasing function of ε.
- $g(0) = 0$ and $g(\infty) = \sqrt{\frac{\sigma^2}{4\pi\theta}}$ are constants irrelevant to ε.

D.2 Intuitive Explanations of the One-Step Policy

Given (D.2) and (D.4), the one-step look-ahead policy ϕ' can be illustrated by Fig. 5.3, where we plot $g(t)$ when $\varepsilon = 0.1$, 1, and 5, respectively.

Let us focus on $\varepsilon = 1$ case, the optimal sampling interval T^* given by the one-step look-ahead policy is the T that makes the shaded area equal to the sampling cost η. This implies, the one-step policy can instruct us to "never sample" if

$$\eta > \int_0^{\infty} g(\infty) - g(t)\,dt. \tag{D.5}$$

In this case, the shaded area will always be less than η as $t \to \infty$, and the system will never sample again. Moreover, since $g(t)$ is a decreasing function of ε, the policy basically tells us to never sample again if the absolute value of one sample is less or equal to ε.

As time goes, the cost rate of the one-step look-ahead policy converges to $g(\infty) = \sqrt{\frac{\sigma^2}{4\pi\theta}}$.

Equation (D.5) gives us the lower bound for η. The next subsection gives an upper bound for η to guarantee that "never sample" is suboptimal. We will show that, if we sample once at the equilibrium state long time later than $t = 0$, and then never sample again, the achieved cost rate can beat the one-step look-ahead policy.

D.3 Sample at the Equilibrium State

When comes to the equilibrium state, an OU process

Assuming we sample at $t = \tau$ in the equilibrium state, and obtain $X(\tau) = x_\tau$, then $x_\tau \sim \mathcal{N}(0, \sigma_\tau^2)$, where $\sigma_\tau^2 = \frac{\sigma^2}{2\theta}$. Given observation $X(\tau) = x_\tau$, the value of $X(t)$ at epoch $\tau + t$ $(t > 0)$ follows (D.3) and is a Gaussian random variable. We have

- $X(\tau + t) \sim \mathcal{N}(\mu_{\tau+t}, \sigma_{\tau+t}^2)$, $t > 0$, where $\mu_{\tau+t} = x_\tau e^{-\theta t}$ and $\sigma_{\tau+t}^2 = \frac{\sigma^2}{2\theta}(1 - e^{-2\theta t})$.
- The instantaneous cost of error $\varphi(\tau + t) = -X(\tau + t)\mathbb{1}_{(-\infty,0)}(X(\tau + t))$.

Then, for $x_\tau \geq 0$, we have

$$\begin{aligned}
\mathbb{E}[\varphi(\tau + t)|X(\tau) = x_\tau] &= \int_{-\infty}^{0} -x f_{X(\tau+t)}(X(\tau + t) = x)\,dx \\
&= \frac{\sigma_{\tau+t}}{\sqrt{2\pi}} e^{-\frac{\mu_{\tau+t}^2}{2\sigma_{\tau+t}^2}} - \frac{\mu_{\tau+t}}{2}\operatorname{erfc}\left(\frac{\mu_{\tau+t}}{\sqrt{2}\sigma_{\tau+t}}\right)
\end{aligned}$$

Averaged over all possible x_τ, $x_\tau \sim \mathcal{N}(0, \sigma_\tau^2)$,

$$\begin{aligned}
\mathbb{E}[\varphi(\tau + t)] &= \int_{-\infty}^{\infty} \mathbb{E}[\varphi(\tau + t)|X(\tau) = x_\tau] f_{X(\tau)}(X(\tau) = x_\tau)\,dx_\tau \\
&= \sqrt{\frac{\sigma^2}{4\pi\theta}(1 - e^{-\theta t})}
\end{aligned}$$

Thus, if we sample at $t = \tau$, and the never sample again, the overall cost from $t = \tau$ to $t = \infty$ will be $\eta + \int_0^\infty \mathbb{E}[\varphi(\tau + t)]dt$. Let $\eta + \int_0^\infty \mathbb{E}[\varphi(\tau + t)]dt < \int_0^\infty g(\infty)dt$, we have

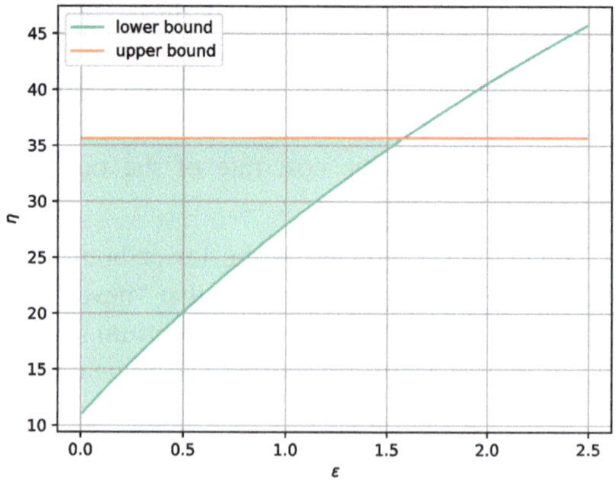

Fig. D.1 An example of the lower and upper bounds for η when $\{\mu = 0, \theta = 0.025, \sigma = 0.5\}$

$$\eta < \int_0^{\infty} [g(\infty) - \mathbb{E}[\varphi(\tau + t)]]dt = \sqrt{\frac{\sigma^2}{4\pi\theta}} \int_0^{\infty} e^{-\theta t} dt = \sqrt{\frac{\sigma^2}{4\pi\theta}} \frac{1}{\theta}. \tag{D.6}$$

Combing (D.5) and (D.6) gives us Proposition 5.2.

We use the example in Fig. 5.2 to show the set of $\{g(t), \eta\}$ for which (5.7) holds is nonempty. Let $\{\mu = 0, \theta = 0.025, \sigma = 0.5\}$. The upper and lower bounds in (5.7) are plotted in Fig. D.1 for different ε. As can be seen, policy ϕ' is suboptimal when the sampling cost η falls into the shaded area.

References

1. S.C. Liew, Y. Shao, New transceiver designs for interleaved frequency-division multiple access. IEEE Trans. Wirel. Commun. **19**(12), 7765–7778 (2020)
2. Y. Shao, S.C. Liew, Flexible subcarrier allocation for interleaved frequency division multiple access. IEEE Trans. Wirel. Commun. **19**(11), 7139–7152 (2020)
3. Y. Du, S.C. Liew, Y. Shao, Efficient FFT computation in IFDMA transceivers. IEEE Trans. Wirel. Commun. **22**(10), 6594–6607 (2023)
4. R. Dinis, D. Falconer, C.T. Lam, M. Sabbaghian, A multiple access scheme for the uplink of broadband wireless systems, in *GLOBECOM*, vol. 6 (2004), pp. 3808–3812
5. T. Frank, A. Klein, E. Costa, E. Schulz, Interleaved orthogonal frequency division multiple access with variable data rates, in *Proc. Int. OFDM Workshop* (2005), pp. 179–183
6. R.A. Al Na'mneh, K.A. Darabkh, An efficient bit reversal permutation algorithm, in *Int. Conf. Robotics, Biomimetics, Intell. Comput. Syst.* (2013)
7. J.-L. Gauvain, C.-H. Lee, Maximum a posteriori estimation for multivariate Gaussian mixture observations of Markov chains. IEEE Trans. Speech Audio Process. **2**(2), 291–298 (1994)
8. P. Ahrendt, The multivariate Gaussian probability distribution. Technical University of Denmark, Tech. Rep., 2005
9. C.B. Do, More on multivariate Gaussians. Tech Report, Stanford CS229, 2008

MIX
Papier aus verantwortungsvollen Quellen
Paper from responsible sources
FSC® C105338

If you have any concerns about our products,
you can contact us on
ProductSafety@springernature.com

In case Publisher is established outside the EU,
the EU authorized representative is:
Springer Nature Customer Service Center GmbH
Europaplatz 3, 69115 Heidelberg, Germany

Printed by Libri Plureos GmbH
in Hamburg, Germany